Energy Supply to the Year 2000 Global and National Studies

Energy Supply to the Year 2000 Global and National Studies

Second Technical Report of the Workshop on Alternative Energy Strategies (WAES)

William F. Martin
Editor

Carroll L. Wilson
WAES Project Director

The MIT Press
Cambridge, Massachusetts, and London, England

This material is based upon research supported in part by the National Science Foundation under Grant No. OEP74-13902 AO1. Any opinions, findings, and conclusions or recommendations expressed in this publication are those of the authors and do not necessarily reflect the views of the National Science Foundation.

This book was set in IBM Composer Century by Eastern Composition, Inc. and printed and bound by Halliday Lithograph Corp. in the United States of America.

Library of Congress Cataloging in Publication Data

Workshop on Alternative Energy Strategies.
Energy supply to the year 2000.

Includes bibliographical references.
1. Power resources. 2. Energy consumption.
3. Economic forecasting. I. Martin, William Flynn, 1950- II. Title.
HD9502. A2W66 1977 333.7 77-5549
ISBN 0-262-23082-8

Contents

Preface

In October 1974 leaders of business, industry, government, and academia from a dozen countries convened the first session of the Workshop on Alternative Energy Strategies (WAES). During its two-and-a-half year lifetime the Workshop studied energy demand and supply, resource availability, and the critical factors in alternative energy futures, both nationally and globally. These studies provided the essential ingredients for what we call "energy supply-demand integrations"—the comparing and contrasting of separate energy demands and supplies. This book reports the results of our supply studies. It is the second volume in a series of three WAES technical reports.

The Workshop was an ad hoc international project involving over seventy-five people from fifteen countries. It set itself the goal of assembling a set of coherent global energy pictures of the future. To this end it developed methods for estimating energy supply and demand through the year 2000 and for "integrating" them.

This goal was set against the recognition that according to the best estimates, world crude oil production will peak and begin to decline within the next twenty-five years. The transition from oil to other forms of energy is coming more rapidly than most people realize. WAES evolved from the common concern of a number of influential people who believe that this transition needs to be widely understood and effectively managed to avoid major national and international dislocations.

This book describes our global and national supply assessments in three major parts. Part I describes the methodology and major conclusions of the supply studies. Part II includes individual global overviews for oil, natural gas, coal, nuclear energy, other fossil fuels, and renewables such as hydroelectricity and geothermal and solar energy. Part III consists of individual national supply studies provided by WAES national teams.

The overall methodology developed by the Workshop is described in Chapter 1. This novel approach to global energy analysis draws partially on traditional analytical techniques and tools that have been adapted by the Workshop Associates for our purposes with the guidance and assistance of the Participants. Following a discussion of our methodology is a summary of our major global supply assessments, including our global supply data base in Chapter 2.

Part II, which includes Chapters 3 to 7, is our assessment of the reserve base and future production potentials of the major contemporary fuels—oil, natural gas, coal, nuclear, other fossil fuels, and renewables.

In Chapter 3 we examine a range of alternative assumptions that will influence future oil production. Given only technical constraints, oil production peaks and declines during the 1990s. But oil production could plateau much earlier (as soon as the early 1980s) if the oil-exporting countries set their production levels substantially below their technical capacity.

In Chapter 4 we assess potential production and transport of natural gas. Meeting future increasing natural gas demand in major consuming centers will require building large liquefied natural gas (LNG) systems and pipelines to connect the major potential producers (primarily OPEC and the U.S.S.R.) to the major consuming countries in Western Europe, North America, and Japan. This chapter discusses the transportation and distribution infrastructure necessary to meet natural gas demands to the year 2000.

Coal resources are abundant, as Chapter 5 points out, but desired demand is less than potential supply because consumers prefer cleaner and more convenient fuels such as oil and gas. Coal may become an important replacement fuel if demand can be generated and the right environmental standards maintained.

Chapter 6 shows that in the highest nuclear case, nuclear energy could contribute as much as 25 percent of primary energy in the year 2000. This is the energy equivalent of all the oil produced today in the world outside Communist areas.

Chapter 7 discusses other fossil fuels and renewables. Resources such as oil sands, heavy oil, and oil shales are more abundant than conventional oil reserves. Their production to date has been small due to high costs and technological and environmental problems. Renewables—especially solar—have recently captured the imagination of the public. Our national teams have estimated the level of energy that might be obtained from solar in the year 2000. The solar contribution may be smaller than most people expect; however, it is expected to play an increasingly important role as we enter the twenty-first century.

Part III includes thirteen national supply studies, from Canada, Denmark, Finland, France, the German Federal Republic, Italy, Japan, Mexico, The Netherlands, Norway, Sweden, the United Kingdom, and the United States. Each national study is attributable solely to the Workshop members from that country, Although all studies were done in a generally consistent manner, each has its own distinctive features.

The WAES Final Report, *Energy: Global Prospects, 1985-2000* (McGraw-Hill, 1977), summarizes the global character of the WAES findings. It draws heavily on the substance of the three-volume technical report series. It identifies the major long-term energy problems that emerge from our studies and reveals some implications for nations acting together and individually.

The Workshop on Alternative Energy Strategies was an independent organization, with no official ties to any governments or private firms.[1] It was organized into three groups:

- The *Participants* were thirty-five senior decision makers from around the world. Experienced in dealing with problems involving the interplay of technology, economics, finance, and government policy, they were well suited to take an active part in the Workshop. Members included officials from organizations that are suppliers and users of fuels or producers of electricity, government officials with energy-related responsibilities, and others whose insights enriched the Workshop. The Participants took part in this study as individuals, not as official representatives of their organization or country. They provided or arranged for financial support for WAES activities in their own country and paid for the expenses involved in attending and sponsoring meetings.
- The *Associates* were chosen by the Participants from their own or cooperating organizations. Scientists, engineers, economists, and managers with expertise in many energy-related fields, they brought a wealth of broad support to the Workshop. They conducted the technical studies that are the basis for the Workshop's findings and conclusions.

To expand their capabilities and influence, Workshop members in some countries recruited *Reference Groups*—people from government and industry who served as advisors, provided financial support, and did some of the analytical work.

- The *Secretariat* included the Project Director and a small staff at the Massachusetts Institute of Technology. The Director provided executive leadership, and the Secretariat provided project coordination. Secretariat expenses were supported by foundation grants.

The Workshop's distinctive character allowed it to have two important effects. Through its technical analyses it provided a unique and necessary quantitative basis for policy decisions. More directly, involvement in such an in-depth study influenced the members themselves who will, it is hoped, assist in the process of turning the options identified by the Workshop into realities.

The Workshop objectives were to establish a quantitative basis for analyzing probable future energy developments through the year 2000 and then to identify feasible alternative global and national energy strategies to deal with these developments. The analyses were designed to provide information that would increase public understanding and assist governments and others in choosing and formulating the strategies necessary to assure a proper balance of energy supply and demand during the period from 1985 to 2000 and beyond.

Few nations acting alone have the resources to ensure their own energy supplies. Therefore, it is in the common interest to avoid placing insupportable demands on specific energy sources. For example, it makes no sense for one country to build power stations that will depend on coal from another country if the latter does not develop the needed capacity for coal production, transport, and export. The international nature of WAES enables it to approach such issues on a global basis.

In looking at the period to the end of the century, one must try to determine what can be achieved in the available time. Economic growth and energy "savings" will be the main determinants of energy demand up to 1985, with oil remaining the principal fuel. Beyond 1985 alternative energy sources

1. Further information about the Workshop can be found in "A Global Assessment 1985-2000," a brochure available from any of the Workshop members.

will be needed on an increasing scale. Because of the long lead times required, however, the size of their contributions will be determined by actions taken now or in the very near future.

Any future study involves uncertainties. In our projections we tried to define possible ranges of energy demand and supply, and their possible global balance or imbalance, in WAES member countries and in the world outside Communist areas.

Many differences in national style and decision making are revealed by the studies in this book. These differences add richness to the studies and illustrate some of the potentials for change in present-day patterns of energy use. Such potentials should be carefully assessed by the many analysts, scientists, and research groups outside the Workshop who have the ability and time to do so.

Carroll L. Wilson
Project Director

Workshop Participants, Associates, and Staff

PROJECT DIRECTOR
Carroll L. Wilson
Massachusetts Institute of Technology

PARTICIPANTS

Canada
Mr. Marshall A. Crowe
Chairman, National Energy Board

Mr. Maurice F. Strong
Chairman, Petro-Canada

Denmark
Professor Bent Elbek
Niels Bohr Institute

Finland
Professor Jorma Routti
Helsinki University of Technology

France
M. Jean Couture
Conseiller du Président
Société Générale

M. Jean-Marie Martin
Director, Institut Economique et Juridique de l'Energie
Centre National de la Recherche Scientifique

Germany
Director Dr. Hans Detzer
Head of Central Planning
Badische Anilin & Soda Fabrik AG

Professor Heinrich Mandel
Member of the Board of Directors
Rheinisch-Westfälisches Elektrizitätswerk AG

Professor Hans K. Schneider
Director, Institute of Energy Economics
University of Cologne

Iran
Dr. Khodadad Farmanfarmaian

Italy
Professor Umberto Colombo
Director, Research & Development Division
Montedison

Professor Sergio Vaccà
Director, Istituto di Economia delle Fonti de Energia
Università L. Bocconi, Milano

Japan
Mr. Toshio Doko
President, Japan Federation of Economic Organizations (Keidanren)

Mr. Shuzo Inaba
President, Industrial Research Institute

Mr. Soichi Matsune
Chairman, Committee on Energy
Japan Federation of Economic Organizations (Keidanren)

Mr. Tatsuzo Mizukami
President, Japan Foreign Trade Council, Inc.

Dr. Saburo Okita
President, The Overseas Economic Cooperation Fund

Mr. Masao Sakisaka
President, National Institute for Research Advancement

Mr. Shigefumi Tamiya
Advisor to the Chairman
Enrichment and Reprocessing Group

Mexico
Ing. Juan Eibenschutz
Executive Secretary
Mexican National Energy Commission

Dr. Victor Urqùidi
President, El Colégio de México

The Netherlands
Dr. A.A. T van Rhijn
Deputy Director General for Energy
Ministry of Economic Affairs

Netherlands/U.K.
Mr. J.C. Davidson
Director, Shell International Petroleum Company, Ltd.
Royal Dutch/Shell Group

Norway
Mr. Christian Sommerfelt
Chairman
Elkem-Spigerverket A/S

Sweden
Mr. Erland Waldenström
Chairman
Granges AB

United Kingdom
Mr. Robert Belgrave
Policy Adviser to the Board
British Petroeum Company, Ltd.

Professor Sir William Hawthorne
Master, Churchill College
University of Cambridge

United States
Mr. Thornton F. Bradshaw
President, Atlantic Richfield Company

Mr. Walker L. Cisler
Retired Chairman of the Board
Detroit Edison Company

Mr. John T. Connor
Chairman of the Board
Allied Chemical Corporation

Mr. Richard C. Gerstenberg
Director, General Motors Corporation

Dr. H. Guyford Stever[1]
Director, National Science Foundation

Venezuela
Dr. José A. Mayobre
Banco Central de Venezuela

Ing. Ulises Ramírez
Executive Secretary
National Energy Council

1. From October 1974 to October 1976.

ASSOCIATES

Canada
Mr. Marc LeClerc
Director General, Special Projects
National Energy Board

Dr. John Ralston Saul
Special Assistant to the Chairman
Petro-Canada

Finland
Mr. Seppo Hannus
Ministry of Trade and Industry
Energy Department

France
M. Bertrand Chateau
Institut Economique et Juridique de l'Energie
Centre National de la Recherche Scientifique

Dr. Maxime Kleinpeter
Eléctricité de France
Service Etudes Economique Générales

Germany
Dr. Georg Klotmann
Strategic Planning Department
Badische Anilin & Soda Fabrik AG

Dr. Dieter Schmitt
Head, Institute of Energy Economics
University of Cologne

Mr. Paul H. Suding
Economist, Institute of Energy Economics
University of Cologne

Italy
Dr. Oliviero Bernardini
Department of Technology Assessment
Research & Development Division
Montedison

Dr. Riccardo Galli
Director, Department of Technology Assessment
Research & Development Division
Montedison

Mr. William Mebane
Department of Technology Assessment
Research & Development Division
Montedison

Japan

Mr. Shinichiro Aoyama
Senior Staff Research, National Institute for Research Advancement

Mr. Kenichi Matsui
Senior Staff Economist
The Institute of Energy Economics

Mr. Yasuhiro Murota
Staff Economist
The Japan Economic Research Center

Mr. Mitsuo Takei
Director, Research Affairs
The Institute of Energy Economics

Mr. Hisashi Watanabe
Executive Director, Japan Alternative Energy Strategies Organization

Mexico

Ing. Gerardo Bazán
Assessor
Mexican National Energy Commission

Ing. Alberto Escofet
Assistant to the Executive Secretary
Mexican National Energy Commission

The Netherlands

Dr. André C. Sjoerdsma
Director, Future Shape of Technology Foundation

Netherlands/U.K.

Mr. Alan W. Clarke
Energy & Oil Economics Division
Shell International Petroleum Co., Ltd.
Royal Dutch/Shell Group

Dr. Hans DuMoulin
Head of Energy & Oil Economics Division
Shell International Petroleum Co., Ltd.
Royal Dutch/Shell Group

Dr. Gareth Price
Group Planning, Long Term Future
Shell International Petroleum Co., Ltd.
Royal Dutch/Shell Group

Norway

Mr. Henrik Ager-Hanssen
Deputy Managing Director
Statoil, Den Norske Stats Oljeselskap A/S

Mr. Kai Killerud
Manager, Power Plant Engineering
Scandpower A/S

Sweden

Dr. Harry Albinsson
Energy Secretary
Federation of Swedish Industries

Mr. Bertil Eneroth
Special Assignments
Skandinaviska Enskilda Banken

United Kingdom

Mr. Michael Clegg
Manager, Systems Group, Corporate Planning
British Petroleum Company, Ltd.

Dr. Edmund Crouch
Energy Research Group
Cavendish Laboratory
University of Cambridge

Dr. Richard J. Eden
Head, Energy Research Group
Cavendish Laboratory
University of Cambridge

Mr. Andrew R. Flower
Assistant Policy Analyst
Policy Review Unit
British Petroleum Company, Ltd.

United States

Mr. Walter F. Allaire
Director, Energy Resources
Allied Chemical Corporation

Mr. Steven Carhart[2]
Assistant Scientist
National Center for Analysis of Energy Systems
Brookhaven National Laboratory

Dr. Paul P. Craig[3]
Director, Energy & Resources Council
University of California

Dr. Henry L. Duncombe, Jr.
Chief Economist
General Motors Corporation

2. Consultant on the Demand Studies, October 1974 to December 1975; Advisor on U.S. Studies from January 1976 to February 1977.

3. Associate from October 1974 to April 1976.

Ms. Sandra Fucigna[4]
Policy Analyst
National Science Foundation

Mr. Edward D. Griffith
Senior Consultant—Policy Analysis and Forecasting
Atlantic Richfield Company

Dr. Kenneth Hoffman
Associate Chairman for Energy Programs
Head, National Center for Analysis of Energy Systems
Brookhaven National Laboratory

Dr. H. Paul Root
Director of Economic Studies
General Motors Corporation

Dr. David Sternlight
Chief Economist
Atlantic Richfield Company

Venezuela

Dr. Felix Rossi-Guerrero
Minister Counselor for Petroleum Affairs
Venezuelan Embassy, Washington, D.C.

STAFF

MIT Program Staff

Mr. Robert P. Greene
Program/Administrative Officer

Mr. Paul S. Basile
Program Officer

Mr. William F. Martin
Program Officer

MIT Research Assistants

Mr. Richard Cheston
Mr. Miles Harbur

MIT Support Staff

Ms. Elaine R. Goldberg
Ms. Susan M. Leland
Ms. Hedy Walsh

European Support Staff

Ms. Karin Berntsen, Norway
Ms. Adriana Cavagna, Italy
Ms. Dalia Jackbo, Norway
Ms. Margaret Martirosi, United Kingdom

Acknowledgments

CONTRIBUTORS TO THIS REPORT

The studies and analyses upon which this report is based result from substantial contributions by many WAES members. The national reports, Chapters 8 through 20, represent extensive efforts by all national teams over the course of the Workshop. Each of these chapters is the responsibility of the Associates from the given country.

The chapters in Part II were prepared by several individuals although the chapters represent the product of contributions by all WAES members. Andrew Flower is the author of Chapter 3 on oil, and he joined William Martin in preparing Chapter 4 on natural gas. Edward Griffith and Alan Clarke prepared Chapter 5 on coal. The nuclear analysis (Chapter 6) was written by Carroll Wilson with revisions agreed on at the final meeting of Participants. The writing of Chapter 7 on fossil fuels and renewable energy sources was coordinated by William Martin, with assistance from Marc LeClerc, Sandra Fucigna, Edward Griffith, John Page, Leonard Topper, and John Ralston Saul.

INSTITUTIONS AND SPONSORS

Many organizations and institutions have provided direct financial support and other services for WAES. The Workshop gratefully acknowledges the following sponsors for their generous support of the Secretariat and related activities:

The Allied Chemical Foundation

The Atlantic Richfield Foundation

The Edna McConnell Clark Foundation

The General Motors Corporation

The Ford Foundation

The German Marshall Fund of the United States

The Andrew W. Mellon Foundation

The National Science Foundation

The Rockefeller Brothers Fund

The Rockefeller Foundation

The Alfred P. Sloan Foundation

The Workshop also gratefully acknowledges the many individuals and institutions in each country that contributed financial and/or professional support for the Workshop and/or the WAES national studies, including:

Canada
National Energy Board
Petro-Canada

Denmark
Danish Research Council for Science
Niels Bohr Institute

Finland
Academy of Finland
Helsinki University of Technology
Ministry of Trade and Industry

France
Centre National de la Recherche Scientifique
Eléctricité de France
Société Générale

Germany
Badische Anilin & Soda Fabrik AG
Energiewirtschaftliches Institut, University of Cologne
Rheinisch-Westfälisches Elektrizitätswerk AG

Italy
Azienda Municipale Nettezza Urbana, Milano
Azienda Servizi Municipalizzati, Brescia
CISE
Confindustria
EGAM
ENI
FIAT
Istituto Economia Fonti di Energia (IEFE), Università L. Bocconi, Milano
Istituto di Economia e Politica Industriale, Università di Bologna
Istituto di Fisica, Università di Napoli
Istituto Internazionale per le Ricerche Geotermiche, Pisa
IRI
Montedison
SNAM
Zanussi

Japan
Electric Machinery Industry Federation
Electric Power Industry Federation
Industrial Research Institute
Industrial Research Institute, Japan
Institute of Energy Economics
Institute for Future Technology
Japan Economic Research Center
Japan EXPO Fund
Japan Foreign Trade Council, Inc.
Japan Iron and Steel Federation
Japan Techno-Economic Society
Mitsubishi Research Institute
National Institute for Research Advancement
Nomura Research Institute
Petroleum Association of Japan

Mexico
Mexican National Energy Commission

The Netherlands
AKZO Zoutchemie
Centraal Planbureau
ESTEL Hoesch-Hoogovens
Ministerie van Economische Zaken
OGEM Holding
Paktank
Stichting Toekomstbeeld der Techniek
Verenigde Machinefabrieken

Netherlands/U.K.
Shell International Petroleum Company, Ltd.
Royal Dutch/Shell Group

Norway
Royal Norwegian Council for Scientific and Industrial Research
Elkem-Spigerverket A/S
Scandpower A/S
Statoil, Den norske stats oljeselskap A/S
A reference committee with observers from the Ministries and other institutions

Sweden
Gränges AB
Skandinaviska Enskilda Banken
Federation of Swedish Industries
Kraangede AB
National Board of Industry
Secretariat for Future Studies
Royal Academy of Engineering Sciences
State Power Board

United Kingdom
British Petroleum Company, Ltd.
Department of Energy
University of Cambridge:
—Churchill College
—Cavendish Laboratory
—Engineering Department

United States
Allied Chemical Corporation
Atlantic Richfield Company
Brookhaven National Laboratory
General Motors Corporation
National Science Foundation
Overseas Advisory Associates, Inc.

Venezuela
Ministry of Energy and Mines

Part I
Introduction

1 The Workshop Methodology

1.1 THE CONCEPTUAL FRAMEWORK

1.1.1 Introduction

No widely accepted methodology exists for projecting energy supply and energy demand for periods of ten to twenty-five years into the future. Therefore, the first major tasks of the Workshop were to identify and agree on the major determinants of future energy supply and demand, to select a range of likely values for these determinants, and to develop an internally consistent and comprehensible framework for synthesizing the various national and global studies.

The period under study was divided into two time frames: (1) the period from 1972 to 1985, in which the contribution of each energy source is largely determined by existing infrastructure and projects already under development; and (2) the period from 1985 to 2000 when pressures on oil and natural gas supplies will increase, alternative energy supply systems could make major contributions, and energy conservation measures now only beginning could reduce energy demand—provided that these systems are started and developed soon enough and implemented on a sustained basis.

No single forecast can encompass all of the uncertainties of the next twenty-five years. The Workshop therefore identified a range of different but plausible futures; that is, a "scenario" approach was adopted. "Scenario" is the term used to represent a plausible future—for our purposes, with particular reference to energy. Several scenarios were defined and analyzed by the Workshop. For each scenario, values of key variables were specified and the consequent energy picture evaluated. Scenarios are not forecasts; they are chosen to span a wide range of possible futures that lead to different estimates of energy demand and supply.

1.1.2 The Overall Framework[1]

The real world[2] of energy decisions is highly interactive and too complex to be simulated precisely. So WAES constructed a model—a simplified conceptual framework—of the real energy world (Figure 1.1). Its essential feature is that it separates those global factors not under control of individual, energy-consuming nations from those factors under national control. The method used in WAES involves seven basic steps, shown in Figure 1.1 from top to bottom.

Step 1 defines in broad terms the sorts of "energy worlds" that might evolve between now and 2000 by making various assumptions about world economic growth and the world price of oil or energy, the key determinants of energy supply and demand outside the control of individual national decision-making processes. The values we selected are not predictions but plausible values that adequately span the WAES assessment of the likely future states of the world—given the assumption that the

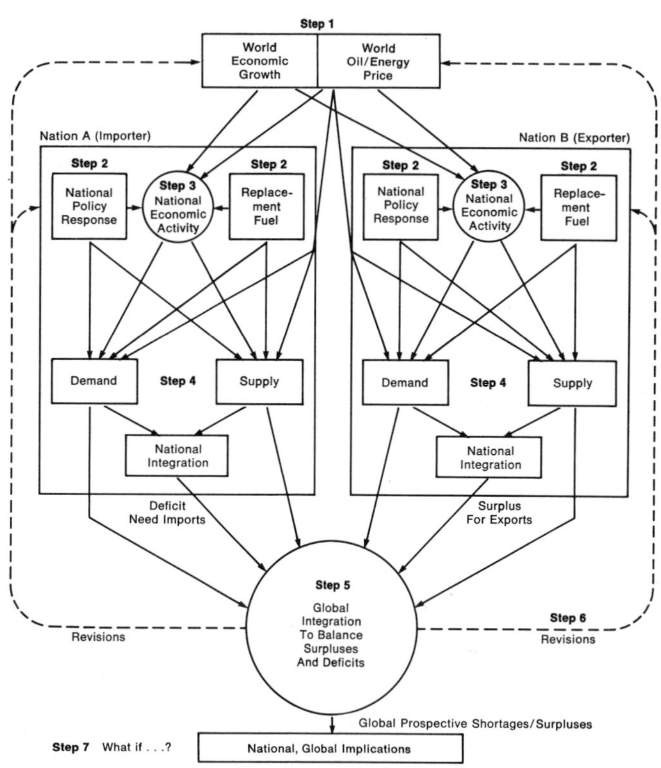

Figure 1.1
The WAES conceptual framework for world energy futures.

1. A discussion of the WAES approach in a general, global context can be found in the WAES report, *ENERGY: Global Prospects 1985-2000* (McGraw-Hill, 1977).

2. Members of the Workshop come from energy-consuming nations in the world outside Communist areas. Although the U.S.S.R. and China are major world energy producers and consumers, their trade in fuels with the countries in the world outside Communist areas (WOCA) has to date been relatively small. In the WAES projections, this situation was assumed to continue through the end of the century.

past trends of world economic growth may be extrapolated into the future.

Step 2 introduces, as a third global scenario variable, a factor that nations can control—the national public policy response to the emerging energy situation. It is assumed that until 1985 each nation has either a vigorous or a restrained national energy policy response: for example, by enacting policies to bring in new supplies and to promote energy conservation. From 1985 to 2000 energy policies of all nations are assumed to be vigorous. Also during the 1985 to 2000 period, a major energy policy question facing nations could well be framed as: What will be the principal replacement fuel for oil as it declines in availability? This is the fourth global scenario variable for this period: i.e., whether coal or nuclear will be the principal replacement fuel. Nations may make different choices but such choices can be added to obtain global (WOCA) totals.

Step 3 moves from these global scenarios to corresponding regional and national scenarios. Each national team individually translated and applied the global assumptions to its own national economic assumptions, taking account of such factors as trade capability and industrial infrastructure.

Step 4 analyzes in detail national energy supply and demand under world, regional, and national assumptions. Each WAES team calculated independent estimates of desired demand and potential supply. Desired demand is what each nation would like to use (assuming that supplies for each kind of fuel are available) at the given price, economic growth, and policy assumptions. Maximum potential supply is the amount that could be produced within a country given the scenario assumptions. These demand and supply estimates do not take into account the availability or unavailability of fuel imports or markets for potential exports. The demand and supply studies are done independently and are then matched in the national supply-demand integration process.

Many existing national energy forecasts stop at step 4 and simply assume that any gap between demand and domestic fuel production can be filled by imports. *Step 5* goes on to compare the fuel imports and exports of the national supply-demand integrations with global fuel availability by means of an international balancing calculation.

The global integration technique will be described later in this chapter. For now it is only necessary to realize that WAES has used two kinds of integration: *unconstrained* and *constrained.*

From the global integrations, an overall energy balance (or imbalance) is obtained—the total energy supplies and demands, including shortfalls and surpluses, for each major fuel and for each scenario case under both the unconstrained and constrained alternatives. The feasibility, consistency, and likelihood of each scenario can then be evaluated for each country.

Step 6 in the WAES framework allows a major choice. Either the analysis can end here, presenting the results as a final projection indicative of prospective shortages or surpluses of fuels in the future, or, if the projection points to severe prospective energy shortages or other major problems—in fact, to an impossible or highly undesirable future—assumptions can be altered or new scenarios calculated. This revision process is shown by the dashed lines in Figure 1.1. Such revisions might be motivated by, for example, balance-of-payments problems in oil-importing countries—causing increased development of domestic resources and greater fuel switching. WAES has not studied balance-of-payments questions in depth.

Step 7 lies outside the main conceptual framework. Here, WAES could ask, in effect, what would happen if the main assumptions proved incorrect. What if Saudi Arabian oil production were to be held near present levels? What if there were a moratorium on nuclear power? What if solar energy rapidly became cheaper than expected? And so on. Such questions are partly explorations of the need for strategies, but they also test the impact of major uncertainties on WAES scenarios.

The next two subsections define more explicitly the scenario variables used by WAES and their combinations into the scenario cases selected for analysis.

1.1.3 The Scenario Variables

For both demand and supply projections, WAES defined the following global variables.

World Energy Price
To 1985 energy price assumptions were based on the price of oil (in constant 1975 U.S. dollars) because oil will continue to be the dominant as well

as the balancing fuel and other internationally traded fuels will therefore be priced in relation to oil. Three alternative price assumptions were made.

- *Constant* real price: No change from the January 1, 1976, price of $11.50 per barrel.[3] This reflects the possibility that the oil price, adjusted for inflation, will remain constant to 1985.[4]
- *Rising* real price: Prices rise steadily to reach $17.25 per barrel by 1985. This price was chosen arbitrarily to be 50 percent above the constant price.
- *Falling* real price: Prices fall steadily to reach $7.66 per barrel in 1985. This price increased by 50 percent equals the constant price.

In both the rising and falling price cases, prices were assumed to change linearly with time; the actual price changes might more realistically be a step function with increases (or decreases) every one or two years.

For 1985 to 2000, only the constant and rising price cases were used because in our study of five cases to 1985, the effect of the falling price assumption proved to be major, with the low oil price helping to drive up energy demand and hold back the development of all fuel supplies. As a result, demand exceeded available supplies. In the 1985 to 2000 period an energy price is specified, as oil may no longer be the balancing source of supply. Two energy prices for the 1985 to 2000 period were taken.

- *Constant* energy price: This is equivalent to $11.50 per barrel of oil (in constant 1975 U.S. dollars).
- *Rising* energy price: This is equivalent in 2000 to $17.25 (in constant 1975 U.S. dollars) per barrel of oil.

World Economic Growth

To 1985, world economic growth was assumed to have two values, high and low. In the high case, an average growth rate of 6 percent per year (1977 to 1985) was assumed for the world outside Communist areas. This is an upper boundary of what is likely, with the world returning to the high economic growth and flourishing international trade of the 1960s and having resolved most of its recent economic problems.

In the low case, an average economic growth rate of 3.5 percent per year (1977 to 1985) was assumed. This growth rate corresponds to the world we have seen during 1974 to 1976 and is sufficiently above population growth to allow an advance in real global GNP per capita. Growth significantly lower than this could result in periods of stagnation or recession more severe than usual in some countries and some measure of political and social instability. In the Third World, where material hopes are closely linked to world trade levels and to economic activity in the developed, industrialized nations, the economic impact could be particularly severe. WAES national studies done under these two assumptions of 6 percent and 3.5 percent global economic growth, when aggregated, actually resulted in global rates of 5.2 percent and 3.4 percent for the high and low cases, respectively, from 1977 to 1985 (see Table 1.1). WAES did not attempt to recalculate national growth rates in light of the modified aggregated global rate. If there had been time to do so, the results of such lower projections would have been taken into account. Their impact on the conclusions of the study, however, would not have been significant.

For 1985 to 2000, the growth rates were reduced slightly. In the high growth case they dropped from 6 to 5 percent because it seemed unlikely that the developed nations could maintain the high rates of the 1960s and 1970s into a period of declining population growth rates.

Similarly, the rate in the low growth case was reduced from 3.5 to 3 percent per year for 1985 to 2000. This rate is just enough above the population growth estimates to allow some increase in the real GNP per capita in most countries. WAES national studies done under these assumptions, when aggregated, actually result in global economic growth rates of 4.0 percent and 2.8 percent for the high and low cases, respectively, from 1985 to 2000. Again, WAES did not attempt to recalculate national growth rates under these lower global rates,

3. All oil or energy prices are in constant 1975 U.S. dollars per barrel of Arabian light crude oil, fob Persian Gulf.
4. The $11.50 per barrel price in October of 1975 represents the actual world price at that time. The $11.50 per barrel price in 1975 U.S. dollars is roughly equivalent to a $9 per barrel price in 1972 U.S. dollars, and the $7.66 per barrel price in 1975 U.S. dollars is roughly equivalent to a $6 per barrel price in 1972 U.S. dollars. These WAES price assumptions are therefore quite similar to the price assumptions used in the OECD study, *Energy Prospects to 1985,* and in several national energy studies such as the U.S. Federal Energy Administration's *1976 National Energy Outlook.*

Table 1.1
Summary of economic growth rate (EGR) assumptions by major region

	Historical 1960–1976					1977–1985				1985–2000			
Major regions	1960 GNP	1960–1972 EGR	1972 GNP	1972–1976 EGR	1976 GNP	1977–1985 EGR High	1985 GNP High	1977–1985 EGR Low	1985 GNP Low	1985–2000 EGR High	2000 GNP High	1985–2000 EGR Low	2000 GNP Low
North America	764	4.3	1264	1.6	1345	4.3	1970	3.0	1760	3.7	3388	2.6	2576
Western Europe	533	4.7	924	2.0	1003	4.6	1500	2.7	1278	3.3	2425	2.2	1774
Japan	75	10.5	248	3.3	284	7.9	563	4.8	434	4.5	1091	2.7	646
Australia and New Zealand	26	4.5	46	3.6	53	5.0	82	3.6	73	4.3	155	2.8	111
Industrialized Countries	1398	4.9	2482	2.0	2685	4.9	4115	3.1	3545	3.7	7059	2.5	5107
Lower-income countries	61	3.7	94	2.3	103	4.4	152	2.8	132	3.1	240	2.5	191
Middle- and upper-income countries	158	6.2	325	5.9	409	6.6	727	4.5	608	4.9	1490	3.9	1079
All non-OPEC developing countries	219	5.6	419	5.1	512	6.2	879	4.2	740	4.6	1730	3.7	1270
OPEC	34	7.2	78	12.5	125	7.2	235	5.5	202	6.5	604	4.3	380
Total WOCA	1651	5.0	2979	2.8	3322	5.2	5229	3.4	4487	4.0	9393	2.8	6757

Notes:
1. All GNP values are given in billions (10^9) of constant 1972 U.S. dollars.
2. All growth rates are given as percent per year increases in constant dollar value of GNP.
3. All data (except historical) for North America, Western Europe, and Japan result from WAES national team estimates, based on WAES global economic scenario assumptions. All data for other regions result from special studies, performed by the World Bank, based on WAES global economic scenario assumptions. Data for 1976 are preliminary.
4. The major regions are defined as follows:

WOCA: World outside Communist areas.

North America: Canada, United States.

Western Europe: WAES-Europe (Denmark, Finland, France, Federal Republic of Germany, Italy, The Netherlands, Norway, Sweden, United Kingdom) plus non-WAES Europe (Austria, Belgium, Greece, Iceland, Ireland, Luxembourg, Portugal, Spain, Switzerland).

OPEC (Organization of Petroleum Exporting Countries): Algeria, Equador, Gabon, Indonesia, Iran, Iraq, Kuwait, Libya, Nigeria, Qatar, Saudi Arabia, United Arab Emirates (Abu Dhabi, Dubai, and Sharjah), Venezuela.

Non-OPEC Rest of WOCA: All other countries outside of Communist areas.

but if they had, the main results would have been only marginally affected.

As explained above, within these global boundary assumptions, each WAES national team determined corresponding national high and low growth projections. National interpretations varied widely, as shown in Table 1.1, with the most significant changes occurring in Japan and the OPEC countries. The WAES approach is designed so as not to lose sight of these important national and regional differences.

National Policy Response

Two types of national energy policy response are assumed as the third scenario variable. For 1972 to 1985 either *restrained* or *vigorous* policies are assumed. Only vigorous policy is assumed after 1985 and until the year 2000. A restrained policy assumes the least possible government action on conservation and supply that can be expected under the oil price and economic growth assumptions of the case. Market forces are assumed to act appropriately with the price and economic growth assumptions. Global stability is assumed. Some changes in social and individual requirements may be induced by national policy responses, but these should not extend beyond what is likely.

The definitions of vigorous and restrained policy actions are flexible, thus permitting WAES national teams to adopt individual approaches. While recognizing that price and policy effects are related, the national studies attempt to consider independently the influence of each. Explicit policies that have an effect beyond the results of price changes can be considered. Such policies have been designed in the national studies to improve the efficiency of energy use or expand supply potentials without altering the economic activities already established for any case. "Conservation" in these studies refers primarily to improvements in the efficiency of energy use rather than to reductions of energy-using activities. The potential environmental desirability of an energy policy and the costs of meeting environmental standards have also been included by each WAES team in their definitions of national policy response.

Oil Discoveries and Production Limits
The methodology used to develop the oil production profile is described in Chapter 3 of this report. Ultimately recoverable world oil resources were estimated to be 2000 billion barrels (of which about 75 to 80 percent is assumed to be in WOCA), but the possibility of higher or lower resources was also considered. However, more critical than the assessment of the total potential resources is the rate at which oil reserves are actually expanded by new discoveries, extensions to known fields, and better recovery techniques in the period to the year 2000. Here we assumed two average rates of "gross additions" to oil reserves—a high average of 20 billion barrels per year and a low average of 10 billion barrels per year. In both cases we have somewhat arbitrarily assumed that 50 percent of these additions will be in OPEC and 50 percent in non-OPEC countries.

Theoretical oil production curves were developed, limited only by the assumption that production will increase to meet demand as long as the global reserve-to-production ratio does not fall below 15 to 1. However, since some OPEC countries may be unwilling to see oil production reach maximum theoretical levels determined by such a ratio because of a desire to extend the life of their oil reserves, we assumed that limits on total OPEC production would be 40 MBD and 45 MBD. These levels are some 10-15 MBD higher than OPEC production in 1973 (the highest annual level that OPEC production has reached to date).

Principal Replacement Fuel
The last global variable for the 1985 to 2000 period is the principal replacement fuel for oil—with coal or nuclear power as the possibilities selected. Other replacements such as oil from oil sands and shales and renewable sources such as solar, wind, and geothermal are of course also considered. Indeed, renewable energy forms represent a rapidly growing and important (if small-scale) source in many national scenarios for 2000. But their contribution was left to the analysis and judgment of national teams, and they were not included as overall scenario variables.

1.1.4 The WAES Scenarios

Figure 1.2 summarizes the WAES scenario variables and the cases selected for each period. For each scenario, separate estimates of energy demand and energy supply were made.

Five scenarios were selected to 1985, and two of these (Cases C and D in Figure 1.2) were coupled to the period 1985 to 2000. Each of these has a coal and a nuclear variant in the latter period, thus giving four scenarios for the year 2000. The choice of extending Cases C and D through to 2000 was made after the results of the projections to 1985 became clear. Case E becomes inconsistent before 1985, with demand exceeding supply. Cases A and B show that the high oil price ($17.25 by 1985) is probably higher than needed to maintain sufficient supplies to 1985. Cases C and D, with closely matched supply and demand in 1985, appeared on first analysis to be the most useful to study beyond 1985. Figure 1.2 shows this coupling and the 1985 to 2000 scenarios.

1.2 SUPPLY STUDY METHODOLOGY

The national and global supply studies include an assessment of the potential supply of each major fuel under the assumptions of each WAES scenario. The criteria that these assumptions are based on include the reserve base, expected future resource discoveries, financing requirements, logistics, political and institutional constraints, and environmental standards. On this basis, the *maximum potential supply* of each source in several selected years from 1972 to 2000 is determined. These potential production levels are not forecasts but estimates of the maximum that could be developed if needed, given the specific scenario assumptions such as economic growth, energy prices, and national policy

	Factors That Influence Future Economy	Variables		
1977-1985	World economic growth rate*: high (6%) or low (3.5%)	High		Low
	Oil price: rising ($17.25) constant ($11.50) or falling ($7.66)	17.25	11.50	7.66
	National policy response: vigorous or restrained	Vig		Res
1985-2000	World economic growth rate: high (5%) or low (3%)	High		Low
	Energy price: rising ($17.25/BOE) or constant ($11.50)	17.25	11.50	
	Gross additions to oil reserves: 20 BB/YR or 10 BB/YR	20		10
	OPEC oil ceiling: 45 MBD or 40 MBD	45		40
	Principal replacement fuel: coal or nuclear	Coal		Nuc

		A	B	C	D	E
1977-1985	Growth rate	High	Low	High	Low	High
	Oil price	17.25	17.25	11.50	11.50	7.66
	National policy	Vig	Vig	Vig	Res	Res

1985-2000	Growth rate	High	High	Low	Low	Low
	Energy price	17.25	17.25	17.25	11.50	11.50
	Reserve additions	20	20	20	10	10
	OPEC ceiling	45	45	45	40	40
	Replacement fuels	Coal	Nuc	Coal	Coal	Nuc
		C-1	C-2	D-3	D-7	D-8

* The period from 1973 to the end of 1975 is assumed to correspond to actual world economic conditions, with recovery to 1973 levels by the end of 1976 postulated.

National economic studies done under these assumptions, when summed, actually result in global rates of 5.2 and 3.4% to 1985 and 4.0 and 2.8% from 1985 to 2000.

We assume that the scenario variables are approximately independent of each other, within a certain range of values. Any combination of values *may*, then, be possible and no combination is automatically excluded from consideration.

Figure 1.2
The WAES scenario cases, 1972–1985–2000.

response. The amount actually developed will depend on demand, long-term purchase contracts, prices, policy inducements and constraints, consumer fuel preferences, lead times, and so on. Environmental factors may be critical in some cases, and difficult choices must be made when weighing energy/environmental priorities. Air quality must be maintained, disturbed land reclaimed, waterways preserved, and sufficient energy supplied. The WAES supply assessments assume compliance with current and expected environmental standards.

Each national team studied supply for their country. In addition special global analyses were prepared for oil, gas, coal, and nuclear energy; these studies summarized the national team data and also estimated supplies for the non-WAES countries, in particular the OPEC nations and the non-OPEC developing countries. Individual supply estimates were then totaled to produce a supply profile for each region and for the world as a whole.

Because energy sources differ by nature, three different types of worksheets were developed for use in the WAES national supply studies—one for fossil fuels, one for nuclear (and uranium), and one for nonfossil fuel and conversion processes. The components of these worksheets are described in Section 1.2.1, along with the definitions developed by WAES for each of the terms used in the worksheets.

The fossil fuel worksheet is designed to aid calculation of potential production from data on proven reserves, the rate of additions to reserves, and a minimum reserve-to-production ratio (R/P). The level of proven reserves for each year is determined by adding to the proven reserves at the end of one year the assumed additions to reserves during the next year and then subtracting production during that year. The maximum potential production from the year-end reserves is determined by demand or the assumed ratio of annual proven reserves to annual production (R/P ratio), whichever gives the lower result.

One of the most important factors in this method is the rate at which reserves are added. This rate depends on (1) undiscovered potential resources and (2) the rate at which such resources might be discovered and added to proven reserves. The rate of discovery is heavily influenced by the scenario assumptions. Higher prices and more vigorous government policy to promote supply will bring about greater additions to reserves than lower price and restrained government policy. Further, additions to reserves may be influenced by technological developments in exploration or resource recovery.

The worksheet for nonfossil fuels and conversion processes is used for all sources other than nuclear, including the nonfossil fuel sources such as solar, geothermal, and hydroelectricity as well as those converted from coal, oil sands, and oil shale. An assessment of the "maximum technological potential" of these sources in our scenarios was based on inputs from each of the national teams and special studies prepared by Associates.

The worksheet for nuclear is similar to that for nonfossil fuels and conversion processes; the only difference is that data for uranium are included on the worksheets.

For our estimates of uranium reserves we have relied on the definitions and estimates used by the OECD/IAEA for reasonably assured reserves (RAR) and extracted additional resources (EAR).[1] The estimates of levels of nuclear electric generating capacity on the national worksheets determine annual and cumulative uranium requirements.

It takes about ten years from decision to completion to build a nuclear plant, and from three to ten years to open a major new coal mine and build associated transport and handling systems. Our studies explicitly recognize these lead times. In some cases 1985 fuel supplies are essentially determined by decisions already made. Thus our 1985 estimates, which are based on present output capacities and announced projects to expand them, are made with a fair degree of confidence given the economic, price, and policy assumptions in each scenario.

In our supply estimates to 2000, the reserves and costs for some sources, such as coal, onshore oil, and hydroelectricity, are fairly well defined, and we therefore have a high degree of confidence in these estimates as well. However, the further the projection, even for well-known activities, the less reliable the results. Thus our estimates to 2000 of speculative sources—offshore oil and gas, and sources requiring new technology such as nuclear, solar, and synthetic fuels—are highly uncertain.

Part II contains a more detailed description of the

1. The OECD Energy Agency and the International Atomic Energy Agency. *Uranium: Resources, Production and Demand, Including Other Nuclear Fuel Cycle Data* (Paris: Organization for Economic Cooperation and Development), 1976.

methodology and assumptions used in making projections for the major fuels. The national team reports in Part III also describe in more detail the assumptions and methodology used to derive the national estimates, especially the interpretation of the WAES scenario assumptions and their effect on domestic supply.

1.2.1 Guide to the Supply Worksheets

Indigenous Supply Assessment Worksheets for Fossil Fuels

These worksheets cover oil, natural gas, and coal. The following definitions have been used for the line entries in the worksheets.

Total resource in place: total quantity of the resource physically in existence.

Ultimate recoverable resource: total quantity of the resource that could be extracted under the assumed economic conditions for the relevant time frame and using assumed technologies.

Reserves proved to date: quantity of measured and identified reserves recoverable under the assumed economic conditions for the relevant time frame and using assumed technologies.

Potential production: level of production that would result from optimal reservoir engineering practices based on assumed prices and costs and the reserves proved to date, but not on political factors.

Expected annual production: production expected under the scenario assumptions; can be lower or higher than potential production depending on whether governments choose to implement conservation measures or to produce beyond optimal recovery practices.

Cumulative production: total production to date based on the expected production to date.

Remaining proved reserves: reserves proved to date minus cumulative production.

Reserve/production ratio: remaining proved reserves divided by current annual production.

Capital cost: total capital investment (1975 U.S.$) required to explore for the resource and to develop, extract, and transport (to the collection center) the expected production, expressed in dollars per unit of daily production capacity; includes all relevant costs, such as interest during construction, site acquisition, and "owner's costs" (these numbers are necessarily approximate, but are based on current best estimates).

Host government take: net tax receipts to governments of producing countries from resources produced, in dollars per units of output.

Operating cost: total cost (1975 U.S.$) required for producing the fuel, including extraction and transportation to the collection center (or to the first processing stage) less the capital cost of the producing facilities, expressed per unit of output.

Nonfossil Sources and Conversion Process Worksheets

These worksheets cover hydroelectricity, coal gasification, coal-to-syncrude conversion, gas liquefaction, regasification, and other relevant nonfossil sources and conversion processes (oil shales, oil sands, solar, geothermal). A separate sheet is provided for nuclear that includes information on uranium reserves. The following definitions have been used for the line entries.

Potential production: same as for indigenous supply assessment worksheets.

Expected production: same as for indigenous supply assessment worksheets.

Yield: amount of output energy as a percentage of the raw energy materials extracted.

Maximum potential output: quantity of energy delivered to the energy distribution system upon completion of processing or conversion to final energy form.

Capital cost: total cost of acquiring and constructing facilities needed to produce energy, expressed in dollars per unit of daily production capacity; includes construction costs and interest during the construction stage (because capital accounting practices differ among countries, footnotes explaining assumptions about capital may appear).

Operating cost: total cost (1975 U.S.$), excluding cost of primary energy output (for nuclear electric generation the cost of uranium is included), required for processing and conversion of primary (input) energy, less the capital costs of processing and conversion facilities, all divided by the output energy.

1.2.2 Energy Units

Energy and fuel quantities are measured by a bewildering variety of units. Attempts to achieve wide acceptance and use of a common convention—for example, the International System of Units (SI)—have met with some degree of success. But different people are accustomed to different units of measure, and changes in training, habit, and usage in these matters come slowly.

Not only are different systems of units in use, but there is also little consensus on precisely how to convert from one system to another. Given the vast variety in the qualities and characteristics of energy and fuels in the world, perhaps this is to be expected. WAES has used several units of measure—each national team has done its study in the units most familiar to it—and has adopted one conversion system, shown in Table 1.2.

Two broad classes of units are in use today: energy units and physical units. Energy units are, for example, calories, joules, and British thermal units (Btu's). One Btu is the energy required to raise the temperature of one pound (0.454 kg) of water from 64.5° F to 65.5° F. It is equivalent to 1055 joules; one calorie is 4.2 joules. Physical units are sizes and weights—barrels, gallons, tons, etc. A barrel of crude oil, for example, has a certain energy content; so does a ton of coal. Often energy units are expressed as the energy equivalent of a barrel of oil, or ton of oil, or ton of coal.

In this report the rate of energy use (both for fuels and electricity delivered to consumers and for supply and demand of primary energy) is generally expressed in terms of millions of barrels per day of oil equivalent (MBDOE). This measure is based upon the conventional unit, barrel of oil equivalent, with a gross calorific value of 5.8 million Btu.

In many places throughout the text we have also employed other units of measurement, for example exajoules (10^{18} joules), quadrillion Btu (10^{15} Btu or "Quads"), or units applicable to particular energy sources, such as tons of coal or cubic meters of gas.

An electrical output of 620×10^9 kWh per year (1 MBDOE) would require power stations of 100 GWe (100,000 megawatts) installed capacity, given an average load factor of 70 percent. If the average efficiency of the power stations was 35 percent, one would need a fuel input of 1/0.35 (approximately 2.8) MBDOE. The difference between the input of energy as fuel and the output of energy as electricity (in this example, 1.8 MBDOE) is the transformation loss in electricity generation, which is included in the WAES data under the heading of "processing losses," along with the losses that occur in all other energy conversion processes.

When presenting data for primary energy supply and demand, WAES adopted the convention of expressing electricity from primary sources (nuclear, hydro, geothermal, etc.) in terms of the fuel input (generally expressed in MBDOE) that would be required to produce the equivalent amount of electricity output in fossil-fuel power stations.

Table 1.2
Standard energy conversions (equivalent values lie in vertical columns)

Unit									
Barrels per day of oil equivalent[a]	—	—	—	—	—	—	—	—	—
Tons of oil equivalent[b]	—	—	—	—	—	—	—	0.022	0.023
Metric tons of coal equivalent[c]	—	—	—	—	—	—	—	0.034	0.036
Barrels of oil equivalent[d]	—	—	—	—	—	0.0064	0.02	0.16	0.17
Cubic meters of "average" natural gas[e]	—	—	—	0.027	0.09	1	2.7	25	27
Kilowatthours	—	—	—	0.3	1	11	29	280	293
Cubic feet of "average" natural gas[e]	—	—	—	1	3.4	37.3	100	950	1000
Kilocalories	0.24	0.25	1	252	860	9400	25,200	0.24 million	0.25 million
British thermal units	0.95	1	4.0	1000	3400	37,300	1 therm	0.95 million	1 million
Kilojoules	1	1.06	4.2	1055	3600	39,400	105,500	1 million	1.06 million

Calorific values are measured gross. Rounded equivalents only are given.
[a]Equivalents in other units are shown on a per annum basis.
[b]of 43 million Btu (≈ 10,000 kcal/kg net cal. val.).
[c]of 12,000 Btu/lb = 7000 kcal/kg.
[d]of 5.8 million Btu.
[e]of 1000 Btu/ft^3 or 9400 kcal/m^3.
Source: Energy Conversion Equivalents Table, Shell International Corporation.

								Abbreviation
--	0.003	0.013	0.02	1	2.7	18,000	470 million	BDOE
0.09	0.13	0.65	1	50	135	0.9 million	23 $\times 10^9$	TOE
0.14	0.21	1	1.5	76	209	1.3 million	36 $\times 10^9$	MTCE
0.68	1	4.8	7.4	365	1 TBOE	6.4 million	170 $\times 10^9$	BOE
106	155	745	1150	57,000	0.155 million	1 $\times 10^9$	27 $\times 10^{12}$	Nm^3 NG
1160	1700	8140	12,600	0.62 million	1.7 million	11 $\times 10^9$	290 $\times 10^{12}$	kWh
4000	5800	27,800	43,000	2.1 million	5.8 million	37.3 $\times 10^9$	1 $\times 10^{15}$	ft^3 NG
1 million	1.5 million	7 million	10.8 million	530 million	1.5 $\times 10^9$	9.4 $\times 10^{12}$	250 $\times 10^{15}$	kcal
4 million	5.8 million	27.8 million	43 million	2.1 $\times 10^9$	5.8 $\times 10^9$	37.3 $\times 10^{12}$	1 Q	Btu
4.2 million	6.1 million	29.3 million	45.4 million	2.2 $\times 10^9$	6.1 $\times 10^9$	39.4 $\times 10^{12}$	1.06 $\times 10^{18}$	kJ

— = insignificant
1 therm = 100,000 Btu
1 MBDOE = 10^6 BDOE
1 TBOE = 1000 BOE
1 Q = 10^{18} Btu
1 Quad = 10^{15} Btu

Summary of the Supply Studies

2.1 BASE YEAR (1972)

In 1972 energy production in the world outside Communist areas (WOCA) was 82 million barrels per day of oil equivalent. About 50 percent of the energy produced was oil, 20 percent gas, 17 percent coal, and the remainder hydro and nuclear. This is a dramatic change from the turn of the last century, when coal provided about 95 percent of all energy. The transition from a coal era to an oil and gas era has spanned the last seventy-five years. In this volume we look ahead twenty-five years and attempt to answer the following questions:

1. To what degree can we rely on oil in the future? To what degree may oil production be constrained by resource availability or by political decision?

2. Gas reserves are estimated by various experts to be two-thirds as large as oil reserves. What are the problems and potentials of increasing gas production and transporting it to the major energy-consuming regions?

3. What can we expect nuclear energy to contribute? On what scale? How soon? How much will it cost? What are the uncertainties?

4. Coal reserves are large. What are the costs and potential problems of increasing coal production? Will coal reverse its historic trend toward lower relative use?

5. What will the contribution of renewable energy sources such as solar, wind, tidal, and nuclear fusions be by the end of the century?

Base year (1972) supply data by region are shown in Table 2.1. Most of the energy was produced in North America and the rest of WOCA (which includes the OPEC countries). The percentage of demand met by indigenous supply varies from region to region. North America supplied about 90 percent of its own needs, Western Europe 38 percent, and Japan 14 percent.

Although the major consuming countries of WOCA (North America, Western Europe, and Japan) consume nearly 80 percent of the WOCA energy, they produce only about 50 percent of WOCA energy.

Our supply analyses can be viewed in two parts. The first are supply options and expectations for North America, Japan, and most countries of Western Europe. The supply reports prepared by the national teams from these regions that participated in WAES can be found in Part III of this volume. The second part includes the supply prospects for non-WAES countries—largely OPEC and non-OPEC developing countries. Estimates of future supply potential in these countries are much more difficult and we have considered a wide range of production possibilities. The energy supply prospects for non-WAES countries are discussed in the global resource studies in Part II of this volume.

2.2 THE RESOURCE BASE

Figure 2.1 shows and estimate of proven reserves for the major fossil fuels in WOCA. Proven reserves are those that have been measured and are considered exploitable under present economic conditions. The most important question is, At what rate can additional reserves be proven and brought into production? Production potential, although constrained by resource availability, depends in the nearer term on discovery rates, improvements in recovery techniques, and the development of infrastructure for exploiting, transporting, and using a particular fuel. Finally,

Table 2.1
WOCA supply of major fuels 1972 (MBDOE)

	Oil	Natural gas	Coal	Nuclear	Hydro	Other	Total
North America	12.68	12.03	6.9	0.33	2.42	0.01	34.37
Western Europe	0.34	2.12	3.8	0.32	1.63	0.14	8.35
Japan	0.02	0.05	0.35	0.04	0.41	—	0.87
Rest of WOCA	30.60	1.62[a]	3.14	0.02	1.13	—	36.51
Total	43.64	15.82	14.19	0.71	5.59	0.15	80.10

[a]Not including gas flared or vented.

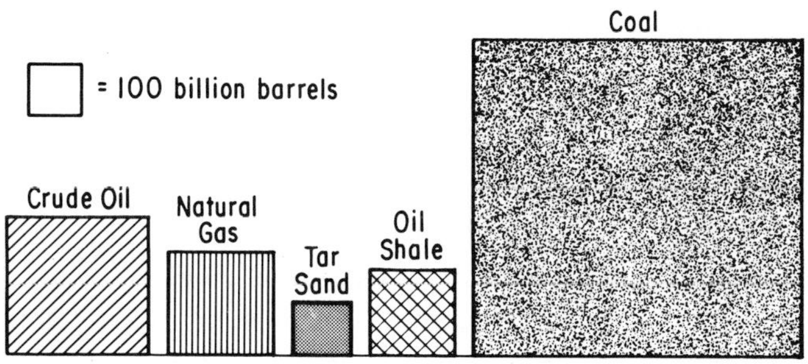

Figure 2.1
Comparison of proven recoverable world reserves of fossil fuels (including Communist areas).

potential also depends on political decisions regarding both production levels and prices for exported energy.

2.3 SUMMARY OF MAJOR RESOURCES

2.3.1 Oil

Two of the most critical questions facing the world are, how long will the production of oil continue to increase, and how soon can dependence on it be reduced. WAES has given much attention to these questions, and the results are reported in Chapter 3.

The principal factors affecting future oil availability are:

- The minimum R/P ratio.
- The magnitude of proven reserves.
- The rate at which these reserves increase as a result of new discoveries or improved production techniques.
- The future level of oil consumption.
- The rate of oil production that exporting countries permit.

Alternative assumptions are made about each of these factors, and oil production profiles that meet oil demand as long as possible are developed.

The main WAES conclusion about oil is that if all the oil-producing countries of WOCA allow production to increase to meet world demand, limited only by technical factors, oil production will peak and decline before the end of the century. This holds true over a wide range of assumptions. However, limitations on production may be imposed, in particular the level at which OPEC sets their production (reflecting their views on acceptable depletion rates). If recent government pronouncements about production levels are adopted, then production could peak early in the 1980s and stay at that level until the end of the century and beyond. Peak production could be delayed until the late 1980s if these production limits are higher. In this case, oil production would probably remain on a plateau through the 1990s.

These results of our analysis are illustrated in Figures 2.2 (Cases C, C-1) and 2.3 (Cases D, D-8) for a wide range of assumptions. In all our projections oil continues to be an important fuel in global energy supply. Yet in every case a production

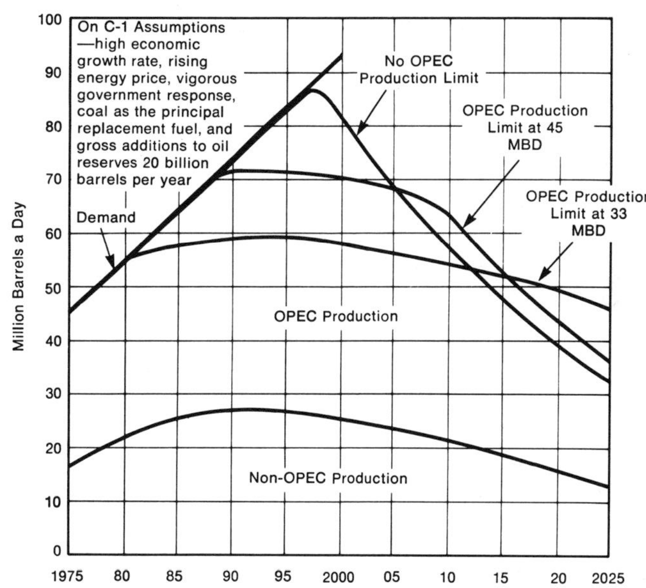

Figure 2.2
WOCA oil production, Case C-1.

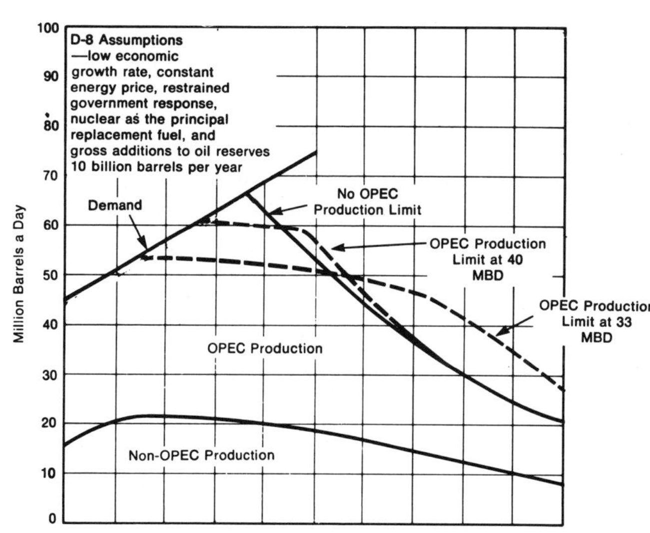

Figure 2.3
WOCA oil production, Case D-8.

peak or plateau occurs before the end of the century—in some cases well before. If technical considerations alone determine production levels, and economic and policy responses are ignored, the peak of oil production may be unrealistically sharp and the decline may be very steep. Even under such assumptions the supply of oil is not sufficient to meet the desired demand.

The oil demand estimates used for these projections are much lower than the 6.2 percent annual increase that prevailed during the 1960s and early 1970s. Our analyses show that annual growth rates for oil demand are likely to decline to between 2.5 and 3.4 percent in the 1975 to 1985 period and to between 1.5 and 2.6 percent in the period 1985 to 2000.

In 1975 proven reserves of oil in WOCA were estimated to be 555 billion barrels. A critical factor is the rate at which these reserves are increased or decreased. In Figure 2.2 the level of gross additions to oil reserves has been assumed to be 20 billion barrels per year, slightly above the historical rate of 18 billion barrels per year from 1940 to 1975 (based on backdating field extensions to the year in which a field was discovered). On this basis, and with production eventually constrained by a 15 to 1 ratio of remaining reserves to production, production would peak around 1997 at about 85 MBD (including OPEC production of 57 MBD). Such a rapid increase and decline of production are unrealistic, on both technical and economic grounds.

The figure for gross additions to oil reserves of 20 billion barrels per year includes an allowance for enhanced recovery. Such improved recovery could add significantly to oil reserves, but extensive and costly research and development will be necessary. Improved recovery is more likely to permit extension of the oil production plateau further into the twenty-first century than to increase the height of the peak of the "oil mountain."

Some OPEC countries may decide for economic, conservation, or other reasons to limit their production to a level below the theoretical maximum. Such action might result in a maximum production of 40 to 45 million barrels per day from the OPEC countries. These levels could probably be maintained for a number of years without running into technical limits if reserve additions continue to be large and steady. The resulting production profile is shown in Figures 2.2 and 2.3. This profile shows not a global peak but a plateau (at around 70 MBD in Case C-1) that continues from 1989 to well beyond 2000, when technical limits (reserve/production ratios) force a reduction in production.

In the C-1 profile (OPEC production at 45 MBD) Saudi Arabia, which has more than 25 percent of known world oil reserves, is assumed to be prepared to increase output from its present rate of about 9 MBD to as much as 20 MBD; but the level to which Saudi Arabia will wish to increase its rate of oil production is uncertain because of the financial and investment implications of revenues from such production levels.

The third curve in Figures 2.2 and 2.3 shows a projection of production levels based on the assumption that Saudi Arabia and some other OPEC countries restrict production to slightly above present production levels. This would mean a maximum OPEC potential of 33.0 MBD. Such a policy, if taken, could lead to a leveling of world oil production in the early 1980s and thus a failure to meet the projected demand for oil much earlier than expected.

The growth era in oil production will therefore end in no more than fifteen years. However, there may be a decade of constant oil production after 1990 when consumers can adjust to declining oil supply provided they have prepared for such a transition in the 1970s and 1980s. There may be additional major oil discoveries or much higher recovery factors than have been assumed, but such favorable developments would only delay, not remove, the problem. Nations that continue to increase their oil consumption in the hope that more optimistic estimates will prove correct risk losing time to adjust their energy consumption patterns if their optimistic expectations are not met.

2.3.2 Natural Gas

Natural gas is a clean and convenient fuel for heating and a valuable raw material for the large and growing petrochemicals industry. World reserves are large and unlikely to limit production within the next twenty-five years, but the future role of gas as an energy source is more likely to be determined by the problems of transporting and distributing gas from the wellhead to the consumer and by the attitudes of producers toward the export of gas.

In contrast to oil, which supplies over half of the energy in almost every major energy-consuming country, natural gas supplies varying amounts of energy—from nearly zero in Sweden, Denmark, and Japan to about 30 percent in the United States and 47 percent in The Netherlands. In North America, where large reserves of gas were discovered with the early development of the oil industry and where prices have been kept low compared with oil, natural gas has become a major source of energy. In Western Europe the natural gas industry was of minor importance until the early 1960s when the discovery and development of the Groningen field stimulated gas consumption in The Netherlands and in surrounding countries. Only negligible amounts of natural gas were consumed in Japan until the late 1960s when the development of liquefied natural gas (LNG) transportation systems permitted imports from Alaska and Brunei. Although the OPEC countries have about 55 percent of WOCA gas reserves, development of natural gas for internal use or export has been very limited. Often gas has to be flared, although its reinjection into oil wells to improve oil recovery is increasingly important.

Much of the world's major undeveloped gas reserves are from present and potential markets. Figure 2.4 shows the contrast between the location of natural gas reserves and consumption of gas in 1975. North America has the most prominent role, accounting for some 68 percent of WOCA consumption out of its own reserves, which are now being depleted more rapidly then new reserves are developed. In Western Europe gas consumption has increased rapidly in the last decade. However, the major Groningen and southern North Sea fields are likely to decline in the early 1980s, although reserves in the northern North Sea could postpone the decline in total Western European production to the 1990s. Japan, which lacks any significant reserves, is developing a natural gas industry based on imported LNG to diversify its energy sources and to take advantage of this relatively pollution-free fuel.

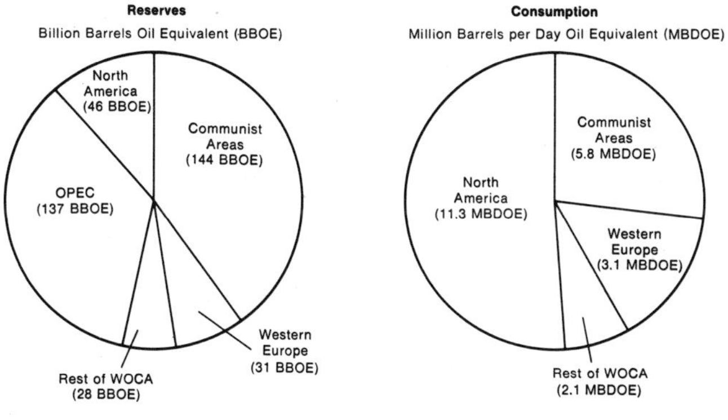

Figure 2.4
Proven reserves and world consumption of natural gas in 1975 (including Communist areas).

By the year 2000 natural gas production in the major consuming countries will probably decline to about two-thirds of the 1972 level. However, demand for natural gas in Western Europe, Japan, and North America is expected to increase, but at a fairly slow rate compared with the growth over the last fifteen years. North America, Western Europe, and Japan together might desire total natural gas imports of about 3.3 MBDOE in 1985 and 8.5 MBDOE by the year 2000. The resources to fill these demands are available and lie primarily in the U.S.S.R. and OPEC. The key question is whether the gas can be moved from the major resource areas to the major consumers.

In 1975 intercontinental gas trade was only 0.3 MBDOE. To achieve a level of 8.5 MBDOE by the year 2000 would require a 30-fold increase over today's levels.

Natural gas has historically moved directly from supplier to consumer through pipelines. Gas has been used on a large scale only in markets that could be economically connected by pipeline to gas reserves.

An alternative to pipelines is transport by tanker as liquefied natural gas. The liquefied gas is carried in refrigerated tankers at $-161°C$ and regasified at receiving terminals. A drawback to LNG is that about 25 percent of the primary energy is lost in processing and transport. Moreover, LNG requires large capital investments (over $10,000/BDOE) at each stage of production and transport. Large numbers of LNG shipments also pose a special environmental problem arising from the risk of tanker collisions in port areas. Such concerns could slow the growth of LNG developments.

The conversion of natural gas to methanol may be more economical and convenient than LNG for transport over long distances because capital costs are lower and conventional tankers and terminals can be used. However, the energy efficiency of the entire system is substantially lower.

If all potential projects now planned are completed, LNG trade could increase from about 0.3

MBDOE in 1975 to as much as 3.1 MBDOE in 1985. Projects currently under construction would furnish about 25 percent of that increase. The remainder would have to come from projects now classified as potential—many only in the early stages of planning. About 80 percent of such LNG shipments would need to come from OPEC countries.

In addition, there are plans to import gas into Western Europe by pipeline from the Middle East and North Africa. By 1985, if all planned projects are completed, about 0.3 MBDOE of natural gas would be available. Lastly, there is the potential importation of 0.5 MBDOE into Western Europe of gas from the U.S.S.R. by 1985. If all such potential gas trade projects are realized, international trade in 1985 could be as high as 4.1 MBDOE, more than enough to meet consuming countries' import requirements.

To satisfy import needs of 8.5 MBDOE in year 2000, major commitments must be made to increase the LNG system capacity beyond the maximum levels planned for 1985. We have assumed that imports from the U.S.S.R. into Western Europe could reach 1.0 MBDOE by 2000. Further imports will be limited by likely increases in demand for natural gas within the U.S.S.R. The remaining import requirements of about 7.5 MBDOE would have to come largely from OPEC countries.

Proven reserves of natural gas are sufficient to support a large expansion in international trade in gas, principally as LNG, and because it is more desirable as a fuel than oil and coal, it may command a premium price. However, some uncertainties about the growth of world gas trade arise from (1) the attitudes of the governments of OPEC countries toward export of gas versus domestic use, including use as a chemical feedstock, (2) availability of capital for investment in LNG systems, and (3) the risks of LNG tanker accidents.

2.3.3 Coal

Today's leading coal producers are the major industrialized countries of the Western World and the Communist bloc. Three countries (the United States, the U.S.S.R., and China) account for nearly 60 percent of world output. In the developing countries, India is the only large producer and consumer. The single largest producer and consumer of coal in the non-Communist world is the United States, with nearly half the production and over half the known reserves.

In 1974, 1200 million metric tons, or 15.2 MBDOE of coal were consumed in WOCA, representing about 18 percent of the primary energy. This is a sharp decline from 34 percent in 1960, although total coal consumption was marginally higher in 1974 than it was in 1960. Coal use has been declining in all market sectors except electricity generation and the iron and steel industry.

The world possesses vast reserves of coal, far greater than those of any other fossil fuel. Economically recoverable world proven reserves of coal are about 600 billion metric tons of hard coal equivalent, or 3000 billion BOE. Potential reserves are much greater, possibly as much as 12,000 billion BOE. This resource base is sufficient to support massive development of coal well into the next century. The amount of coal actually produced, however, will be governed by the level of coal demand, the development of infrastructure for transportation, handling, storage, and use, and the resolution of environmental, social, and economic policy issues related to coal production and use.

The analysis in Chapter 5 shows that under the WAES assumptions about energy prices, economic growth, and national policies, WOCA coal production could expand threefold to 3.2 billion metric tons a year by 2000, with 2.0 billion tons in North America.

But such figures are for potential production. The amount that will actually be produced depends on whether, when, and on what scale decisions and commitments are made. To reach such high production levels, major investments will be needed for mines, coal handling equipment, and transportation systems; consumers will have to adapt to a new fuel. A detailed study of U.S. coal production potential, outlined in Chapter 5, shows that by year 2000 significant new and expanded transportation facilities (unit trains, conventional trains, rail lines, barges, trucks, and slurry pipelines) would have to be built to move the coal from mines to markets and to ports for exports.

One problem that would have to be solved if coal is to be a major alternative to oil is the maintenance of environmental standards. A better understanding of the long-term climatic effects of fossil fuel combustion is essential. Surface-mined areas will have to be reclaimed, and sufficient water must be made available for mining operation and land restoration. Many environmental, economic, and social issues require resolution before a policy of mine expansion and increased domestic use and export would be acceptable.

Our studies reveal a major conflict between desire and reality in the case of coal just as with oil, but with opposite effects. National preferences for oil, as we have seen, produce prospective shortages. National distaste for coal produces a potential surplus. Our projections of "desired demand"—the mix of fuels each nation would like to have (at given price, growth, and policy assumptions)—show a total coal demand well below levels needed to avoid an overall energy supply gap.

Large increases in coal demand will require action to encourage its use, ways to simplify its handling, and means to prevent excessive air pollution. Distribution systems must be built, new coal-fired equipment designed and installed, and techniques for controlling air pollution substantially improved. The price of coal in the market must be sufficiently below that of competing fuels to justify the additional costs of new coal handling and using equipment as well as pollution control facilities.

The alternative of converting coal to synthetic oil or gas may also be considered, although this option costs more than using coal directly. It avoids the conversion of existing oil and gas distribution and consuming systems and may simplify pollution control by concentrating coal use at a small number of sites.

The scale of the coal challenge is large and so are the potential benefits. It challenges our ability to anticipate and collaborate. Coal will not be produced in time, in sufficient quantities, unless potential exporting countries make firm commitments very soon. It will not be exported in sufficient quantities, in time, unless consumer countries recognize the need for coal imports well in advance and place firm long-term contracts. This recognition may not occur unless potential users believe that the coal will be available, that it can be burned cleanly, and that the governments and people of coal-producing/exporting countries will not interrupt the flow of coal exports.

2.3.4 Nuclear

Some people believe that nuclear power, by supplying a large fraction of the world's electricity, will substantially ease the pressure on fossil fuel resources by the end of the century. Their confidence is based on the relatively low cost of nuclear electricity and the outstanding safety record of operation of nuclear reactors and related facilities over the last twenty-five years. Yet the unique character of nuclear energy as a source of radioactivity and its potential for destruction have led to resistance in many countries to the growth of the nuclear power industry. The public debate on such issues is widespread and goes on in many countries at different levels of intensity. Debate has focused around several serious concerns. Probably the most serious is containing radioactivity. The safe transport, storage, and treatment of spent fuel elements and the resulting highly radioactive wastes, which remain active for hundreds and thousands of years, is a matter of great public concern. These concerns and others are discussed in Chapter 6.

WAES did not attempt to seek agreement on how and when the nuclear debate would be resolved in various countries. Instead, we decided to show the scale of potential contributions to the world's primary energy needs in 1985 and year 2000 that could come from an intensive and continuing nuclear power program. In this way, we could show how the pace of nuclear development would affect the total supply of primary energy required to meet the world's future energy needs.

To estimate the maximum contribution from nuclear energy we asked each WAES national team to determine, in the general context of our scenarios to 1985 and 2000, the "maximum likely" and "minimum likely" levels of installed nuclear electric capacity. The high nuclear level assumes that nuclear will be the principal replacement for fossil fuels and that present political and technical obstacles will be quickly overcome. The lower nuclear level also assumes early resolution of obstacles that now delay nuclear growth in many parts of the world, but is based on coal as the principal replacement fuel for oil. In many cases these high and low projections are based on national forecasts. After adding the national projections we estimated growth in the rest of the world to determine what nuclear energy might contribute to world primary energy in the year 2000. Such estimates also allowed us to show quantitatively how much energy from other fuels would be required if the supply of nuclear energy fell short of such WAES projections.

The reader can form his or her own conclusions about the likely outcomes and their impacts on the probability that the WAES nuclear projections will be achieved. If nuclear energy supply turns out to be lower than the WAES projections, then it will be necessary to consider carefully how the shortfall might be met—from what other fuels, or with what additional conservation measures.

The maximum likely and minimum likely WAES projections of future installed capacity to 2000 are shown in Table 2.2; more detailed national and regional estimates of future installed nuclear capacity are contained in Chapter 6. The maximum projection would require an annual growth rate of 14 percent per year for twenty-five years. The minimum projection would require an annual growth rate of 11 percent. While these numbers may seem high, it must be recalled that the potential growth is from a comparatively low base.

Such rapidly rising levels of nuclear energy would permit this source to supply as much as 21 percent of the world's primary energy in 2000, an amount equivalent to 43 MBDOE or the energy contained in total non-Communist world oil production in 1975 or in 3 billion tons of coal per year. Even with this sizable nuclear contribution, over three times as much primary energy in 2000 would have to be met by other fuels.

The analyses in Chapter 6 suggest that with political will and firm, prompt actions on the many technical problems, WAES projected nuclear expansion is technically feasible in terms of uranium, physical facilities, and other critical resources. To achieve such nuclear growth requires action soon. Reactors would have to be built, and uranium exploration and mining would have to be rapidly expanded, as would uranium enrichment and fuel reprocessing facilities. Further research and demonstration are required before the most acceptable reprocessing methods and the most suitable sites for waste disposal can be selected. Without such research and construction the projected nuclear expansion might be delayed several years.

The nuclear issue does not have to involve an all or nothing choice. Chapter 6 calls attention to three stages of choice in the nuclear fuel cycle, pointing out the possibility of considering separately (a) reactor operation involving a single use of uranium without fuel reprocessing, (b) the processing of used fuel to extract and recycle plutonium and uranium, and (c) operation of fast breeder reactors. Much of the nuclear debate centers on stages (b) and (c). There is very little debate centering on stage (a), yet in some places approval to start construction of nuclear power plants to implement stage (a) is withheld because issues in stages (b) and (c) are unresolved. Separation of these stages of choice could clarify issues and might allow nuclear power plant construction to continue (with its six to ten year lead time) while the issues of stages (b) and (c) are being resolved. Some countries have already selected a course of action; others will have to choose.

The period beyond 2000 involves other challenges that must be faced in the next few years. High nuclear expansion can be maintained only if fast

Table 2.2
Summary of WAES estimates of installed WOCA nuclear power capacity

	1974	1985		2000	
	Installed capacity	Maximum likely	Minimum likely	Maximum likely	Minimum likely
Nuclear capacity (GWe)	66.9	412	291	1772	913
Share of total primary energy consumed in WOCA (%)	2	9	6	21	14
Oil equivalent (MBDOE)	1.7	10	7	43	22

Note: We have adopted the convention of expressing nuclear electricity production in terms of the fuel input that would be required to produce the equivalent amount of electricity output from fossil-fueled power stations. The assumed generating efficiency is 0.35 in 2000. The formula used to convert GWe to MBDOE, assuming 0.35 efficiency and a 0.6 load factor, is

$$\text{GWe} \times \frac{8700}{620{,}000} \times \frac{1}{0.35} \times 0.6 = \text{MBDOE primary energy input.}$$

breeder reactors—which both "burn" and produce nuclear fuels—become commercial in the period 1990 to 2000. At the same time very large additional reserves of uranium must be found and developed. France, the United Kingdom, the United States, Japan, the Debenelux venture (Germany, Belgium, The Netherlands, and Luxemburg), and an international venture comprising electricity companies from France, Italy, and Germany are developing fast breeder reactors. In the United Kingdom and the United States progress is very slow and surrounded by controversy. The technology is new and the lead times long; the scale of expansion by 1990 is thus uncertain.

Overall, it seems unlikely that breeder reactors will provide more than 5 percent of nuclear energy in the world outside Communist areas by the year 2000. Most people have little hope that nuclear fusion will provide any energy in the twentieth century.

Uncertainties surround all of our estimates of demand and supply to 2000. Because different countries may choose different nuclear policies, the range of uncertainty in our nuclear projection is greater than for other fuels. On the one hand, extended delays in nuclear programs in some countries could hold nuclear power to the levels projected for 1985; these levels are based on commitments and construction already underway in most cases. On the other hand, a new sense of awareness of the imminence of a deeper and continuing energy shortfall arising from reduced oil supplies might lead to a public reappraisal of the risks and benefits of nuclear energy and a decision to accept the risks. All that we can do in this report is to show the scale of the contribution nuclear could make in 2000 and also to describe the issues in the public debate that will influence each country's political decision on nuclear risks.

2.3.5 Hydroelectricity and Geothermal

Hydroelectric power is a major energy source in many countries today and will continue to play an important role in the future. Major hydroelectric expansion is most likely to occur in the developing areas of the world, where it is estimated that only about 4 percent of the potential is currently developed. Although the potential for hydroelectric expansion is vast in the developing countries, growth is somewhat constrained by the potential sites and by the long lead times necessary for construction. We have therefore made conservative estimates of the future growth of hydroelectric power in the developing countries, assuming primary electricity production to increase from about 1 MBDOE in 1972 to a maximum production of about 4.5 MBDOE in 2000. The estimate could be higher if a number of developed countries, because of energy shortages or for environmental reasons, choose to transfer some energy-intensive industrial activities (such as aluminum production) to developing countries that have abundant hydro resources, the access to required raw materials, and wish to develop such industries.

Growth of hydroelectricity in the developed world—which presently accounts for over 80 percent of WOCA hydroelectric capacity—will be lower than in the developing countries because the most favorable dam sites are already developed and because environmental concerns may limit development of some potential sites. New technologies such as axial flow turbines might extend the number of developable sites. Our projections show a primary hydroelectric growth in Western Europe, Japan, and North America from 5 MBDOE in 1972 to a maximum of 7.5 MBDOE in the year 2000.

Total installed global geothermal electricity generating capacity was only 1400 MWe in 1975. Geothermal energy can be tapped from either natural steam resources or hot dry rocks. The first are presently economically competitive (as exemplified by the Geysers in the United States and the Larderello fields in Italy), but have a limited resource base since they require the relatively rare geologic combination of hot rocks, an underground water system, and an impermeable cap rock for trapping the steam and providing pressure. It has been estimated that exploitation of energy from hot dry rocks could greatly increase geothermal resources, although the technology is still at an early stage of development.

2.3.6 Other Fossil Fuels

The other fossil fuels—heavy oil, oil sands, and oil from shale—are of immediate interest because they can be converted into liquid fuels similar to those obtained from conventional crude oil and can be fed into the existing energy infrastructure. The resources of such fossil fuels are large compared with resources of conventional oil, but current production is less than 0.2 MBDOE. The analyses of Chapter 7 show that they may contribute modest amounts of energy—perhaps 3 MBDOE—by the turn of the century. The capital and operating expenses to produce such oil are greater than for conventional oil, and developing some of these sources presents significant environmental problems that must be resolved before production is greatly expanded. World oil prices higher than at present may be needed to stimulate large increases in production of these fuels.

Heavy oil and oil sands, the two major "other fossil fuel" reserves, are found principally in Venezuela and Canada. The resource base in Canada and Venezuela totals about 2000 billion barrels of oil, but only 700 to 800 billion barrels are considered ultimately recoverable. In Canada about 28 billion barrels of oil are recoverable using surface mining methods. Most of this vast resource will be available only through in situ recovery methods. Before there is large-scale production, serious social, financial, technical, and environmental problems must be solved. Projections in the WAES scenario indicate a possible Canadian production of 0.8 MBDOE by 2000. Venezuelan estimates of production in the year 2000 are of similar magnitude.

The resource base in these countries implies that production could be many times these levels, but this would require vigorous technical and financial action on a scale not yet in evidence.

The largest known oil shale reserves are found in the United States, with significant amounts also in Brazil, the U.S.S.R., and China. Smaller quantities are found elsewhere, for example in Sweden. Development activities to date suggest that various technologies could be used to extract oil from oil shale at a substantially higher cost than the current price of imported oil. Government subsidies for at least the early stages of development appear necessary. Environmental problems must be solved before large-scale production can occur. A maximum worldwide production of oil from shale of about 2 million barrels a day in year 2000, mainly in the United States, is projected in our rising price scenarios.

2.3.7 Solar and Other Renewables

Solar heat, solar electric, wind, tidal, and other renewable energy sources are beginning to attract research and development efforts because of their obvious longer-term attraction as clean and sustainable energy sources of considerable size. For example, solar water heating and space heating are economically competitive now in Israel. Applications like these could become more competitive as other forms of energy become scarcer. In other alternative technologies investment costs are still unattractive, operating costs and maintenance costs are uncertain, and difficult technical problems must be resolved (particularly storage of energy for sunless or windless periods). Furthermore, there are marked differences between countries, even between regions only one hundred miles or so apart, in solar or wind "climates" and therefore in the cost-effectiveness of these sources.

These factors make it extremely difficult to assess how rapidly renewable sources will expand. The WAES national studies show a maximum contribution from solar energy (both thermal and electric) in the developed countries of about 2 MBDOE by the year 2000. This figure is uncertain; if the price of oil were to double in real terms, the contribution of renewable sources could be much higher. Balance-of-payments problems might accelerate action. Major technical and cost breakthroughs could also bring on these sources more rapidly than we have assumed.

Although the aggregate amounts of energy available from these sources may be relatively small in the near future, they can be important in human terms because of the large number of people, particularly in the developing world, who could benefit from them. Other advantages may lie in the decentralization of energy production and use offered by such systems.

Renewable sources should be researched, developed, and demonstrated with a high priority beginning as soon as possible. They have a critically important role to play beyond 2000 as oil and natural gas decline further, particularly if nuclear and coal are limited by resource and/or environmental and safety constraints. The potential for solar and other renewables is discussed further in Chapter 7.

2.4 GLOBAL ENERGY SUPPLY DATA BASE (TABLES 2.3 TO 2.11)

The global energy supply data base is the accumulation of the national supply data submitted by WAES national teams. It also includes the results of special WAES assessments of energy supply in the non-WAES countries and regions. These supply estimates represent the *maximum potential supply* given the specific scenario assumptions. Five 1985 scenario cases, labeled A-E, plus four 2000 scenario cases, C-1, C-2, D-7, and D-8, are reported. The data are presented in million barrels per day of oil equivalent. Only data for the major fuels—oil, natural gas, coal, nuclear, and hydroelectricity—are shown. Data for other fuels such as geothermal and renewables are reported in individual country worksheets.

There are minor inconsistencies between some of the data in this data base and other WAES reports. However, all discrepancies are the result of rounding, energy unit conversions, or differing load factors assumed in the generation of electricity. In no case do these discrepancies result in material differences in the results of the analysis or have an impact on the conclusions of the Workshop.

Table 2.3
Global data supply base, 1985, Case A (MBDOE)

	Oil	Natural Gas	Coal	Nuclear (primary)	Hydro (primary)
Canada	3.44	1.79	0.66	0.29	1.38
United States	12.16	9.62	12.27	4.02	2.18
Total North America	15.60	11.41	12.93	4.31	3.56
Denmark	0.01			0.02	
Finland			0.02	0.07	0.06
France		0.18	0.26	1.09	0.28
Federal Republic of Germany	0.08	0.28	1.69	0.80	0.08
Italy	0.06	0.23	0.02	0.23	0.15
The Netherlands	0.04	1.51		0.08	
Norway	0.77	0.38	0.02		0.46
Sweden				0.18	0.30
United Kingdom	3.15	0.95	1.50	0.36	0.02
Rest of Western Europe			0.80	0.75	0.70
Total Western Europe	4.11	3.53	4.31	3.58	2.05
Japan	0.08	0.07	0.26	0.97	0.54
Australia, New Zealand, and South Africa	0.50	0.63	2.63	0.07	0.25
Mexico[a]	1.87	0.55	0.13	0.07	0.09
OPEC Countries	40.00	5.47	0.42	0.10	0.10
Non-OPEC Developing Countries	6.70	2.40	2.67	1.10	1.70
Total World	66.99	23.51	23.22	10.13	8.20

Note: Blanks denote insignificant production.
[a]Mexico's total is included in the non-OPEC developing country total.

Table 2.4
Global data supply base, 1985, Case B (MBDOE)

	Oil	Natural Gas	Coal	Nuclear (primary)	Hydro (primary)
Canada	3.20	1.79	0.66	0.29	1.38
United States	12.40	9.62	11.21	4.02	2.18
Total North America	15.60	11.41	11.87	4.31	3.56
Denmark	0.01			0.02	
Finland			0.02	0.07	0.06
France			0.26	1.09	0.28
Federal Republic of Germany	0.08	0.41	1.69	0.80	0.08
Italy	0.06	0.23	0.02	0.23	0.15
The Netherlands	0.04	1.51		0.08	
Norway	0.77	0.38	0.02		0.46
Sweden				0.18	0.30
United Kingdom	3.15	0.87	1.50	0.36	0.02
Rest of Western Europe			0.80	0.75	0.70
Total Western Europe	4.11	3.40	4.31	3.58	2.05
Japan	0.08	0.07	0.26	0.97	0.54
Australia, New Zealand, and South Africa	0.50	0.59	2.37	0.06	0.23
Mexico[a]	1.33	0.43	0.13	0.07	0.09
OPEC Countries	40.00	5.08	0.30	0.10	0.10
Non-OPEC Developing Countries	6.70	2.10	2.24	1.10	1.70
Total World	66.99	22.65	21.35	10.12	8.18

Note: Blanks denote insignificant production.
[a]Mexico's total is included in the non-OPEC developing country total.

Table 2.5
Global data supply base, 1985, Case C (MBDOE)

	Oil	Natural Gas	Coal	Nuclear (primary)	Hydro (primary)
Canada	2.84	1.31	0.53	0.29	1.38
United States	11.20	9.14	11.21	4.02	2.18
Total North America	14.04	10.45	11.74	4.31	3.56
Denmark	0.01			0.02	
Finland			0.02	0.07	0.06
France		0.13	0.26	1.09	0.28
Federal Republic of Germany	0.08	0.28	1.69	0.80	0.08
Italy	0.06	0.23	0.02	0.23	0.15
The Netherlands	0.04	1.51		0.08	
Norway	0.77	0.38	0.02		0.46
Sweden				0.18	0.30
United Kingdom	3.15	0.95	1.50	0.36	0.02
Rest of Western Europe			0.80	0.75	0.70
Total Western Europe	4.11	3.48	4.31	3.58	2.05
Japan	0.08	0.07	0.26	0.97	0.55
Australia, New Zealand, and South Africa	0.50	0.63	2.63	0.07	0.25
Mexico[a]	1.52	0.52	0.15	0.07	0.09
OPEC Countries	40.00	5.47	0.42	0.10	0.10
Non-OPEC Developing Countries	6.70	2.40	2.67	1.10	1.70
Total World	65.43	22.50	22.03	10.13	8.21

Note: Blanks denote insignificant production.
[a]Mexico's total is included in the non-OPEC developing country total.

Table 2.6
Global data supply base, 1985, Case D (MBDOE)

	Oil	Natural Gas	Coal	Nuclear (primary)	Hydro (primary)
Canada	2.40	1.15	0.45	0.19	1.38
United States	9.50	8.43	9.43	3.07	1.62
Total North America	11.90	9.58	9.88	3.26	3.00
Denmark	0.01			0.02	
Finland			0.02	0.07	0.06
France		0.13	0.26	0.85	0.28
Federal Republic of Germany	0.08	0.28	1.51	0.51	0.08
Italy	0.06	0.23	0.02	0.13	0.11
The Netherlands	0.04	1.51		0.05	
Norway	0.77	0.38	0.02		0.46
Sweden				0.09	0.30
United Kingdom	3.15	0.87	1.50	0.34	0.02
Rest of Western Europe			0.80	0.56	0.70
Total Western Europe	4.11	3.40	4.13	2.62	2.01
Japan	0.08	0.07	0.26	0.61	0.54
Australia, New Zealand, and South Africa	0.50	0.59	2.37	0.06	0.23
Mexico[a]	1.38	0.47	0.13	0.05	0.09
OPEC Countries	36.00	5.08	0.30	0.10	0.10
Non-OPEC Developing Countries	5.10	2.10	2.24	0.40	1.80
Total World	57.69	20.82	19.18	7.05	7.68

Note: Blanks denote insignificant production.
[a]Mexico's total is included in the non-OPEC developing country total.

Table 2.7
Global data supply base, 1985, Case E (MBDOE)

	Oil	Natural Gas	Coal	Nuclear (primary)	Hydro (primary)
Canada	1.58	1.11	0.39	0.19	1.38
United States	8.00	6.74	9.43	3.07	1.62
Total North America	9.58	7.85	9.82	3.26	3.00
Denmark	0.01			0.02	
Finland			0.02	0.07	0.06
France		0.13	0.26	0.85	0.28
Federal Republic of Germany	0.08	0.28	1.51	0.51	0.08
Italy	0.06	0.23	0.02	0.13	0.11
The Netherlands	0.04	1.51		0.05	
Norway	0.77	0.38	0.02		0.46
Sweden				0.09	0.30
United Kingdom	3.15	0.95	1.50	0.34	0.02
Rest of Western Europe			0.80	0.56	0.70
Total Western Europe	4.11	3.48	4.13	2.62	2.01
Japan	0.08	0.07	0.26	0.61	0.54
Australia, New Zealand, and South Africa	0.50	0.63	2.63	0.07	0.25
Mexico[a]	1.35	0.47	0.13	0.05	0.09
OPEC Countries	40.00	4.00	0.42	0.10	0.10
Non-OPEC Developing Countries	5.10	2.40	2.67	0.40	1.80
Total World	59.37	18.43	19.93	7.06	7.70

Note: Blanks denote insignificant production.
[a]Mexico's total is included in the non-OPEC developing country total.

Table 2.8
Global data supply base, 2000, Case C-1 (MBDOE)

	Oil	Natural Gas	Coal	Nuclear (primary)	Hydro (primary)
Canada	3.06	0.86	1.43	1.17	1.84
United States	9.00	7.22	22.69	12.26	2.17
Total North America	12.06	8.08	24.12	13.43	4.01
Denmark				0.05	
Finland			0.09	0.16	0.06
France		0.13	0.26	2.36	0.32
Federal Republic of Germany	0.05	0.13	1.93	1.31	0.09
Italy	0.03	0.15	0.04	0.52	0.61
The Netherlands	0.04	0.35		0.31	
Norway	1.18	0.59	0.02	0.10	0.47
Sweden				0.26	0.35
United Kingdom	1.54	0.50	1.62	0.87	0.02
Rest of Western Europe			0.80	1.89	0.70
Total Western Europe	2.84	1.85	4.76	7.83	2.62
Japan	0.38	0.08	0.26	1.96	0.83
Australia, New Zealand, and South Africa	0.50	0.93	5.66	0.09	0.35
Mexico[a]	3.53	1.11	0.50	1.05	0.41
OPEC Countries	45.00	13.60	1.46	0.46	0.40
Non-OPEC Developing Countries	11.50	4.56	5.26	4.00	3.80
Total World	72.28	29.10	41.52	27.77	12.01

Note: Blanks denote insignificant production.
[a]Mexico's total is included in the non-OPEC developing country total.

Table 2.9
Global data supply base, 2000, Case C-2 (MBDOE)

	Oil	Natural Gas	Coal	Nuclear (primary)	Hydro (primary)
Canada	3.06	0.86	1.19	1.94	1.84
United States	9.00	7.22	16.19	20.01	2.17
Total North America	12.06	8.08	17.38	21.95	4.01
Denmark				0.13	
Finland			0.08	0.22	0.06
France		0.13	0.26	3.67	0.32
Federal Republic of Germany	0.05	0.13	1.69	3.15	0.09
Italy	0.03	0.15	0.02	1.57	0.68
The Netherlands	0.04	0.35	0.07	0.50	
Norway	1.18	0.59	0.02	0.10	0.47
Sweden				0.52	0.35
United Kingdom	1.54	0.50	1.04	1.52	0.02
Rest of Western Europe			0.08	2.46	0.70
Total Western Europe	2.84	1.85	3.26	13.84	2.69
Japan	0.38	0.08	0.26	3.15	0.83
Australia, New Zealand, and South Africa	0.50	0.93	3.95	0.09	0.35
Mexico[a]	3.53	1.11	0.39	2.10	0.41
OPEC Countries	45.00	11.60	1.30	1.00	0.40
Non-OPEC Developing Countries	11.50	4.50	5.00	8.30	3.10
Total World	72.28	27.04	31.15	48.33	11.38

Note: Blanks denote insignificant production.
[a]Mexico's total is included in the non-OPEC developing country total.

Table 2.10
Global data supply base, 2000, Case D-7 (MBDOE)

	Oil	Natural Gas	Coal	Nuclear (primary)	Hydro (primary)
Canada	2.57	0.73	1.43	1.17	1.49
United States	6.00	5.42	16.91	12.26	1.74
Total North America	8.57	6.15	18.34	13.43	3.23
Denmark				0.05	
Finland			0.04	0.11	0.06
France		0.13	0.26	1.84	0.32
Federal Republic of Germany	0.03	0.13	1.77	1.31	0.09
Italy	0.03	0.15	0.04	0.52	0.51
The Netherlands	0.04	0.35		0.20	
Norway	0.91	0.45	0.02	0.06	0.47
Sweden				0.26	0.35
United Kingdom	1.54	0.80	1.40	0.76	0.02
Rest of Western Europe			0.80	1.89	0.70
Total Western Europe	2.55	2.01	4.33	7.00	2.52
Japan	0.38	0.08	0.26	1.97	0.79
Australia, New Zealand, and South Africa	0.40	0.77	5.66	0.07	0.27
Mexico[a]	3.29	1.09	0.50	0.52	0.41
OPEC Countries	39.00	12.30	0.80	0.30	0.30
Non-OPEC Developing Countries	7.00	3.88	4.10	2.80	2.70
Total World	57.90	25.19	33.49	25.57	9.81

Note: Blanks denote insignificant production.
[a]Mexico's total is included in the non-OPEC developing country total.

Table 2.11
Global data supply base, 2000, Case D-8 (MBDOE)

	Oil	Natural Gas	Coal	Nuclear (primary)	Hydro (primary)
Canada	2.57	0.73	1.19	1.79	1.49
United States	6.00	5.42	13.09	20.01	1.74
Total North America	8.57	6.15	14.28	21.80	3.23
Denmark				0.13	
Finland			0.04	0.11	0.06
France		0.13	0.26	2.62	0.32
Federal Republic of Germany	0.03	0.13	1.51	2.10	0.09
Italy	0.03	0.15	0.02	1.57	0.56
The Netherlands	0.04	0.35		0.31	
Norway	0.91	0.45	0.02	0.08	0.47
Sweden				0.52	0.35
United Kingdom	1.54	0.44	0.90	1.31	0.02
Rest of Western Europe			0.80	2.46	0.70
Total Western Europe	2.55	1.65	3.55	11.21	2.57
Japan	0.38	0.08	0.26	2.81	0.79
Australia, New Zealand, and South Africa	0.40	0.77	3.95	0.07	0.27
Mexico[a]	3.29	1.09	0.48	0.79	
OPEC Countries	39.00	10.20	0.80	0.60	0.30
Non-OPEC Developing Countries	7.00	3.40	3.60	5.90	2.10
Total World	57.90	22.25	26.44	42.39	9.26

Note: Blanks denote insignificant production.
[a]Mexico's total is included in the non-OPEC developing country total.

Part II
Global Resource Studies

3 Oil

3.1 INTRODUCTION

Oil is now the major fuel in the world outside Communist areas (WOCA); it accounts for over half the energy supply in many countries. Therefore, an analysis of the future supply of oil is critical in any long-term energy study.

Our analysis identifies the following principal factors that will affect future oil production:

- Known reserves.
- The rate at which new discoveries or improved production techniques will add to them.
- The level of oil demand and the rate of production that OPEC countries might allow.

Alternative assumptions are made about each of these factors, and oil production profiles that meet oil demand for as long as possible are developed.

The main conclusion is that if all the oil-producing countries of WOCA increase production to meet demand, limited only by technical factors, oil production will peak and decline before the end of the century. This holds true over a wide range of assumptions. However, an even more important limitation on production is a governmental one—the rate at which OPEC countries are willing to produce. If recent government announcements of possible production limits take effect, production could reach a peak early in the 1980s and stay at that lower level until the end of the century and beyond. Peak production could be delayed until the late 1980s if OPEC production limits are higher. Even in this case, oil production would at best remain on a plateau through the 1990s.

Oil production beyond 2000 is also projected. The assumptions used beyond 2000 are highly speculative, and the production profiles should be viewed as such. Oil production figures for the period 2001 to 2025 are not used elsewhere in the WAES energy analysis because our time frame is 1975 to 2000.

3.2 PRODUCING OIL

A popular misconception is that oil is found in vast underground pools or lakes that, once discovered, can be pumped dry with relative ease. Nothing could be further from the truth. Oil is found trapped in the small spaces or pores between individual rock grains—rather like water in a sponge. Over time oil and water seep through porous rocks until impervious rock is reached. Figure 3.1 shows one example of an oil field. The porous rock is capped by an impervious layer of rock or cap rock. If this cap rock is in the form of a dome, as shown in the figure, then fluids (mainly water but often including oil) gradually accumulate under it. Oil gradually separates from the water and finds its way to the top of the structure, accumulating in the porous rock under the cap rock and above the layer of water in the porous rock. This is the oil field or reservoir. Gas is also often found in the reservoir, either dissolved in the oil or as a gas cap above the oil, as shown in Figure 3.1.

The existence of underground structures that might contain oil can be determined from the surface by seismic techniques. Although these techniques have improved in the past twenty years, a well must still be dug to determine whether a structure actually contains oil in commercial quantities.

Drilling a well into an oil-bearing structure releases the natural pressure in the reservoir, forcing oil into the well. Oil produced by this natural pres-

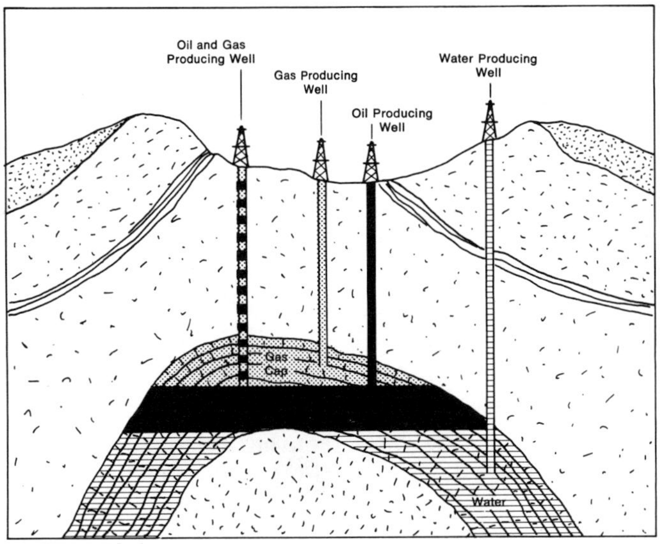

Figure 3.1
The geology of oil.

sure is known as *primary* production. The portion of oil in the reservoir that can be produced in this way depends on such factors as the porosity of the rock and the viscosity of the oil. Faults in the rock structure may also affect primary production. It is difficult to state a global recovery average for primary production, but in the United States, for example, it yields an average of about 25 percent of the oil in place.

The amount of oil recovered can be enhanced by pumping water or gas into the reservoir to increase or maintain the reservoir pressure. The success of these *secondary recovery* techniques varies. Some fields respond well because of their physical characteristics; others do not. In the United States secondary recovery is mainly responsible for increasing recovery from 25 percent of the oil in place in the 1940s to about 32 percent by 1975. In some Middle East countries such as Iran, where large reserves of natural gas are available, gas is already being injected into oil fields to improve recovery.

A third method of improving recovery rates from oil fields is to lower the viscosity of the oil so that it flows more easily through the pores of the rock. This can be done, for example, by injecting steam to heat the oil or by injecting chemicals to dilute the oil. This is called *tertiary recovery*. It is not widely used yet since the technology is costly and not well developed. In tertiary recovery energy is injected into the oil field in the form of heat or chemicals, reducing the net energy recovered in the oil by the amount used in the heat or chemicals.

Most of the world's oil is produced from these conventional oil fields. Other current, potential sources of oil are synthetic oil produced from coal, nonconventional sources, and natural gas liquids.

In 1976 coal liquefaction provided a significant amount of oil only in South Africa. And even these supplies represented only a tiny fraction of world oil supply. The potential of coal liquefication is discussed in Chapter 5. Nonconventional sources of oil such as tar sands, heavy oil, and oil shales are excluded from this analysis, but are discussed in Chapter 7. Currently they supply less than 0.2 MBD.

Natural gas liquids (NGL) are by-products of natural gas production. They broadly comprise hydrocarbons that can be extracted in liquid form from natural gas when it is produced. They are generally blended with crude oil and its products and contribute to oil supply rather than to natural gas supply. In 1975 about 5 percent of non-Communist world oil production was from NGL. In this analysis NGL will be implicitly included with oil production, and oil supply or production includes crude oil from conventional sources plus NGL.

Proven reserves of oil are usually defined as oil recoverable from known reserves with today's technology and prices. Therefore, in addition to primary recovery, proven reserves include potential production based on using secondary and tertiary techniques—where such techniques have been evaluated and are expected to be used in the fields.

Ultimately recoverable reserves are estimates of how much oil will eventually be produced. They usually include new discoveries plus an allowance for enhanced recovery as secondary recovery becomes more widely used and as tertiary recovery techniques are developed.

3.3 FACTORS THAT DETERMINE OIL SUPPLY

Estimates of ultimately recoverable reserves of crude oil for the total world have been made by many oil geologists. Table 3.1 shows how estimates

Table 3.1
Estimates of total world ultimately recoverable reserves of crude oil

Year	Source	Billion Barrels
1942	Pratt, Weeks, and Stebinger	600
1946	Duce	400
1946	Pogue	555
1948	Weeks	610
1949	Levorsen	1500
1949	Weeks	1010
1953	MacNaughton	1000
1956	Hubbert	1250
1958	Weeks	1500
1959	Weeks	2000
1965	Hendricks (USGS)	2480
1967	Ryman (Esso)	2090
1968	Shell	1800
1968	Weeks	2200
1969	Hubbert	1350–2100
1970	Moody (Mobil)	1800
1971	Warman (BP)	1200–2000
1971	Weeks	2290
1975	Geiger and Moody	2000

have increased from around 500 billion barrels in the early 1940s to about 2000 billion barrels in the 1960s. Since 1960, estimates by geologists have tended to converge to around 2000 billion barrels, although some estimates are well above this figure. Although geologists might be using the figure because it is convenient, their agreement is probably not coincidental.

It therefore seems reasonable to take 2000 billion barrels as one estimate for the ultimately recoverable reserves of oil from conventional sources and NGLs. Because some geologists are less optimistic, we shall also look at a lower figure of 1600 billion barrels. Finally, we look at a more optimistic case in which ultimately recoverable reserves are assumed to be 3000 billion barrels.

These estimates may include reserves in the deep sea and Antarctica. Such reserves will be produced only at high cost and with advanced technology, and they are unlikely to provide significant volumes of oil before the end of the century.

These estimates are for the total world, including Communist countries. Since the WAES analysis is for WOCA, we must subtract estimated reserves in the Communist area. Geologists do not agree on oil reserves in the Communist world, but a figure of 20 percent of the total world reserves will be assumed for this analysis. Using this percentage for all our cases gives estimates of ultimately recoverable reserves in WOCA of 1300, 1600, and 2400 billion barrels.

However, these numbers are in one sense academic. Oil can only be produced from fields that have been discovered and for which production facilities have been installed. Oil production is based on proven reserves, the rate at which they are added to, and the rate at which production facilities are developed. Ultimately recoverable reserves are important because they determine how long a rate of additions to reserves can be maintained.

Each field has a *potential* production rate that depends on the size of the field, its geological characteristics, and its installed facilities, such as secondary recovery facilities. In some areas governmental controls on production may also determine the production rate. Together these factors set an upper limit to the annual production from an oil field.

Primary recovery relies on natural pressure within the reservoir, and the maximum yield is obtained by releasing this pressure gradually. In general this means that no more than 10 percent of the recoverable reserves can be produced in any one year without reducing the amount of oil that can be eventually recovered. In some fields it may be possible to produce at a faster rate. In others the rate may be lower, but overall a proven reserves-to-production (R/P) ratio of 10 to 1 is probably the minimum feasible for the world's oil reserves.

Applied to the world, however, an R/P ratio of 10 to 1 would imply that all known oil fields are producing at the maximum rate. As discoveries continue to be made, some fields will be under development, so although they are included in the world's proven oil reserves it will be some years before they produce any oil. This was the position in 1976 of the Prudhoe Bay field in Alaska and many North Sea fields. Thus an R/P ratio of 15 to 1 is probably a more justifiable estimate of the maximum rate of production from the aggregate of the world's proven oil reserves. As discoveries decline, the R/P ratio may decrease toward 10 to 1.

These factors form the technical basis of the WAES estimates of future oil production. The level of proven oil reserves for each year is equal to the proven reserves at the end of one year plus the gross additions to the reserves during the year less the *actual* production during that year. The maximum *potential* production from these reserves is determined by the annual proven reserves and the the limiting R/P ratio. As long as oil demand is less than this *potential* production, *actual* production will equal demand. When demand exceeds *potential* production, demand will be limited to this potential, which becomes *actual* production.

Thus the following factors can be used to project an oil supply profile:

- An estimate of proven reserves in a base year.
- An estimate of future annual rate of gross additions to reserves.
- A limiting R/P ratio, which, for the reasons discussed above, is assumed to be 15 to 1.
- An oil demand curve.

3.4 PROVEN RESERVES AT THE END OF 1975

Proven oil reserve figures published at the end of each year in journals such as the *Oil and Gas Journal* and *World Oil* are derived from estimates made by oil companies and governments of oil-producing countries. The estimates are stated to be reserves recoverable at current prices and with current technology. Therefore they probably include oil recoverable by primary production plus oil recoverable by secondary or tertiary recovery where the potential has been evaluated and facilities are planned.

Estimates of recoverable reserves are subject to much uncertainty and can therefore change—up or down—from year to year as new assessments are made. In this analysis we have used the *Oil and Gas Journal* figures published at the end of 1975. (At the end of 1976 these estimates were revised downward by 40 billion barrels.) The figures are broken down by region in Table 3.2. Remaining proven reserves in the world at the end of 1975 were estimated at 658 billion barrels, of which 555 billion barrels were in WOCA. The uncertainty in this figure is illustrated by comparing these estimates with those given by *World Oil* for the end of 1975—579 billion barrels, of which 500 billion barrels are in WOCA.

Table 3.2 shows that about 80 percent of the remaining proven reserves are in OPEC countries, especially the Middle East. Saudi Arabia is in a class by itself with 152 billion barrels—about 27 percent of the proven oil reserves in the non-Communist world. Much of the difference between *World Oil* and *Oil and Gas Journal* estimates is found in the latter's higher estimates for proven reserves in Saudi Arabia. The non-OPEC countries in WOCA have only 105 billion barrels distributed fairly evenly among North America, Western Europe, and the rest of WOCA.

Table 3.2 also shows oil production to date. In contrast to proven reserves, the total of 291 billion barrels is divided almost equally between OPEC and non-OPEC. The largest portion (46 percent) of past production has been in North America. Adding proven reserves and cumulative production, some 846 billion barrels of oil had been discovered in the non-Communist world by 1975, and about one-third had been consumed.

Table 3.2
Total world reserves and cumulative production to the end of 1975

	Remaining Proven Reserves (billion barrels)	Cumulative Production to End of 1975 (billion barrels)
Saudi Arabia	152	23
Other Middle East	208	61
Other OPEC	90	55
Total OPEC	450	139
North America	40	133
Western Europe	25	2
Rest of WOCA	40	17
Total Non-OPEC	105	152
Total WOCA	555	291
Communist Countries	103	50
Total World	658	341

Source: *Oil and Gas Journal,* December 29, 1975.

3.5 ADDITIONS TO PROVEN OIL RESERVES 1950 TO 1975

Before considering the rate at which proven reserves might be increased over the next twenty-five years, we must look at what has happened in the past twenty-five years. Two methods of considering past additions to reserves are advocated: backdating estimates of reserves to the year in which they were discovered, and comparing year-end proven reserve figures as published in consecutive annual surveys of proven reserves in the *Oil and Gas Journal* or *World Oil.* Each method has advantages and disadvantages.

The backdating method attributes the estimated proven reserves in an oil field to the year in which the field was discovered. For example, a field discovered in 1960 might originally be estimated to contain 1 billion barrels of recoverable reserves. Five years later, information obtained in developing the field might result in a new estimate of recoverable reserves as 1.2 billion barrels. The backdating method would attribute the extra 0.2 billion barrels to discoveries in 1960. Comparing year-end estimates of proven reserves would attribute this extra 0.2 billion barrels to additions to reserves in 1965.

Figure 3.2 shows the additions to reserves between 1930 and 1975 given by the backdating procedure. It also shows how these additions are divided between the Middle East and areas outside the Middle East. The rate at which reserves were discovered outside the Middle East was a relatively constant 5 billion barrels per year between 1930 and 1955; most discoveries were made in the United States, Venezuela, and Canada. The rate increased slowly to around 10 billion barrels per year by

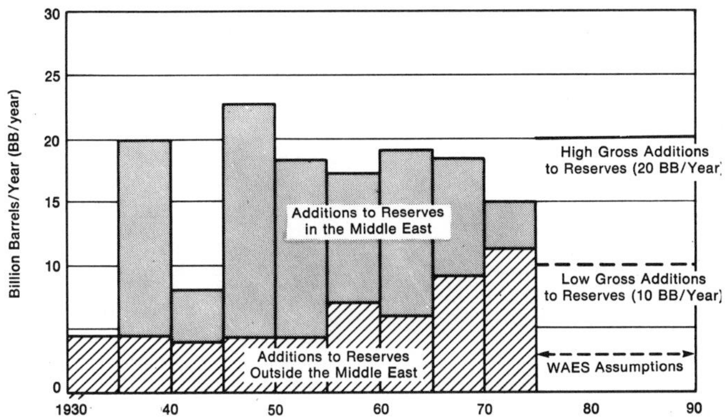

Source: EXXON Press Briefing, World Energy Outlook, Dec. 7, 1975.

Figure 3.2
Oil discovery rate in WOCA from 1930 (five-year average obtained by backdating discoveries to the year of field discovery).

1975 as reserves were found in such areas as Libya, Algeria, Nigeria, Indonesia, and in Alaska and the North Sea during the past decade.

These discoveries are overshadowed by those in the Middle East, which now contains around 60 percent of the non-Communist world's proven oil reserves. Between 1950 and 1970 the average discovery rate in the non-Communist world was about 18 billion barrels per year. Between 1970 and 1975 this rate fell to 15 billion barrels per year. This rate may be revised as further information on these recent discoveries becomes available.

Figure 3.3 shows a similar picture prepared by comparing annual estimates of proven reserves as published by *Oil and Gas Journal.* As in Figure 3.2, five-year averages are used to smooth out the large annual fluctuations that can occur. The figure covers the period 1950 to 1975, since data are not available before that. It shows a picture different from that obtained by backdating. Additions to reserves averaged about 22 billion barrels per year between 1950 and 1965, but they increased to over 50 billion barrels per year in the five years between 1965 and 1970. Though the fields had actually been discovered before 1950, this was the period when the magnitude of the Middle East reserves, especially in Saudi Arabia, was being realized. Since 1970 additions to reserves have been much lower—around 25 billion barrels per year. Overall, additions to reserves averaged about 27 billion barrels per year.

In 1950 the *Oil and Gas Journal* estimated proven reserves of WOCA at 72 billion barrels. Yet, as Figure 3.2 indicates, we now know that the reserves in oil fields discovered before 1950 were about 300 billion barrels. Thus about 230 billion barrels of oil, which backdating attributes to pre-1950 discoveries, are included in the additions to reserves between 1950 and 1975 if comparison of annual proven reserve estimates is used.

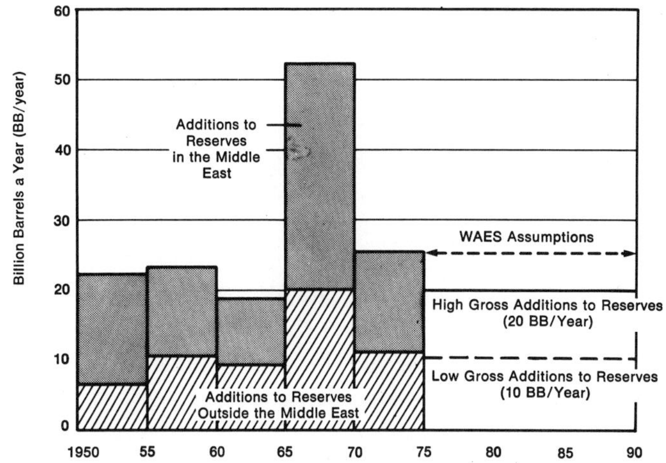

Figure 3.3
Oil discovery rate in WOCA from 1950 to 1975 (five-year average obtained by comparing year-end estimates of proven reserves as published in *Oil and Gas Journal*).

The increase is not surprising. Oil exploration and production techniques improved rapidly in the period. New techniques increased the average rate of recovery from 25 percent to 32 percent in the United States, for example. Seismic techniques made tremendous progress in the 1960s, resulting in a much earlier and more accurate estimate of recoverable oil in a new field. Also, the surplus in oil production capacity in the 1950s provided oil explorers little incentive to speed the evaluation of new reserves, especially in the Middle East where oil production was already being rapidly developed and was fully able to meet rising demand.

Which of these two methods is best? Warman (reference 1) believes that "the most meaningful analysis is obtained by taking the presently known reserves and backdating them to the year of discovery . . . . [Nowadays] the size of reserves is known fairly accurately to the operating company within a short time of the discovery of a field, and the size of the reserves fairly rapidly becomes generally known." Yet, even in the estimates of additions to reserves between 1970 and 1975 we see that new discoveries would be estimated at 15 billion barrels per year by the backdating method and 25 billion barrels per year by comparing annual year-end reserve estimates.

In looking at future discovery rates, then, we should look at results from both these methods. Backdating suggests that the rate of new discoveries has remained remarkably constant since 1950. Comparing annual proven reserves estimates suggests there were important additions as reserves in fields discovered before 1950 were reassessed.

In making estimates for the period after 1975 we should consider gross additions to reserves. By this we mean the total of (1) reserves in genuine new discoveries and (2) additions to reserves in fields discovered before 1975, either through reassessment of reserves or enhanced recovery of the oil in place.

3.6 FUTURE GROSS ADDITIONS TO RESERVES

3.6.1 Genuine New Discoveries

Genuine new discoveries in the past would have been made at a much lower rate without the discovery of the massive oil reserves of the Middle East. About 60 percent of WOCA oil reserves have been found in an area some 800 miles by 500 miles. The chances that such a prolific oil-bearing region will be found again appear small. Many of the remaining possible areas have been evaluated by seismic techniques or exploration wells, but no evidence of a new Middle East has been found.

More oil is likely to be discovered in the Middle East, as Figure 3.2 shows, but the rate at which new reserves are being found is already beginning to decline despite continued drilling activity there. With only a small chance of discovering either a new Middle East or new reserves in the Middle East as large as those already discovered, the past rate of genuine new discoveries can be achieved only if a large number of smaller producing areas are found.

The major uncertainties about where new producing areas might be found are well illustrated by two examples. In the early 1960s few people would have forecast the extent of the North Sea reserves; by 1975 proven reserves of about 23 billion barrels had been discovered. Conversely, after Prudhoe Bay, Arctic Canada was forecast as a possible major oil province; despite much seismic work and numerous exploration wells, however, no commercial oil has been found.

Table 3.3
Approximate size of world oil fields

	Number of Fields Discovered	Estimated % of Non-Communist World's Reserves
All fields	30,000	100
Fields greater than 0.5 billion bbl recoverable reserves	240	73
Fields greater than 10.0 billion bbl recoverable reserves	15	34
Four largest fields[a]	4	21

[a]Ghawar (Saudi Arabia), Greater Burgan (Kuwait), Bolivar Coastal (Venezuela), and Safaniya-Khafji (Saudi Arabia/Neutral Zone).

Since the resource base is finite, the oil reserves remaining to be found must decline as oil is discovered. New techniques of assessing prospective regions make it possible to identify structures that might contain large oil accumulations. These areas will be drilled first; as the search moves into less likely areas, the discovery rate will probably decline. The importance of finding large fields is shown in Table 3.3. Over 30,000 oil fields have been discovered, but about 75 percent of the oil lies within 240 large fields, each with over 500 million barrels of recoverable reserves. All the effort put into oil exploration around the world over the past hundred years has yielded only 240 large oil fields. Yet in hostile environments such as the North Sea, about 500 million barrels of recoverable reserves are necessary to justify the costly production platforms and undersea pipelines needed to bring the oil ashore.

It therefore seems unlikely that genuine new discoveries will be maintained at even the 15 million barrels per year rate achieved over the past five years.

3.6.2 Additions to Reserves in Fields Discovered before 1975

Estimating recoverable oil reserves is much more accurate today than in the 1950s, and the corrections of the past twenty-five years resulting in large additions to known fields are unlikely to be repeated on the same scale.

Future additions to reserves are more likely to be obtained by improved recovery techniques. The percentage of oil that can be recovered varies from field to field. Where the oil is heavy (viscous), recovery rates may be under 10 percent (with over 90 percent left in the ground). Conversely, in some rare cases where tertiary recovery is successful, recovery rates have reached over 80 percent. Estimating recovery factors on a global basis is difficult because published information relates to recoverable reserves without stating recovery factors. However, in the United States recovery currently averages 32 percent, and the figure for the non-Communist world is probably between 30 and 35 percent.

It is uncertain how the recovery rate might increase in the future. Secondary recovery techniques have been used for several years, and in many cases oil from secondary recovery is now included when estimates of recoverable reserves are made. Tertiary recovery is a newer technique that has been used with varying success in the United States. Its effect to date has probably been small. The increase in oil prices in the winter of 1973-1974 certainly increased the incentive to use secondary and tertiary techniques and may continue to increase that incentive in the future.

If we conservatively assume that global recovery rates are currently 30 percent, then each 1 percent increase in the recovery rate would increase proven reserves by a thirtieth. Proven reserves were 555 billion barrels at the end of 1975, and cumulative production was 291 billion barrels. Applying a 1 percent increase in the recovery rate to all the proven reserves plus half the cumulative production would add reserves of about 25 billion barrels. If achieved before 2000, this increase would add reserves at an annual rate of 1 billion barrels per year.

Moody and Geiger's figure of ultimately recoverable reserves of 1500 billion barrels in WOCA assumes that global recovery factors will reach 40 percent (reference 2). Such an improvement might eventually add 250 billion barrels to reserves in known fields and, if achieved before 2000, would give average additions to reserves of 10 billion barrels per year between 1975 and 2000.

However, such an increase is unlikely to occur before 2000 without a major breakthrough in technology. More likely, recovery rates will improve gradually, and annual additions to reserves from enhanced recovery will initially be small but will increase over the next twenty-five years.

3.7 WAES ASSUMPTIONS FOR GROSS ADDITIONS TO RESERVES

Based on this analysis the WAES projections use two future rates of gross additions to reserves: 20 billion barrels per year and 10 billion barrels per year.[1] The higher rate assumes that future exploration will be relatively successful, but that the discovery rate will gradually decline from the average of 18 billion barrels per year as yet undiscovered reserves are reduced. To offset this effect the contribution from enhanced recovery is projected to increase, giving an overall average rate of additions of 20 billion barrels per year. Although exact figures cannot be given, between 1975 and 1985 possibly 75 percent of the additions will come from genuine new discoveries and 25 percent from known reserves. By the end of the century the proportions could be 50 percent from new discoveries and 50 percent from enhanced recovery from known reserves.

The lower rate of 10 billion barrels assumes that the rate of new discoveries will decrease rapidly and that enhanced recovery from fields discovered before 1975 will be relatively small. The division between genuine new additions and additions from enhanced recovery from known reserves might be similar to that assumed for the higher rate.

We also looked at the effect of an unlikely case—average gross additions to reserves of 30 billion barrels per year, which is slightly above the rate observed by comparing historical annual estimates as given by the *Oil and Gas Journal*.

These average rates of additions to reserves are assumed for the period 1975 to 2000. Looking beyond 2000, we have to consider how they might decline as the figure for total ultimately recoverable reserves is approached. The rate of additions to reserves after 2000 must be highly speculative, but Figure 3.4 shows what might happen. Cumulative discoveries could exceed 1600 billion barrels

1. These rates of gross additions to reserves in April 1975 were used as averages for the twenty-five years from 1975-2000. Since then, new estimates of proven reserves made by *Oil and Gas Journal* indicate gross reductions in 1976 of 40 billion barrels. Thus to achieve the average of 20 and 10 billion barrels for the period 1975 to 2000 will, in fact, require additions to reserves of about 23 and 12 billion barrels per year over the twenty-four years between 1976 and 2000.

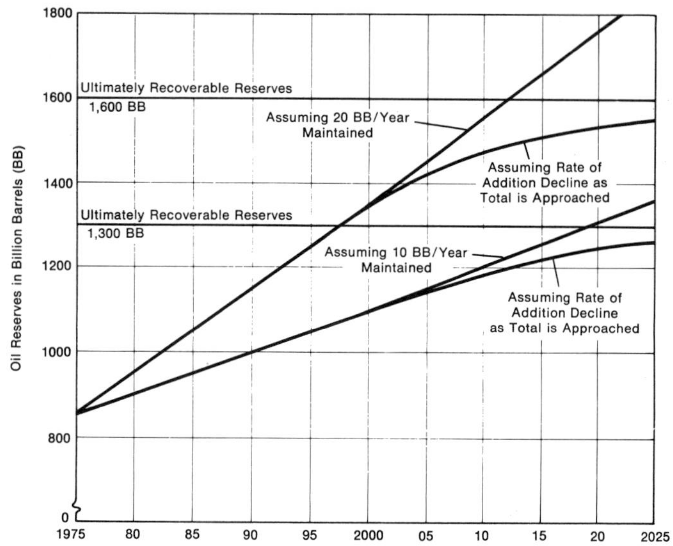

Figure 3.4
Cumulative oil discoveries in WOCA.

Table 3.4
Annual gross additions to reserves (in billion barrels per year)

	High	Low
1975–2000	20	10
2000–2010	12	8
2010–2020	7	6
2020–2025	4	3

from conventional sources by 2010 if additions to reserves continued at 20 billion barrels per year after 2000. Thus the actual rate would probably decline after 2000 so that cumulative discoveries slowly tend toward ultimately recoverable reserves.

The lower rate for gross additions to reserves—10 billion barrels per year—could be maintained until 2050 before cumulative discoveries passed 1600 billion barrels. However, this lower rate should perhaps be associated with a lower ultimate recoverable reserve of, say, 1300 billion barrels in WOCA (or 1600 billion barrels in the total world). In this case the annual rate of gross additions to reserves might start to decline around 2000 as shown in Figure 3.4.

Therefore we assumed that annual gross additions to reserves will follow the curves shown in Figure 3.4. The average annual rates in billion barrels per year are given in Table 3.4.

Although the WAES study is for the period 1975 to 2000, for oil we have extended the study beyond 2000 to obtain a picture of the oil profile over a fifty-year period.

3.8 OIL DEMAND

In this study of oil supply, we have assumed that WOCA oil demand, unconstrained by available supplies, can be met by production for as long as is theoretically possible (limited only by technical factors such as the R/P ratio). When oil production can no longer increase, then oil demand will be constrained to the level of production.

WOCA oil demand as projected by the unconstrained supply-demand integration is given in Table 3.5.

Between 1960 and 1972 oil demand growth in WOCA averaged 6.2 percent per year. The projected growth rates in the WAES cases represent a significant decline from historical rates.

Because there is so little difference in the unconstrained oil demand in 2000 for Cases C-1 and C-2 and for Cases D-7 and D-8, we considered only C-1 and D-8. The results of Case C-1 will be used for Case C-2 and the results of Case D-8 for Case D-7.

3.9 COUPLING OF OIL DEMAND AND OIL DISCOVERY RATES

As described in Chapter 2, Case C-1 assumes high economic growth rates and oil prices that remain constant in real terms until 1985, gradually increasing by 50 percent by 2000. Case D-8 assumes low economic growth and constant oil prices in real terms until 2000.

The assumptions about future gross additions to reserves depend to some extent on the size of the world oil resources. For example, additions to reserves cannot average 20 billion barrels per year over the next twenty-five years if ultimately recoverable reserves are only 1300 billion barrels. This suggests that assumptions about additions to reserves may not be related to particular economic scenarios. However, to make a consistent study of the interaction between oil supply and demand, such a relationship must be assumed.

In the case of high additions to reserves, we assume that enhanced recovery might add as much as 10 billion barrels a year by the end of the century. Enhanced recovery is more likely to occur in the rising price than in a constant price scenario. Therefore, the following couplings are adopted for this study: high gross additions to reserves (20 billion barrels per year) with the high growth/rising price

Table 3.5
WOCA oil demand

Scenario	Million barrels per day 1975	1985	2000	Average annual growth rate (%) 1975–1985	1985–2000
C-1	45.1	63	93	3.4	2.6
C-2	45.1	63	92	3.4	2.5
D-7	45.1	58	76	2.5	1.8
D-8	45.1	58	75	2.5	1.8

Table 3.6
Assumptions for oil supply profile

	Case C-1	Case D-8
Proven reserves in WOCA as of December 1975	555 × 10^9 bbl	555 × 10^9 bbl
WOCA oil demand in 1975	45 MBD (16.5 10^9 bbl/year)	45 MBD (16.5 × 10^9 bbl/year)
WOCA oil demand growth:		
1975–1985	3.4%/year	2.5%/year
1985–2000	2.6%/year	1.8%/year
Gross additions to reserves:		
1975–2000	20 × 10^9 bbl/year	10 × 10^9 bbl/year
2000–2025	declining to 4 × 10^9 bbl/year by 2025	declining to 3 × 10^9 bbl/year by 2025
Limiting R/P ratio	15 to 1	15 to 1

Note: Oil demand growth assumptions should be compared with the rates for 1960–1975: 1960–1972, 6.2%/year; 1972–1975, 0.5%/year.

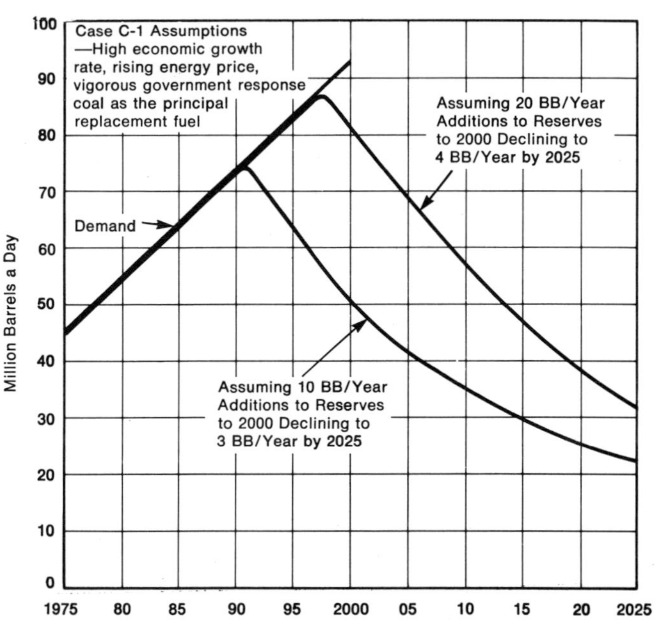

Figure 3.5
WOCA oil production, Case C-1 (no government-imposed limits).

scenario (C-1), and low rate of gross additions to reserves (10 billion barrels per year) with the low growth/constant price scenario (D-8).

The couplings are arbitrary, but they do represent two different views of oil and supply and demand. There are other possibilities. For example, the WOCA oil resource base may be only 1300 billion barrels, and even with high economic growth and rising oil prices, exploration may yield only low gross additions to reserves. On the other hand, a high gross addition to reserves might occur in a low growth/constant price world. In this chapter we shall consider in detail the high additions (C-1) and the low additions (D-8) couplings, but we shall also discuss possible effects of alternative couplings.

Table 3.6 summarizes these basic assumptions for Cases C-1 and D-8. The resulting oil supply curves are shown in Figure 3.5 (Case C-1) and Figure 3.6 (Case D-8).

3.9.1 Case C-1

With the high gross additions to reserves, production ceases to meet demand in 1997 when the R/P ratio falls to 15 to 1. Thereafter production is limited to maintain an R/P ratio of 15 to 1 and falls from the peak of 86 MBD in 1997 to 81 MBD in 2000 and to around 32 MBD in 2025.

Figure 3.5 also shows the effect of a low gross addition to reserves for this scenario. Production first ceases to meet demand in 1990 when it peaks at 72 MBD. Production then declines to 50 MBD in 2000 and to around 22 MBD in 2025.

3.9.2 Case D-8

With the low gross additions to reserves, oil production first ceases to meet demand in 1994. Thereafter production is limited to maintain an R/P ratio of 15 to 1 and falls from the peak of 66 MBD in 1994 to 53 MBD in 2000 and to just over 20 MBD in 2025.

Figure 3.6 also shows the effect of a high gross addition to reserves in this scenario. With such high additions, WOCA demand for oil of 75 MBD in 2000 can be satisfied. If demand continued to increase at 1.8 percent per year after 2000, production could meet demand until 2004. Supply would peak at 81 MBD and then decline to around 30 MBD by 2025.

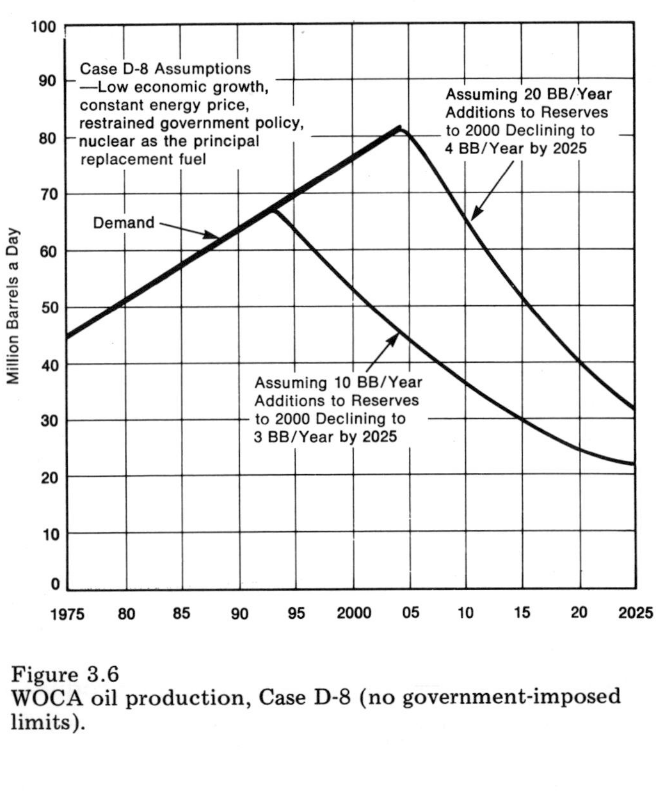

Figure 3.6
WOCA oil production, Case D-8 (no government-imposed limits).

3.10 THE REAL WORLD

These oil supply curves for WOCA are based on assumptions about proven reserves, rates of additions to reserves, oil demand, and R/P ratios. We have assumed that all countries will allow production to increase to meet demand and that the only limit on production would be an R/P ratio of 15 to 1

However, WOCA production is unlikely to actually develop in this way. Some nations will want to reduce dependence on imported oil and will therefore produce domestic oil at a much faster rate than assumed under a 15 to 1 ratio. For example, in 1975 the United States was already moving to a ratio of about 10 to 1.

For strategic and conservation reasons, other nations will restrict oil production well before a 15 to 1 ratio is reached. In Iran, for example, a limiting R/P ratio of 25 to 1 is more likely, while some Saudi Arabian spokesmen have suggested that in certain circumstances production might be limited to 8.5 MBD, even though proven reserves in Saudi Arabia could probably support a physical production rate of over 20 MBD.

Thus the supply profiles in Figures 3.5 and 3.6 represent a most optimistic production profile designed to meet potential oil demand for as long as possible. Because they give an unlikely picture of WOCA oil supply, they have been labeled "WOCA oil production (no government-imposed limits)." To obtain a more likely oil supply profile we must attempt to disaggregate the WOCA picture and look at production by region.

3.11 OPEC AND NON-OPEC

A first step in disaggregating oil producers is to look at OPEC and non-OPEC countries separately. Most oil-exporting countries belong to OPEC while most of the countries outside OPEC are oil importers.

Although all thirteen members of OPEC are developing countries, their attitudes and dependence on oil production differ greatly. Nigeria, Iran, Indonesia, and Venezuela, for instance, have relatively large populations and need high oil revenues to support economic development programs. These countries must balance the need for immediate income against the desire to prolong the flow of oil revenues by limiting production. On the other hand,

Saudi Arabia, the United Arab Emirates, Kuwait, and Libya are currently earning oil revenues well in excess of domestic requirements and therefore have little incentive to increase oil production above present levels.

The non-OPEC group includes countries at all stages of economic development and with various degrees of dependence on imported oil. With the possible exception of a few countries such as the United Kingdom, Norway, and Mexico, they are likely to remain oil importers. Consequently most of them can be expected to maximize domestic oil production to reduce their dependence on OPEC oil.

Because of these important differences, OPEC and non-OPEC aggregations must be considered separately, and where possible individual countries must be considered in detail. Table 3.2 shows how total past discoveries have been divided between OPEC and non-OPEC countries. Although production to date has come almost equally from OPEC and non-OPEC, about 70 percent of WOCA discoveries have been in OPEC, which now contains over 80 percent of remaining reserves.

Warman and Moody and Geiger estimate the potential for future discoveries by region, but they do not divide estimates between OPEC and non-OPEC. We have made the working assumption that undiscovered reserves are equally divided between OPEC and non-OPEC. Within the non-OPEC area are countries where there has been little exploration and where many new oil fields may be found. However, the gross additions to reserves include enhanced recovery and additions to known reserves revealed by further development. These will be largest in OPEC where most of the currently proven reserves are located.

3.12 NON-OPEC OIL PRODUCTION 1975-2000

Each national team in WAES has estimated oil production for their own country. The estimates of oil production for the non-OPEC world outside Communist areas are shown by region in Table 3.7.

The 1985 figures are determined largely by reserves already known and under development. Because long lead times are involved in exploring and developing new oil regions in difficult environments, such as the North Sea and Alaska, discoveries made between now and 1985 will probably contribute only marginally to production in 1985. This is especially true in Western Europe, where most of the production will be from the North Sea. Production in Western Europe will therefore be largely unaffected by economic growth or oil price, and a similar production level is assumed for both Case C-1 and Case D-8. In North America and in the developing world, the hope is that a more favorable economic climate could lead to increased development. This is reflected in the higher production forecasts for Case C-1 compared with Case D-8. Japanese production is expected to remain of marginal importance in both scenarios.

The situation is much more uncertain for 2000. Oil reserves that are being produced now will, to a large extent, be exhausted; an increasing share of production will come from discoveries made after 1975. The rate of discovery becomes increasingly important and results in different production rates for each case, especially in North America and the developing world. In Europe reserves in the North Sea, the one major oil region, are limited, and no further major discoveries are expected. Thus the Western Europe production forecasts are similar for each scenario.

Table 3.7
Estimates of non-OPEC non-Communist world oil production 1975–2000 (MBD)

	Case C-1			Case D-8		
Area	1975	1985	2000	1975	1985	2000
North America	11.7	14.0	9.3	11.7	11.9	8.0
Western Europe	0.6	4.5	2.9	0.6	4.5	2.6
Japan	0.01	0.03	0.4	0.01	0.03	0.4
Rest of non-Communist world	4.0	6.2	12.0	4.0	24.7	12.0
Total non-OPEC	16.3	24.7	24.6	16.3	22.0	18.5

Note: Production estimates for North America, Western Europe, and Japan are taken from the WAES national teams' supply studies.

3.13 NON-OPEC PRODUCTION 2000-2025

Because the WAES national studies have not generally looked beyond 2000, we must resort to the methods used earlier in this chapter to consider the potential production rates from proven reserves given limiting R/P ratios.

The remaining non-OPEC reserves in 2000 depend on the oil production and the gross additions to reserves between 1975 and 2000. In Case C-1, where the additions are assumed to be high with half occurring in the non-OPEC region, the remaining non-OPEC reserves would be about 130 billion barrels in 2000 with an R/P ratio around 14 to 1. The comparable figures for Case D-8 are 76 billion barrels and an R/P ratio of 11 to 1.

After 2000 the rate of addition to reserves is likely to decline as the undiscovered reserves are reduced. The R/P ratio may also decline as demand increases pressure on the remaining reserves. A production forecast based on declining gross additions to reserves and a declining R/P ratio is given in Table 3.8.

3.14 OPEC PRODUCTION 1975-2000

To achieve the non-Communist world oil production profile shown in Figures 3.5 and 3.6 would require the production rates from OPEC shown in Table 3.9.

Table 3.8
Non-OPEC potential oil production rates (MBD)

	Case C-1	Case D-8
2000	25	19
2005	23	17
2010	21	15
2015	19	13
2020	16	11
2025	13	9

Table 3.9
Required OPEC oil production rates (MBD)

	Case C-1	Case D-8
1975 (actual)	27	27
1980	32	29
1985	39	36
1990	47	42
1995	57	44
2000	56	34

These rates assume that all OPEC countries will allow production to increase to meet WOCA oil demand as long as the R/P ratio is at least 15 to 1. The rates required, especially in Case C-1, are well in excess of current capacity of around 37 MBD. They are probably not feasible either on technical or policy grounds. Iran, for example, holds that a 25 to 1 ratio is more realistic, and the expansion in oil production infrastructure in other countries is unlikely to take place on the scale required.

Individual OPEC countries may also be unwilling to see oil production reach the maximum theoretical level because of a desire to extend the life of their oil reserves. Some OPEC governments have already expressed such views by announcing possible limits to oil production as shown in Table 3.10.

If imposed, these government restrictions on production would reduce OPEC's usable capacity by some 4.8 MBD from the estimated usable capacity of 37 MBD in 1975. The majority of these restrictions are in countries that are "low absorbers" of oil revenues: Kuwait, Saudi Arabia, Abu Dhabi, Qatar, and Libya. Venezuela and Ecuador have imposed restrictions to conserve declining reserves and prolong the flow of oil revenues, even though these countries could probably absorb all the oil revenues they could earn.

The real impact of government-imposed restrictions, however, is on future potential production rates. Table 3.11 summarizes the present oil reserves, oil production, and usable capacity in the OPEC countries. It shows that the OPEC countries

Table 3.10
Possible production limits announced by governments (MBD)

Country	Estimated usable capacity at end 1975[a]	Government-announced possible production limits
Venezuela	2.5	2.2
Ecuador	0.22	0.2
Libya	2.5	2.0
Qatar	0.7	0.5
United Arab Emirates	2.3	1.8
Kuwait	3.0	2.0
Saudi Arabia	10.8	8.5
Total	22.0	17.2

[a]As given in *Petroleum Intelligence Weekly*, February 2, 1976.

Table 3.11
OPEC oil production statistics

	Reserves at 1/1/76[a] (10^9 bbl)	Production in 1975[b] (MBD)	Production in 1975[b] (10^9 bbl/year)	R/P ratio	Estimated usable capacity (MBD)
Saudi Arabia	152.1	7.1	2.59	59	10.8
Kuwait	71.3	2.1	0.77	93	3.0
United Arab Emirates	32.2	1.7	0.62	52	2.3
Qatar	5.9	0.4	0.15	40	0.7
Iran	64.5	5.4	1.97	33	6.8
Iraq	34.3	2.3	0.84	41	2.6
Total Middle East	360.3	19.0	6.94	52	26.2
Nigeria	20.2	1.8	0.66	31	2.7
Libya	26.1	1.5	0.55	48	2.5
Algeria	7.4	1.0	0.37	20	1.0
Gabon	2.2	0.2	0.07	30	0.2
Venezuela	17.7	2.3	0.84	21	2.5
Indonesia	14.0	1.3	0.47	30	1.7
Ecuador	2.5	0.2	0.07	34	0.2
Total non-Middle East	90.1	8.3	3.03	30	10.8
Total OPEC	450.4	27.3	9.97	45	37.0

[a] *Oil and Gas Journal,* December 29, 1975.
[b] *Petroleum Intelligence Weekly,* February 2, 1976.

of the Arabian Peninsula (Saudi Arabia, Kuwait, Qatar, and the United Arab Emirates) dominate OPEC reserves with 261 billion barrels, or 58 percent of total OPEC proven reserves. The R/P ratios of these Arabian Peninsula countries are, with the exception of Qatar, in excess of 50 to 1. If a ratio of 15 to 1 were the only limit on production, they could today produce oil at a rate of 47 MBD (equivalent to current total WOCA output). This compares with a possible government-imposed production limit in 1976 of 12.8 MBD or some 4.0 MBD below current installed capacity.

The remaining countries with government-imposed production limits have a much smaller potential to expand production. Currently the restrictions reduce usable capacity by 0.8 MBD. Such limits will have little effect on the conclusions of a long-term study such as ours, in spite of the possible implications for the countries concerned. The effects of any government limits imposed by countries with major potential for expanding oil production—Saudi Arabia, Kuwait, and the United Arab Emirates—will be important and need to be considered in more detail.

The remaining OPEC countries outside the Arabian Peninsula have estimated usable capacity of 20.2 MBD, but the possible government limits shown in Table 3.10 could reduce this by 0.8 million to 19.4 MBD. The reserves of some of these countries are already being reduced fairly rapidly (the R/P ratios in Algeria and Venezuela are already down to 20 to 1, while Iran is nearing its probable limit of 25 to 1). New additions to reserves in these countries will probably enable them to maintain production at present levels but not to significantly increase it. We have therefore assumed a limit of 20 MBD on the oil production from OPEC countries outside the Arabian Peninsula. This limit, similar to present capacity, takes into account both the physical limits imposed by reserves and any other limits that may be imposed by governments. It is an assumption rather than a forecast, and will be used in both the high and low additions to reserves cases.

For the Arabian Peninsula countries (Saudi Arabia, Kuwait, Qatar, and the United Arab Emirates) three alternative assumptions were studied:

1. Government limits are near present production levels. This would limit output to around 13 MBD.

2. Saudi Arabia allows production to increase to about 15 MBD, and as a result, total Arabian Peninsula potential production increases to 20 MBD.

3. Saudi Arabia allows production to increase to about 20 MBD, and as a result, total Arabian Peninsula potential production increases to 25 MBD.

The first assumption will be considered in both the high and low additions to reserves cases. The third assumption would require a rapid increase in production capacity and will be considered only in the high gross additions to reserves case. The second assumption probably requires a smaller increase in infrastructure and in reserves and will be considered only in the low gross additions to reserves case.

We therefore have three assumptions for limits to total OPEC production:

- 33 million barrels per day
- 40 million barrels per day
- 45 million barrels per day

These levels define the maximum potential production assumed available from OPEC. Once the required production from OPEC reaches such a limit it will be assumed to remain at that level as long as the R/P ratio exceeds 15 to 1. From then on maintaining a 15 to 1 ratio will result in declining OPEC production.

The resulting OPEC production rates from 1975 to 2000 are shown in Table 3.12.

In Case C-1 the 45 MBD rate of production is reached in 1989 and the lower limit of 33 MBD in 1981. In Case D-8 the 40 MBD limit is also reached in 1989 and the lower limit in 1983.

3.15 OPEC PRODUCTION 2000-2025

Beyond 2000, the OPEC production rates are more speculative. They depend on the additions to reserves made between 1975 and 2000, the production rates that the countries outside the Arabian Peninsula can maintain, and the attitudes of Arabian Peninsula countries toward restraints. If we assume that OPEC limits of 33, 40, and 45 MBD are maintained—if necessary by some countries relaxing restraints—then the production rates shown in Table 3.13 might be possible.

3.16 TOTAL NON-COMMUNIST WORLD OIL PRODUCTION

Combining the production estimates for non-OPEC countries and the assumed OPEC production levels gives estimates for total WOCA oil production.

Figure 3.7 shows non-Communist oil production for Case C-1 with gross additions to reserves of 20 billion barrels per year. The profiles that assume maximum OPEC production of 33 and 45 MBD are contrasted with the profile assuming no government limits on production. The production profile for non-OPEC is also shown.

An OPEC limit of 45 MBD results in constrained WOCA production by 1989, compared to 1997 for the unlimited case. OPEC production can be maintained at 45 MBD until 2007. For nearly twenty years the total WOCA profile is nearly constant, with peak production of about 71 MBD, or some 15 MBD below the unconstrained peak.

Table 3.12
Projected OPEC production rates 1975–2000 (MBD)

	Case C-1 High gross additions to reserves[a]			Case D-8 Low gross additions to reserves[b]		
	OPEC limit 33 MBD	OPEC limit 45 MBD	No govt. limit	OPEC limit 33 MBD	OPEC limit 40 MBD	No govt. limit
1975 (actual)	27	27	27	27	27	27
1980	32	32	32	29	29	29
1985	33	39	39	33	36	36
1990	33	45	47	33	40	42
1995	33	45	57	33	40	44
2000	33	45	56	33	39	34

[a] 10×10^9 bbl/year in OPEC.
[b] 5×10^9 bbl/year in OPEC.

Table 3.13
OPEC production rates 2000–2025 (MBD)

	C-1 High additions to reserves[a]			D-8 Low additions to reserves[b]		
	OPEC limit 33 MBD	OPEC limit 45 MBD	No govt. limit	OPEC limit 33 MBD	OPEC limit 40 MBD	No govt. limit
2000	33	45	56	33	39	34
2005	33	45	45	33	31	26
2010	33	42	37	33	25	22
2015	33	33	28	32	20	19
2020	33	28	23	25	17	17
2025	33	22	19	20	13	13

[a] 10×10^9 bbl/year in OPEC.
[b] 5×10^9 bbl/year in OPEC.

If OPEC maximum production were limited to 33 MBD, then WOCA oil production would be constrained as early as 1981. The limit of 33 million barrels could be maintained until 2025, and the only variations in WOCA production after 1981 would arise from changes in non-OPEC production. The result is a slow increase in WOCA production to a maximum of about 59 MBD in 1995 followed by a slow decline to 46 MBD by 2025.

Figure 3.7 shows clearly the impact of OPEC constraints. Oil is kept in the ground in the 1990s, increasing the remaining reserves. This oil can then be produced after 2000, making the oil production decline less severe. An OPEC production limit of 45 MBD results in 1997 production 15 MBD lower than the theoretical maximum. However, production in the early part of the next century would be higher than it would be were reserves depleted at the fastest rate possible in the 1990s.

Figure 3.8 shows comparable WOCA oil supply curves for Case D-8, assuming that gross additions to reserves will be 10 billion barrels per year. With an OPEC production limit of 40 MBD, oil supply reaches a peak of 61 MBD in 1989, or some 5 million less than the unconstrained peak which occurs in 1994. OPEC can hold production at this level until 1999. The profile for the non-Communist world between 1989 and 1999 declines slowly because of the decline in non-OPEC production. The impact of keeping the OPEC oil in the ground is already felt by 2000 since the reserves conserved enable production in 2000 to be 58 MBD, or about 5 MBD above the level possible without OPEC constraints.

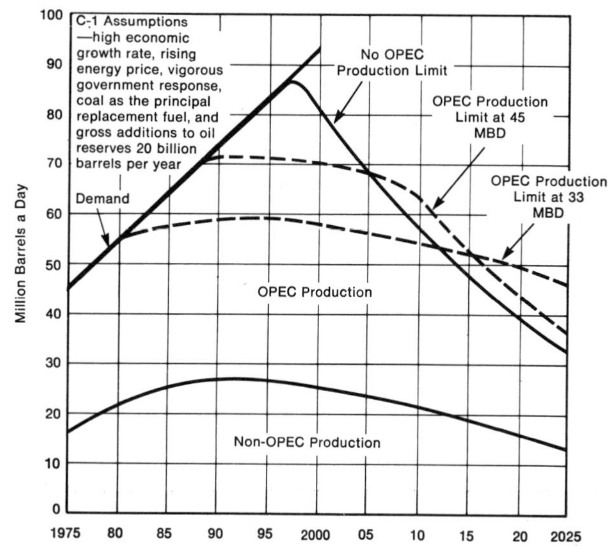

Figure 3.7
WOCA oil production, Case C-1.

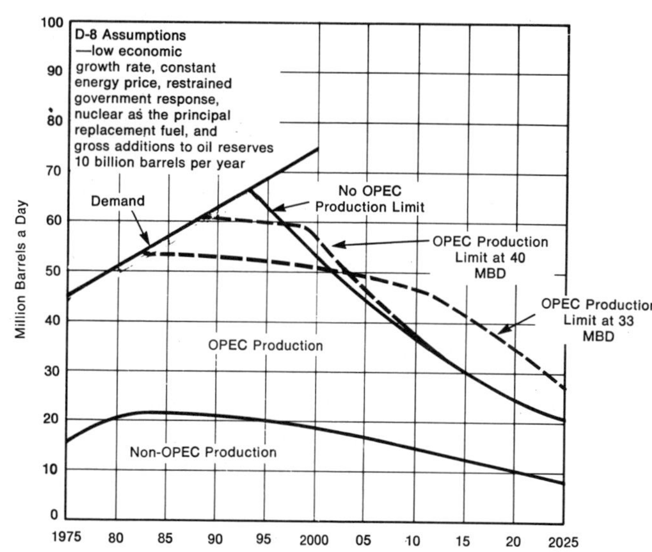

Figure 3.8
WOCA oil production, Case D-8

Table 3.14
Range of peak production

WAES Case	Gross annual addition to reserves (10^9 bbl)	OPEC limits (10^6 bbl)	Year production ceases to meet demand	Production in year when demand first ceases to be met (MBD)	Production in 2000 (MBD)	Potential demand in 2000 (MBD)	Possible shortfall in oil supply (MBD)
C-1	20	none	1997	86	81	93	12
C-1	10	none	1990	72	50	93	43
C-1	20	33	1981	55	58	93	35
C-1	20	45	1989	71	70	93	23
D-8	10	none	1994	66	53	75	22
D-8	20	none	2004	81	75	75	0
D-8	10	33	1983	55	52	75	23
D-8	10	40	1989	61	58	75	17

If OPEC production were to be limited to 33 MBD, then WOCA oil supply would be constrained as early as 1983. OPEC production could be held at this limit until 2013, so that between 1983 and 2013 total WOCA production declines slowly as non-OPEC production declines.

3.17 WOCA OIL SUPPLY-DEMAND BALANCE IN 1985 AND 2000

Potential oil demand in 1985, as projected in Cases C and D, can be met even if OPEC were to impose a limit on production of 40 or 45 MBD. However, if members of OPEC who have recently suggested possible production ceilings were to actually impose them, WOCA oil production could be constrained early in the 1980s. Then there could be a gap between potential oil demand and oil supply in 1985 for both Cases C and D.

Table 3.14 summarizes the various cases considered to 2000. Potential oil demand can be met only if gross additions to reserves are high, oil demand is low, and OPEC does not impose any limits on production. This is an unlikely set of events, and it is more probable that potential oil demand will not be met in 2000 for either Case C-1 or Case D-8.

In the high growth scenario C-1, the shortfall in oil supply in 2000 is about 12 MBD if gross additions to reserves are high and if OPEC countries do not limit production. OPEC limits of 45 MBD would increase this shortfall to about 23 MBD while more severe limits of 33 MBD would increase it to 35 MBD.

The effect of OPEC limitations of 40 MBD in Case D-8 with low gross additions to reserves is the opposite: they tend to reduce the projected shortfall in oil supply by 2000. Oil kept in the ground is by 2000 already helping to defer the decline in production. Thus the shortfall of 22 MBD with OPEC production unconstrained could be lower (17 MBD) if OPEC production were not allowed to exceed 40 MBD. In the case of an OPEC limit of 33 MBD, the shortfall in 2000 could be 23 MBD.

3.18 A MORE OPTIMISTIC FUTURE?

These assumptions for gross additions to oil reserves may be considered conservative by some observers. Estimates that ultimately recoverable reserves may be 3000 billion barrels or even higher have been published. If these estimates prove accurate, then gross addition to reserves may indeed be higher than our high case of 20 billion barrels per year.

More optimistic estimates assume that either genuine new discoveries will be high or that enhanced recovery will make a major contribution. To maintain our high gross additions to reserves assumption of 20 billion barrels will require a major successful exploration and development program. It would probably require new oil regions the size of the North Sea (23 billion barrels proven reserves at the end of 1975) or Nigeria (20 billion barrels) to be discovered at much more frequent intervals than in the recent past. However, some estimates for the developing countries are much higher than ours. If oil exists in these areas in the suggested volumes, then we might see a much higher level of genuine new discoveries than is assumed here.

In our high case, we have assumed that enhanced recovery may contribute up to 50 percent of the gross additions to reserves by 2000. To achieve such a figure would require major investment and success in research and development of secondary and tertiary recovery techniques. However, on average some 65–70 percent of the oil in an oil field is at present left unrecovered. A major breakthrough in secondary and tertiary recovery techniques that substantially reduced this proportion could make a contribution to additions to reserves even higher than we have assumed. Such a breakthrough would have to occur soon for a major contribution to oil supplies to occur in this century. It therefore appears that improvements in enhanced recovery above those assumed in this chapter are more likely to extend the oil plateau into the twenty-first century than to substantially increase oil production before year 2000.

However, the possibility that gross additions to reserves may be higher than 20 billion barrels per year cannot be ruled out. We have therefore considered the possible effects of additions to reserves averaging 30 billion barrels per year over the next twenty-five years, that is, slightly higher than the figure of about 27 billion barrels per year between 1950 and 1975 obtained by comparing successive year-end estimates as published by the *Oil and Gas Journal.* If OPEC production were not constrained, such a rate of additions to reserves would allow production to meet potential demand in Case C-1 in 2000. Demand could continue rising for the first five years of the next century, but sometime before 2010 a limiting R/P ratio of 15 to 1 would be reached and production would peak and decline.

Thus the level at which OPEC countries are prepared to produce their oil is likely to be a much more important factor than the theoretical limits imposed by a 15 to 1 minimum R/P ratio. The year in which an OPEC limit of 45 MBD is reached in this case depends on the level of production in non-OPEC countries. Even with a discovery rate outside OPEC of 15 billion barrels per year, which might allow a rapid increase in non-OPEC production, OPEC production would still have to exceed 45 MBD soon after 1990 for oil production to continue to meet demand. By 2000, with OPEC production at 45 MBD, production might be between 75 and 80 MBD. This is less than 10 MBD above production if additions to reserves were 20 billion barrels per year and still about 15 MBD below potential oil demand.

3.19 CONCLUSION

The future of oil supply is uncertain. The cases considered in this analysis show differing years and different levels of peak production, with dissimilar levels of production in year 2000. However, one conclusion is very clear: potential oil demand in the year 2000 is unlikely to be satisfied by crude oil production from conventional sources. Even without governmental constraints on oil production, oil supply will meet demand only under the most optimistic assumptions about gross additions to reserves. A more important constraint on oil production is likely to be the level of production set by producers with large reserves and limited needs for more revenue. Possible constraints on oil production by members of OPEC are likely to cause oil supply to peak at the latest sometime around 1990, although lower production limits could bring this date forward into the early 1980s.

The end of the growth era in oil production is probably no more than fifteen years away. However, there may be a decade of constant oil production after 1990 in which consumers will have to adjust to a decline in oil supply. Oil discoveries or recovery factors might be higher than we have assumed, but the effect would be to delay the impact for a few years rather than to solve the problem of transition to other fuels. For nations to continue to increase their oil consumption in the hope that more optimistic estimates will prove correct is to risk accelerating the peak in oil production, making the necessary adjustments in energy consumption patterns even more severe.

REFERENCES

1. H.R. Warman. "Future Availability of Oil," presented at the Conference on World Energy Supplies, September 18–20, 1973.

2. John D. Moody and Robert E. Geiger. "Petroleum Resources: How Much Oil and Where?" *Technology Review,* March/April 1975.

4.1 INTRODUCTION

Natural gas is a clean and convenient fuel for cooking and heating and is a valuable raw material for the large and growing petrochemical industry. World reserves are large and unlikely to limit production within the next twenty-five years. However, the future role of gas as an energy source will be determined not so much by supply as by the problem of transporting and distributing gas from the wells to the consumer. Problems of transport have limited intercontinental trade in gas, which amounted to only 0.3 MBDOE in 1975.

Historically, natural gas has moved directly from supplier to consumer through pipelines. The large-scale use of gas has been in markets that could be economically connected by pipeline to natural gas reserves. The expense of constructing costly pipeline networks can only be justified where there are both large reserves and an assured demand. In contrast to oil, which supplies over a third of the energy in every major energy-consuming country, dependence on natural gas varies from virtually zero in Sweden, Denmark, and Japan to around 30 percent in the United States and 47 percent in The Netherlands.

An alternative to pipelines is transport by tanker as liquefied natural gas (LNG). The technology for liquefaction, transportation, and regasification has been available on a commercial scale only during the last decade. The gas is liquefied, then carried in special refrigerated tankers at –161°C, and regasified at receiving terminals. The method presents some problems. Approximately 25 percent of primary energy is lost in the processing of LNG. There is concern that an LNG tanker explosion in a port could severely damage life and property. An LNG system also requires large capital investments for moving gas from the Middle East (where the largest unexploited reserves are) to Europe, Japan, or the United States (the largest potential users). Because of the large sunk investment in infrastructure for gas distribution and use in some countries, LNG may become increasingly desirable to avoid replacements based on other fuels.

In the following sections we describe how gas is produced and the size and distribution of the reserves. This is followed by projections of future demand for and supply of natural gas by region for the WAES scenario cases. The chapter concludes with a discussion of the potentials and problems of future international gas trade.

For consistency with the rest of the report, the units used in this chapter are expressed in MBDOE. Natural gas is usually discussed in trillion (10^{12}) cubic feet (TCF), billion (10^9) cubic feet per day (BCFD), or in billion (10^9) cubic meters (BCM). The standard conversions from MBDOE into the other units are approximately:

1 MBDOE = 2.12 TCF/year
= 5.9 BCF/day
= 57.0 BCM/year

4.2 PRODUCING GAS

Natural gas occurs in underground structures similar to those containing oil (see Chapter 3), and the methods of exploring for and producing natural gas are similar to those for oil. There are varying quantities of gas in most oil reservoirs, either in solution with the oil or in a gas cap above the oil. Such gas is called *associated gas.* When it is dissolved in the oil, gas is necessarily produced along with the oil. On the other hand, gas found in a cap above the oil is rarely produced until after the oil has been extracted. Earlier production of the gas would reduce the yield of oil. In some structures, only natural gas can be produced. This is called *unassociated gas.* Globally, about 40 percent of gas reserves are associated with oil, 60 percent are unassociated.

4.3 PROVEN RESERVES

At the end of 1975, natural gas equivalent to about 520 billion barrels of oil had been discovered—probably between 25 and 50 percent of the ultimately recoverable reserves. Of this total, 123 billion BOE has been consumed, leaving about 400 billion BOE in proven reserves. Estimates of remaining proven reserves as of January 1976 are shown in Table 4.1.

Two-thirds of the proven reserves are in the world outside Communist areas (WOCA) and one-third is in the Communist world. Although the major share of the non-Communist world's natural gas reserves are in OPEC, the distribution of gas reserves is much wider than oil reserves. In WOCA, 45 percent of the natural gas reserves lie outside OPEC countries (compared with only 20 percent for oil).

Table 4.1
Total world proven natural gas reserves

Region	Remaining reserves (10^{12} ft^3)	Remaining reserves (10^9 BOE)
Iran	330	57
Saudi Arabia	103	18
Other Middle East OPEC	96	17
Algeria	126	22
Other Non-Middle East OPEC	133	23
Total OPEC	788	137
North America	268	46
Western Europe	181	31
Rest of WOCA	160	28
Total Non-OPEC	609	105
Total WOCA	1397	242
Communist Areas	835	144
Total World	2232	386

Source: *Oil and Gas Journal,* December 1975.

About 30 percent of the reserves lie in the major energy-consuming areas of North America and Western Europe (compared with 12 percent for oil).

4.4 ULTIMATELY RECOVERABLE RESERVES

Estimates of ultimately recoverable world natural gas reserves are subject to even more uncertainty than oil reserve estimates. This is partly because in the past, much of the gas found was associated with oil and was treated as an unwanted by-product of oil production. Or it was found in areas too remote for it to be economically brought to market. Table 4.2 gives some estimates of ultimately recoverable natural gas reserves made over the last twenty years.

The estimates have been converted into barrels of oil equivalent to facilitate comparison with estimates of oil reserves in Chapter 3. Estimates of ultimate recoverable reserves of oil increased until the mid-1960s. Since then, oil estimates have tended to converge around 2000 billion barrels. No such convergence is seen in the estimates of gas reserves. Estimates made since 1965 have ranged from 1030 to 2640 billion BOE.

Estimates of ultimately recoverable reserves of gas and the rate at which these reserves are increased are less important than the location of the reserves. Currently proven reserves and possible future additions

Table 4.2
Estimates of world ultimately recoverable reserves of natural gas

Year of estimate	Source	Reserves (10^{12} ft^3)	(10^9 BOE)
1956	U.S. Department of the Interior	5000	860
1958	Weeks	5000–6000	860–1035
1959	Weeks	6000	1035
1965	Weeks	7200	1240
1965	Hendricks (USGS)	15300	2640
1967	Ryman (ESSO)	12000	2070
1967	Shell	10200	1760
1968	Weeks	6900	1200
1969	Hubbert	8000–12000	1380–2070
1971	Weeks	7200	1240
1973	Coppack	7500	1300
1973	Hubbert	12000	2070
1973	Linden	10400	1800
1975	Kirkby and Adams	6000	1030
1975	Moody	8150	1400

Source: World Petroleum Congress, Tokyo, May 1975.

(even if ultimately recoverable reserves are as low as 1030 billion barrels) are likely to allow for a substantial increase in natural gas production during the next few decades provided the gas can be moved to market.

Moody and Geiger (reference 2) estimate ultimately recoverable gas at 1400 billion BOE by region (see Table 4.3). They show that only about 25 percent of the world's natural gas reserves are likely to be found near the main energy-consuming countries of the non-Communist world. About 30 percent is expected to be found in the Communist areas, and the remaining 45 percent is located in the Middle East and other developing countries.

4.5 NATURAL GAS SUPPLY AND DEMAND 1950–1972

An analysis by region of natural gas demand shows how the proximity of reserves has stimulated consumption. In North America, where large reserves of gas were discovered in association with oil in the early part of the twentieth century and where prices have been low compared to oil, natural gas has become a major source of energy. In 1950 natural gas consumption in North America was 3.2

Table 4.3
Ultimately recoverable reserves by region

	Reserves (10^9 BOE)	Share of total (%)
North America	280	20.2
West Europe	77	5.5
Middle East	270	19.3
Rest of WOCA	348	24.8
Communist Areas	425	30.3
Total World	1400	100

Source: John D. Moody and Robert E. Geiger, "Petroleum Resources: How Much Oil and Where?" *Technology Review,* March 1975.

MBDOE, or 9 percent of total energy consumption. This was over 91 percent of world natural gas consumption in 1950. By 1975 gas consumption had grown to 11.3 MBDOE, providing a third of North American total primary energy and about 68 percent of the gas consumed in WOCA, as shown in Table 4.4.

In the United States, over 40 million households and 3 million businesses rely on gas. Industry relies on gas both as an energy source and as a petrochemical feedstock. It has also been used for electricity generation in areas close to abundant gas reserves.

In Western Europe, the natural gas industry was of minor importance until the early 1960s when the discovery and development of the Groningen field stimulated gas consumption in The Netherlands and surrounding countries. Consumption was further stimulated by discoveries of gas in the southern North Sea in the mid 1960s, which provided the United Kingdom with a source of natural gas to replace gas manufactured from coal. Production from these gas fields allowed an increase in gas consumption in the United Kingdom from 2 percent of total energy in 1960 to 15 percent in 1975. Between 1969 and 1972 natural gas consumption in Western Europe increased at a rate of 30 percent per year when total energy was increasing only 5 - 6 percent per year. As Table 4.4 shows, natural gas consumption was 3.1 MBDOE in Western Europe in 1975—19 percent of the total natural gas consumption in WOCA.

Almost no natural gas was consumed in Japan until the late 1960s when the development of LNG technology made possible imports from

Table 4.4
WOCA natural gas consumption in 1975

	Consumption (MBDOE)	% of WOCA
North America	11.3	68
Western Europe	3.1	19
Rest of WOCA	2.1	13
Total WOCA	16.5	100

Source: *BP Statistical Review,* 1975.

Alaska and Brunei. Domestic production is still very small. In Australia and New Zealand gas consumption is very small, but significant offshore discoveries made in the past decade will help stimulate demand.

In the less developed countries (LDCs), development of natural gas production has been limited. Lack of markets for natural gas has discouraged the construction of gathering and distribution facilities. OPEC countries with large reserves currently have very small gas requirements. There have been some minor exports from Algeria and Libya as well as from the non-OPEC LDCs—Afghanistan, Argentina, and Brunei.

4.6 NATURAL GAS SUPPLY AND DEMAND 1972-2000

4.6.1 Method of Analysis

Our projections for natural gas demand and supply in 1985 and 2000 are made on a national basis. National estimates of gas imports needed to satisfy demand are then combined by region: North America, Japan, Western Europe, and the rest of WOCA. Desired gas imports may appear reasonable on a national basis, but global aggregation is needed to determine whether desired imports for all countries can be met. The total of these regional imports is an estimate of the imports required from OPEC and Communist areas, the two regions with potential surpluses.

4.6.2 North America

In 1975 natural gas production in the United States was 9.5 MBDOE. Since 1966 annual production of natural gas from the lower forty-eight states has been greater than new additions to reserves, resulting in a decline in the reserves-to-production ratio from about 17 to 1 in 1966 to about 11 to 1 in 1974. Even if the large discoveries in Alaska are added, consumption for more than a decade has been increasing appreciably faster than additions to reserves.

For conservation and technical reasons we believe that the R/P ratio of 11 to 1 will be maintained. Therefore, potential production is expected to be constrained by the volume of new discoveries. It is generally agreed that future gas discoveries are likely to be found either in deep onshore wells (below 4500 meters), in offshore structures, or in arctic frontier areas such as Alaska.

The rate at which the offshore continental shelf is leased and whether the price of natural gas is deregulated will both affect the level of exploration and probable discovery in the United States. Potential U.S. production in 1985 could be as high as 9 MBDOE if (1) prices of newly discovered natural gas are deregulated in the near future, (2) the offshore continental shelf leasing program attracts many bidders for leases, and (3) these two actions result in large increases and success in exploration. Alternatively, a continuing low regulated gas price could result in gas supply falling to 7.6 MBDOE by 1985. After 1985 we assume that reserves of gas will be increasingly hard to find, and in all our cases production declines. Between 1985 and 2000, an increasing part of production is expected to come from Alaska (either by pipeline or LNG shipments) as the lower forty-eight states' production continues to decline. By 2000 U.S. natural gas production is expected to fall to between 5.4 and 7.2 MBDOE.

With U.S. natural gas supplies likely to decline between now and 2000, an increase in imports will be needed to maintain present consumption levels. U.S. natural gas demand projections range from 8.6 to 10 MBDOE in 1985. By 2000 demand is expected to fall to between 7.8 and 8.8 MBDOE. At such levels, maximum import requirements for the United States would be about 1.2 MBDOE in 1985 and as much as 2.5 MBDOE in 2000.

There are plans to import quantities of LNG. Table 4.5 shows that these potential LNG projects for 1985 for the United States amount to 1.2 MBDOE. This is sufficient to meet the required U.S. gas imports for the WAES scenarios in 1985.

In 1975 Canadian natural gas production was 1.6 MBDOE; of this about 40 percent was exported to the United States. Since 1971 Canadian annual production has been greater than additions to reserves; as a consequence the R/P ratio has declined from 29 to 1 in 1971 to about 24 to 1 in 1975. Proven gas reserves in Canada are extensive, amounting to some 10 billion BOE in accessible producing areas. Another 3 billion BOE have been found in the frontier areas of the Arctic and in hostile environments off the east coast. We assume that most future discoveries will probably be in similar areas. Production from such areas requires large reserves and

Table 4.5
U.S. LNG imports

Project Exporter	Commencing	Length of Contract (years)	Approximate Annual Plateau Quantities (BDOE)	Status
Algeria	1976	20	8,000	Operational
Algeria	1978	20	175,000	Under Construction
Indonesia	1979	20	100,000	Under Construction
Algeria	1977/78	22	106,400	Under Construction
Algeria	1978	20	22,000	Under Construction
Algeria	1980	20	74,300	Potential
Nigeria	1982/83	20	125,000	Potential
Nigeria	1982/83	20	125,000	Potential
Algeria	1981/82	20	175,000	Potential
Iran	1981/82	20	62,250	Potential
Iran	1983/84	20–25	175,000	Potential

Note: In August 1976 the Energy Resources Council issued a policy statement recommending 2 trillion cubic feet per annum, or about 1 MBDOE, as an acceptable level of LNG imports for the United States.

large amounts of capital. Significant production by 2000 from these areas will probably require an increase in energy prices above the WAES assumptions.

Because of these factors, Canadian production is expected to decline slightly to between 1.1 and 1.3 MBDOE. Canadian demand is expected to increase to 1985. This increase will probably be met by reducing exports to the United States. Between 1985 and 2000, we expect production to decline to between 0.7 and 0.85 MBDOE while gas demand increases. Any increased demand will have to be satisfied by imports, and our analysis shows that by 2000 Canadian import requirements could be as high as 1 MBDOE.

4.6.3 Western Europe

Three countries in Western Europe—Norway, The Netherlands, and the United Kingdom—have large reserves of natural gas. Gas production during the next twenty-five years is expected to come principally from these reserves although there may be some production in Western Germany, France, and Italy (totaling about 0.6 MBDOE in 1985 and 0.4 MBDOE in 2000).

The gas industry in The Netherlands is largely based on the Groningen field, although smaller reserves of gas have been found in the Dutch areas of the North Sea. Current production from Groningen is about 1.6 MBDOE, of which about 50 percent is exported. Government policy is to use this field as a strategic energy reserve. Production will be allowed to peak in 1978. As it begins to decline exports will be phased out, particularly after 1985 when existing contracts expire. Such a policy will enable existing Dutch gas reserves to fulfill domestic demands into the twenty-first century.

Natural gas production in the United Kingdom started in the mid 1960s with discoveries in the southern North Sea. These fields are now within a few years of peak production, but discoveries of gas in the northern North Sea—often in association with oil—will allow production to continue to increase into the late 1980s. Production is likely to be restricted to meet U.K. demand. There will be no exports. It is more likely that the U.K. will be a net importer of gas, since it has already contracted to import gas from the Frigg field in the Norwegian sector of the North Sea.

Natural gas production in Norway has not started, although the proven reserves could support rapid development. Norway has only a limited present need for natural gas and is expected to limit its natural gas exports for domestic fiscal policy reasons. Total production may reach 0.5 MBDOE in 1985 and increase to about 0.6 MBDOE by 2000.

In Western Europe, we project that natural gas production will increase from 2.1 MBDOE in 1972 to around 3.5 MBDOE in 1985 and then decline to around 2 MBDOE by 2000. We project demand to increase from about 2.2 MBDOE in 1972 to 5 MBDOE in 1985 and to between 4.6 and 6 MBDOE in year 2000. This would produce a regional shortfall that would have to be met with imports of 1.3 MBDOE in 1985 and as much as 4 MBDOE in

Table 4.6
Western Europe LNG imports

Project Exporter	Commencing	Length of Contract (years)	Approximate Annual Plateau Quantities (BDOE)	Status
Algeria	1964	15	18,100	Operational
Algeria	1965	25	10,000	Operational
Libya	1971	20	44,200	Operational
Libya	1971	15	20,000	Operational
Algeria	1972	25	62,250	Operational
Algeria	1977	20	62,250	Under Construction
Algeria	1980/81	20	62,250	Potential
Algeria	1980/81	20	62,250	Potential
Algeria	1980/81	20	62,250	Potential
Iran	1983/84	20-25	174,700	Potential
Algeria	1983/84	20	211,000	Potential

2000. The 1985 import requirements can be filled with firmly contracted imports of pipeline gas from the U.S.S.R. and potential LNG and pipeline shipments from Algeria, Iran, and Libya (Table 4.6). Additional import capacity will have to be built if year 2000 demands are to be met.

4.6.4 Japan

Japan's interest in increasing imports of LNG stems partly from its desire to reduce atmospheric pollution in the large towns and cities. Japanese power companies and gas utilities have already entered into long-term LNG contracts and are expected to negotiate a number of others (Table 4.7).

Domestic natural gas production in Japan is expected to be small—less than 0.1 MBDOE by the year 2000. Demand for gas, however, is expected to increase from around 0.15 MBDOE in 1975 to 1.5 MBDOE by the year 2000. Almost all of this demand is expected to be met through LNG imports. The maximum planned LNG import by 1985 is 1.2 MBDOE, more than enough to meet demand. By 2000 another 0.3 MBDOE capacity must be added to meet desired demand.

4.6.5 OPEC

Proven reserves of natural gas in OPEC are estimated to be 140 billion barrels—about 55 percent of WOCA reserves. Much of this is associated gas, a product of oil production that must now be wasted for lack of markets. Currently there is only a small demand within OPEC countries, either for local consumption or for reinjection to maintain oil reservoir pressure. As much as 2 to 3 MBDOE of this gas is now burned off (flared) in OPEC countries, but reinjection of substantial quantities into oil reservoirs is planned for the future.

OPEC countries are expected to use an increasing part of their gas for domestic purposes. The government of Iran, for example, recently announced its intent to increase natural gas as a share of domestic energy from 18 percent to 24 percent by 1978 and 35 percent by 1987. Other OPEC countries may follow a similar policy so that by year 2000 natural gas might contribute 30-40 percent of OPEC's internal primary energy use. If this happens, natural gas consumption within OPEC countries in the year 2000 could be as high as 5 MBDOE. Proven OPEC reserves are already large and will almost certainly be increased. Despite increases in domestic use, gas production would allow substantial exports. The amount depends on the policy of various OPEC countries toward the production of natural gas for export.

4.6.6 Rest of WOCA (Non-OPEC)

Although gas consumption in 1975 was only 1 MBDOE in the rest of WOCA outside OPEC, 28 billion BOE of proven reserves are estimated in this region. Consumption is likely to increase as reserves are developed, as has already occurred in Australia and New Zealand. Total production in the rest of the non-OPEC, WOCA countries might reach 2 to 3 MBDOE by 1985 and 3 to 4.5 MBDOE by year 2000. Most of this gas will probably be consumed

Table 4.7
Japanese LNG imports

Project Exporter	Commencing	Length of Contract (years)	Approximate Annual Plateau Quantities (BDOE)	Status
Alaska	1969	15	24,100	Operational
Brunei	1972	20	124,500	Operational
Abu Dhabi	1977	20	70,300	Under Construction
Indonesia	1977	20	187,000	Under Construction
Sarawak	1981/82	20	149,000	Potential
Australia	1983	20	162,000	Potential
Indonesia	1980/81	20	62,250	Potential
Iran	1981/82	20	62,250	Potential
Qatar	1983/84	20	149,000	Potential
U.S.S.R.	1984/85	20	175,000	Potential

Table 4.8
Domestic supply, demand, and import requirements for North America, Western Europe, and Japan:1985 and 2000 (MBDOE)

	1985			2000		
	Domestic supply	Domestic demand	Import requirements	Domestic supply	Domestic demand	Import requirements
Case C-1						
North America	10.3	11.3	1.0	8.0	10.8	2.8
Western Europe	3.5	5.0	1.5	2.0	6.1	4.1
Japan	0.1	0.9	0.8	0.1	1.6	1.5
Total	13.9	17.2	3.3	10.1	18.5	8.4
Case D-8						
North America	8.7	9.6	0.9	6.2	9.6	3.4
Western Europe	3.4	4.8	1.4	1.9	4.6	2.7
Japan	0.1	0.9	0.8	0.1	1.6	1.5
Total	12.2	15.3	3.1	8.2	15.8	7.6

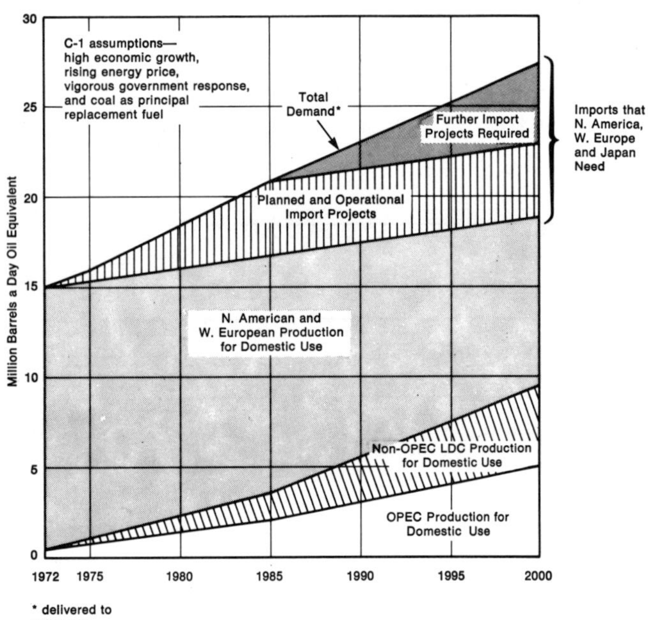

Figure 4.1
Natural gas supply and demand in WOCA, Case C-1.

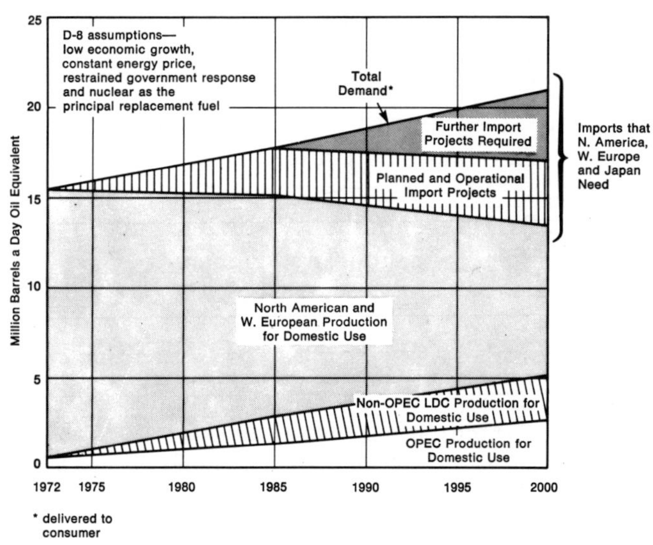

Figure 4.2
Natural gas supply and demand in WOCA, Case D-8.

locally. These countries are unlikely to import significant amounts of natural gas because of the high costs of building a gas distribution infrastructure.

4.7 REQUIRED GAS IMPORTS IN 1985 AND 2000

Table 4.8 is a summary of natural gas supply, demand, and required imports for the major consuming regions of North America, Western Europe, and Japan. By 1985 import requirements amount to about 3.3 MBDOE and are fairly evenly divided among the regions. By 2000 import requirements increase to between 7.6 and 8.4 MBDOE. These figures are shown in Figures 4.1 and 4.2.

4.8 POTENTIAL NATURAL GAS TRADE

Although international trade is currently small, there are plans to expand it. The planned projects can be divided into three groups: (1) exports by the U.S.S.R., (2) LNG exports, mainly from OPEC countries, and (3) pipeline exports from the OPEC countries in the Middle East and North Africa.

4.8.1 Exports from the U.S.S.R.

Table 4.9 shows existing and planned import project(s) for gas pipelines connecting Western Europe with the Soviet Union by 1985. Projects in operation and under contract would allow imports of 0.46 MBDOE into Western Europe by 1985. If additional new projects now under discussion all went ahead and were completed on time, they could add 0.06 MBDOE, making a total of 0.52 MBDOE by 1985. Projects for the export up to 0.3 MBDOE of Soviet natural gas to the United States and Siberian gas to Japan are also under discussion.

Table 4.9
Current and potential imports from the U.S.S.R. into Western Europe (BDOE)

Market	1975	1985
Firm Projects (including supplies under contract or option)		
Austria	36,000	36,000
France		71,000
Finland	18,000	36,000
Italy	36,000	142,000
West Germany	53,000	178,000
Total	143,000	463,000
Possible Additional Projects		
Austria		36,000
Switzerland		18,000
Total		54,000

Although the U.S.S.R. has large resources of natural gas that might allow larger exports, increasing domestic demand will probably limit further expansion. Projects now under discussion are unlikely to be completed before 1985. We have therefore assumed that total gas exports from the U.S.S.R. in 1985 will be 0.5 MBDOE. This includes projects now operating or under contract and allows for the completion of some of the projects now being discussed. For 2000 we have assumed an import figure from the U.S.S.R. of 1.0 MBDOE. This is based on an assumption that projects similar in size to those expected to be on stream by 1985 will be completed between then and 2000.

4.8.2 LNG Trade

Current and planned LNG projects to Western Europe, the United States, and Japan are listed in Tables 4.6–4.8 and summarized in Table 4.10. Table 4.10 shows that LNG trade starts from a very small base in 1975 and could reach 3.1 MBDOE by 1985 if all the current and potential projects are completed. This would mean a tenfold increase over 1975. Projects currently under construction would furnish only about 25 percent of that increase. The remainder would have to come from projects classified as potential, many of which are only in the early stages of planning. About 80 percent of such LNG imports would come from OPEC countries.

An LNG system capable of delivering 3.1 MBDOE from OPEC to Europe, North America, and Japan by 1985 would require about 25 treating and lique-

Table 4.10
Current and potential LNG imports by 1985 (MBDOE)

Importer	Current	Under Construction	Further Potential	Total
United States	0.01	0.4	0.8	1.21
Western Europe	0.15	0.06	0.57	0.78
Japan	0.15	0.26	0.76	1.17
Total	0.31	0.72	2.13	3.16

Source: British Petroleum, *LNG in the Next Ten Years* (1976).

faction plants, 140 LNG tankers, and 25 receiving terminals and regasification plants. The capital cost for such a system is estimated to be approximately $29 billion (1975 U.S.).

4.8.3 Pipeline Gas from OPEC

There are plans to pipe gas into Western Europe from the Middle East and Northern Africa. By 1985, if planned projects are completed, between 0.2 and 0.3 MBDOE of natural gas could be so piped. Iran is also expected to export around 0.2 MBDOE of natural gas to the U.S.S.R. This may be balanced by Soviet exports to Western Europe.

4.9 BALANCING NATURAL GAS SUPPLY AND DEMAND IN 1985

Table 4.11 summarizes potential natural gas imports to the United States, Western Europe, and Japan by 1985 on the basis of firm projects and others now under discussion.

Our studies show that such an amount of gas exceeds the required natural gas imports of 3.3 MBDOE in the main consuming regions. The amount of exports shown in Table 4.11 would require the majority of potential projects to be completed by 1985.

Many projects, especially those for LNG, have not proceeded beyond the planning stage. Some projects may be cancelled, others may be delayed beyond 1985. Projects now in operation or under construction could supply about 1.0 MBDOE of the 3.3 MBDOE needed in 1985. The maximum shortfall would thus be about 2.3 MBDOE.

Table 4.11
Summary of potential international trade in natural gas: 1985

Exports from	MBDOE
U.S.S.R. (pipeline)	0.5
OPEC Countries	
pipeline	0.5
LNG	2.5
LNG from other countries	0.6
Total	4.1

4.10 BALANCING SUPPLY AND DEMAND IN 2000

Our studies show that satisfying the desired demands of consumers in North America, Western Europe, and Japan would require imports of 7.6 to 8.4 MBDOE of natural gas by year 2000. U.S.S.R. exports by year 2000 might be 1.0 MBDOE. The remainder, 6.6 to 7.4 MBDOE, is the amount of gas that would have to be imported from the rest of WOCA. Projects currently planned and in operation could supply about half of this requirement; additional projects would be needed to supply the remainder. Only the OPEC countries have the reserves to supply such volumes of gas.

What is the outlook for such gas exports? Some gas might be moved by pipeline to Western Europe but most would have to be converted to LNG. Assuming that losses between the wellhead and consumer average about 25 percent, imports of 6.6 to 7.4 MBDOE of gas would require producing 8.8 to 10 MBDOE of gas at the wellhead. Domestic consumption of OPEC countries might take 5.0 MBDOE plus an unknown amount of gas for reinjection into oil fields.

Total gas production in OPEC countries would therefore need to be over 15 MBDOE by 2000. In view of the scale of currently proven reserves and possible discoveries over the next twenty-five year, production at 15 MBDOE is probably feasible.

Gas can probably command a premium price over other fuels because of its very desirable properties. However, the price that can be paid at the wellhead may be considerably lower than the price of equivalent oil because of the cost of liquefaction, transport, and regasification. Long-term contracts between producer and consumer that will be needed to secure the necessary capital for the LNG system must reflect such factors. Because the producer's income from gas production will be lower than that from oil production of equivalent energy content, it might be preferable for OPEC to replace oil with gas for domestic use or to reinject gas into oil wells to improve recovery.

Some OPEC countries have expressed concern about the future implications of selling large amounts of gas to the same consuming countries that are now critically dependent on them for oil. On the other hand, part of the gas is produced in association with oil. OPEC countries would undoubtedly prefer to export such excess gas rather than continue to waste it. If the price of energy rises beyond our scenario assumptions, OPEC countries will have additional incentives to export greater quantities of gas.

Uncertainty about continuity of supply and price at the well may discourage importing countries from making the large investments required for LNG. Investments in the LNG system delivering 7.4 MBDOE by 2000 would cost approximately $70 billion (1975 U.S.$) if capital costs are similar to those of today. The safety of LNG systems is also a concern to some. An accident resulting in serious casualties and property damage could affect the acceptability of LNG at the necessary delivery points.

There are some indications of what the future may hold. Iran and Algeria have recently indicated that they are reluctant to consider further natural gas export projects before 1985. These are two of the largest OPEC producers and have substantial reserves. If LNG shipments are to reach the levels we have projected by 2000, production for export will be needed from these countries. With a lead time for LNG projects of about ten years, decisions on new projects must be made soon in order to satisfy a rising demand for gas imports after 1985.

Proven reserves of gas are sufficient to support a large expansion in international trade in gas, principally as LNG. Because of its desirable qualities as a fuel, compared with oil and coal, it may command a premium price. Uncertainties about the growth of world gas trade—perhaps to levels well above our projections—lie in (1) attitudes of the governments of OPEC countries toward export of gas versus domestic use, including use as a chemical feedstock, (2) availability of capital for investment in LNG systems, and (3) possible repercussions of an LNG tanker accident.

Uncertainties about the level of LNG trade in 2000 may lead to the consideration of other alternatives to moving natural gas as LNG. There are at least three other possibilities: (1) converting natural gas into liquid methanol, (2) producing synthetic fuels from coal, and (3) reducing demand.

4.10.1 Methanol

One alternative is to convert natural gas into liquid methanol (alcohol). This would simplify transportation and eliminate the need to build complex LNG tankers. Methanol could act as a blending component for the gasoline used in gas turbines in electric power generation or as a general fuel. It could also be reconverted to synthetic natural gas.

There are disadvantages, however. Converting natural gas to methanol involves a 40 percent processing loss compared to 25 percent for LNG. It is also capital intensive, although it may be more economical than LNG for distances greater than 10,000 km. Transporting methanol removes the need for regasification plants in the importing country, but the direct use of methanol requires investment in new consuming equipment. Finally, as with LNG, it requires OPEC countries to accept the need to produce natural gas for export.

4.10.2 Coal Gasification

Large amounts of coal could be available by the year 2000, as described in Chapter 5. Coal could provide the basis for a synthetic fuels industry for the manufacture of both gas and liquids from coal. In North America, indigenous coal could be used. Western Europe and Japan would have to import coal. Substantial production of syngas from coal could help fill the gap between desired gas demand and natural gas supply in the large gas markets. However, the capital investments for this technology are high, resulting in costly gas, especially compared to the low prices many consumers currently pay for natural gas. Moreover, the technology has not yet been satisfactorily developed. Research, development, and demonstration projects are needed to bring this alternative to the stage of commercial operation with known costs.

4.10.3 Reducing Demand

A third option is to reduce natural gas demand. A first step would be to change the pricing policy for natural gas in countries where government regulations now hold natural gas prices well below the price of alternatives. Another would be to restrict natural gas to its premium uses—domestic heating, chemical feedstocks—and to use in geographical areas where it is needed to reduce atmospheric pollution. In The Netherlands, for instance, even with their large reserves, the government gas policy is designed to stretch reserves into the twenty-first century. These policies include (1) highest priority to small customers, (2) priority for high-grade industrial applications, (3) no new supply contracts with power stations, and (4) no growth in export sales.

4.11 SOME DIFFICULT CHOICES AHEAD

By the year 2000 natural gas production in the major consuming countries will probably decline to about two-thirds of the 1972 level. North American production will reach a plateau and start to decline before 1985. Western Europe, owing largely to expanded production in the northern North Sea, can steadily increase production until the latter part of the 1980s when it will probably start to decline unless reserves are much larger than now estimated.

Demand for natural gas in the major consuming countries is expected to increase, but at a fairly slow rate compared to the growth over the last fifteen years. North America, Western Europe, and Japan together might require total natural gas imports of about 3.3 MBDOE in 1985 and 8.4 MBDOE by the year 2000 if natural gas demands are to be met.

The resources to fill these import demands are available and lie primarily in the U.S.S.R. and OPEC. Pipeline gas from the U.S.S.R. may provide as much as 1 MBDOE by the year 2000. The bulk of the remaining import requirements can come only from OPEC countries. Some countries may choose to build synfuel industries that will enable them to produce gas from imported or domestic coal. Others may decide to reserve gas for high priority uses only.

The policy on natural gas use will vary widely from country to country. Countries that are now large users of natural gas with a large distribution system and gas-user equipment will have difficult economic and political tradeoffs to make if they are to continue to meet demand over the next twenty-five years in the face of declining production from domestic gas fields. Countries will have to decide to what extent they should become dependent on imports to meet their gas requirements. The costs and risks of imported gas will have to be compared with the costs and risks of imported coal and the large capital investment and environmental impacts of converting coal to synthetic gas. Some countries will decide to reduce demand for natural gas. Finally, large gas producers, such as some OPEC countries, will have to compare the advantages of using gas as a fuel or for petrochemicals versus selling gas for export as LNG.

REFERENCES

1. Kirkby and Adams. Presentation at the World Petroleum Congress, Tokyo, May 1975.

2. John D. Moody and Robert E. Geiger. "Petroleum Resources: How Much Oil and Where?" *Technology Review*, March 1975.

3. British Petroleum. *LNG, The Next Ten Years* (1976).

Coal

5.1 INTRODUCTION

Coal was once the world's dominant commercial fuel. For the last decade, however, its production has been level while the production of oil has increased sharply to become the major fuel in the advanced industrial nations. World oil production is expected to peak and decline during the next ten to twenty-five years, and nations must seek alternative fuels. Once again coal may have to sustain industrial production and economic development throughout the world.

Large increases in coal production and use would have a profound impact on people's attitudes toward coal. It would require developing new infrastructures and new technologies for coal mining, processing, and use. Coal could ultimately replace existing energy systems designed for oil and gas. It would force the world to confront the serious environmental issues related to large-scale coal mining and coal burning.

This chapter assesses prospects for future coal development in the WAES scenarios. We shall analyze the world coal resource base, the outlook for coal

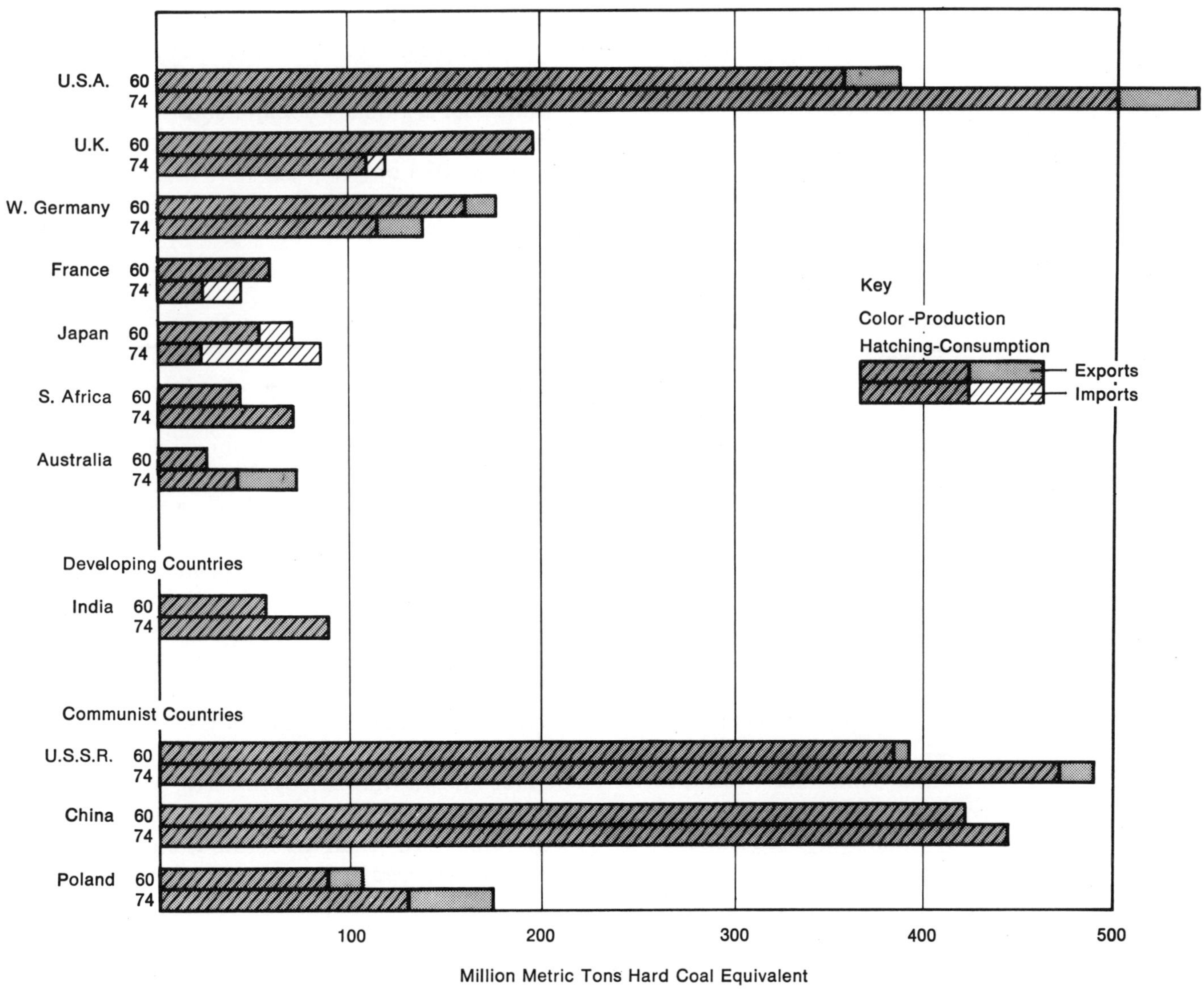

Figure 5.1
Coal production, 1960–1974.

Table 5.1
Coal consumption and WOCA coal markets 1975 (MBDOE)

	North America	OECD Europe	Japan	Rest of WOCA	Total WOCA
Coal consumption	6.4	4.7	1.1	3.0	15.2
Total primary energy consumption	37.4	23.7	6.8	15.3	83.2
Coal's share (%)	17	20	16	20	18

	Rail transport	Industry	Domestic	Electricity generation (input)	Gas manufacture (input)
Coal consumption	0.3	5.2	1.2	7.5	0.8
Total energy consumption	1.0	19.8	17.6	22.7	1.0
Coal's share (%)	35	26	7	33	76

demand, and potential coal production over our time frame, 1985 to 2000. The analysis will deal specifically with such potential constraints on coal production and coal usage as environmental concerns, the need for infrastructure development, and the relative prices of alternative fuels. We shall also consider the production of synthetic fuels from coal as an alternative to the direct use of coal.

The final section outlines implications of large-scale development as an alternative to oil, the options available, and the choices facing national policy makers.

5.2 COAL TODAY

Today's leading coal producers are the major industrialized countries of the Western world and the Communist bloc. Figure 5.1 shows that three countries (the United States, U.S.S.R., and China) account for nearly 60 percent of world output, while the top six (these three plus Poland, West Germany, and the United Kingdom) produce over 75 percent. Coal production in developed countries is similarly concentrated; India is the only large producer/consumer. In the non-Communist world, the single largest producer and consumer of coal is the United States, with nearly half the production and over half the known reserves.

In 1974, 1200 million metric tons of coal, or 15.2 MBDOE, provided 18 percent of the primary energy consumed in the non-Communist world. This is a sharp decline from 34 percent in 1960, although total coal consumption was marginally higher in 1974 than it was in 1960. Coal use has been declining in all market sectors except electricity generation, which now accounts for half of all coal consumption (see Table 5.1).

While coal is globally a major source of energy, most coal is consumed in the country where it is mined. In 1974 only 9 percent of the world's production (233 out of 2513 million metric tons) moved in international trade. A large part of this was coking coal for the iron and steel industry.

5.3 THE RESOURCE BASE

The world possesses vast reserves of coal, far more than any other fossil fuel. This base is sufficient to support massive development of coal well into the next century. The amount of coal actually produced, however, will be governed by the level of coal demand, the development of infrastructure for transportation, handling, storage, and use, and the resolution of environmental, social, and economic policy issues related to coal production and use.

The World Energy Conference *Survey of Energy Resources 1974* estimates the world's total resources of all ranks of coal to be about 11,000 billion metric tons. Of this total, about 1300 billion count as known (measured) reserves, of which approximately 50 percent were deemed economically recoverable at the time the resource studies or surveys were prepared (generally 1972 or 1973).[1]

1. The World Energy Conference uses the following definitions: *Total resources* include all known reserves-in-place plus additional indicated and inferred resources which could exist in unexplored extensions of known deposits or in undiscovered deposits in known fuel-bearing areas. *Known measured reserves* are the estimated quantities of coal in-place in known deposits based on specific sample data, measurement of the deposits, and detailed knowledge of

Table 5.2 indicates the levels of economically recoverable and measured reserves adopted for WAES national studies. Data for non-WAES areas were derived from the World Energy Conference survey.

Estimates of ultimate recoverability vary from one coal field to another; they depend on the accessibility of the coal, which is affected by a number of factors such as seam thickness and depth, type of terrain, and land ownership. Recovery rates also vary widely, from 85 to 95 percent in surface mining and from 25 to 70 percent in deep mines.

The aggregation of reserves on a ton-for-ton basis, as in Table 5.2, is only a rough indication of the extent, location, and energy content of coal resources. The energy content of coal varies. Brown coal, for instance, contains only a third as much energy per ton as bituminous coal. It also contains a good deal of moisture, making it costly to transport. Yet its sulfur content is generally lower than that of bituminous coal. Taking into account the inferior calorific value of the lower ranks of coal, total economically recoverable reserves of 737 tons of all ranks amount to some 600 billion tons of hard coal equivalent—sufficient for over 200 years' consumption at the current rate of coal usage.

Table 5.2
World coal reserves (10^9 metric tons)

	Known (Measured) Reserves	Economically Recoverable Reserves[a]
United States	396	248
Canada	13	6
North America	409	254
West Germany	100	16
United Kingdom	99	4
Rest of Western Europe	26	21
Western Europe	225	41
Japan	3	1
Rest of WOCA	140	53
Non-Communist world	777	349
U.S.S.R. and Eastern Europe	349	287
China	201	101
Total world	1327	737

[a]The majority of these estimates were prepared prior to the oil price increases in 1973–1974. In the new era of higher energy prices, economically recoverable tonnages are likely to be higher.

Expressed in terms of oil equivalent, the amount of coal assessed as economically recoverable from proven reserves is equal to around 3000 billion barrels—between four and five times the current level of proven reserves of crude oil. However, undiscovered reserves of coal could be much greater than the reserves now known.

Most estimates of ultimately recoverable crude oil range around 2000 billion barrels, as described in Chapter 3. The World Energy Conference estimate of total world coal resources of 11,000 billion tons is equivalent to some 50,000 billion barrels of oil. If we take a conservative view that an average of 25 percent is recoverable, a total of 2500 billion tons of coal—12,000 billion BOE—could ultimately be produced; this is about six times the estimated level of recoverable oil. The relative orders of magnitude are illustrated in Figure 5.2.

The picture for coal is less clear than for oil because exploration has been less widespread and generally less intensive. Many estimates of coal reserves were made when readily available, low-cost oil was rapidly displacing coal, resulting in little incentive to look for coal.

While coal deposits are widespread, occurring to some extent in many countries, the really large known reserves are concentrated in just three—

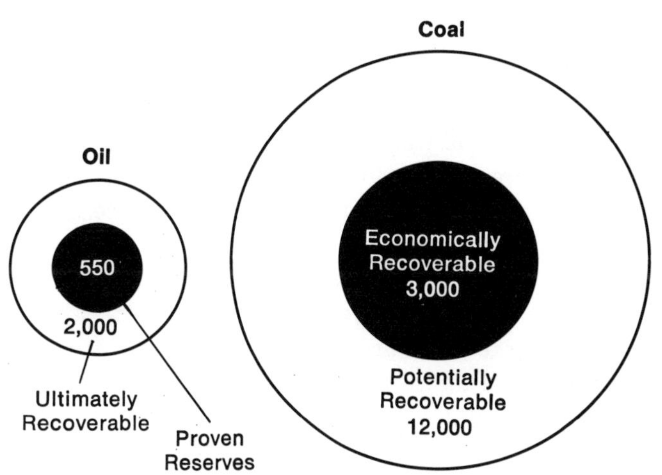

Figure 5.2
Coal versus oil reserves (units: MBOE).

---

the quality or the grade of the deposits, including that part of total solid fuels in-place normally remaining in the ground due to extraction requirements. *Economically recoverable reserves* involve the portion of the total amount of known reserves considered to be actually recoverable under current economic conditions and using current mining technology. The forms of coal are hard coal (anthracite and bituminous) and lower ranks (sub-bituminous, lignite, and brown).

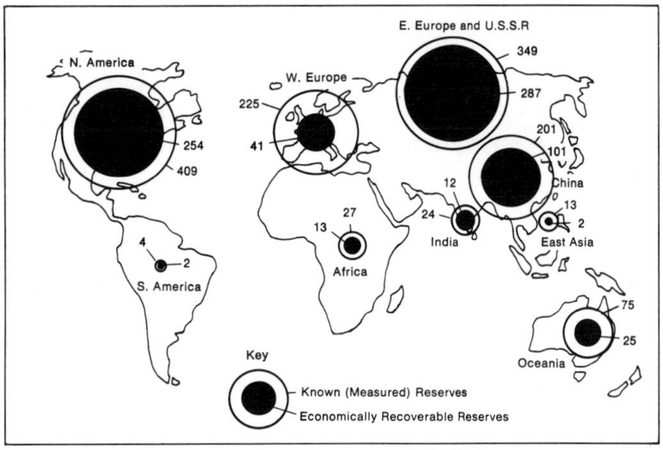

Figure 5.3
Distribution of coal reserves (units: 10^9 metric tons hard coal equivalent.

the United States, the U.S.S.R., and China. Figure 5.3 shows that much of the coal discovered so far lies in the northern temperate zone, perhaps because most coal exploration has been in this area. Early industrialization was based upon locally available coal, and since a sufficient amount was found, there was no need to look further afield. Since World War II, oil and natural gas have dislodged coal from many of its former markets and have met most of the incremental demand for energy throughout the world. Few of the developing countries have ever had to look for coal. However, geology suggests that there are significant portions of the world, including much of the Southern Hemisphere, that have a high coal potential.

Australia is a good example. Less than a hundred years ago coal was being shipped from Britain to Australia. Then an Australian coal industry sufficient for local needs was developed, only to be dampened by oil usage. However, in the last twenty years wider exploration has been carried out in Australia, largely to meet the Japanese demand for coking coal. As a result, output has trebled and measured reserves have been found in New South Wales and Queensland; in calorific terms these reserves are comparable to the proven oil reserves of Saudi Arabia (150 billion barrels).

Similarly, exploration in the southern part of Africa and in Indonesia is now yielding favorable results, and it seems probable that the world's coal resources may be far greater than previously estimated.

But even these projections only consider coal minable from onshore locations. Just as with oil, there are believed to be major coal deposits offshore. Some of the best coal reserves in Europe are thought to lie under the North Sea. The large gas fields in the southern North Sea are in fact gas formed from coal deposits. These coal reserves are not included in even the most optimistic resource estimates.

Certainly, we do not yet have the technology to mine undersea coal, but a device to exploit offshore coal would require only modification of existing systems rather than invention of new systems. Alternatively, a breakthrough in in situ gasification could also change the picture, both onshore and offshore. Such developments are within future engineering capability.

Estimates of eventual coal resources can only be speculative, but the conclusion is not critically influenced by the numbers chosen. By almost any criterion, coal resources can be regarded as ample. What is in doubt is the willingness and ability of the world to accept large increases in coal production and use.

5.4 DEMAND

The key to estimating the future role of coal is determining the potential development of coal demand. The potential coal production estimated for the WAES scenarios will only be realized if the demand for coal develops and is perceived far enough in advance so that the necessary investments in production and transportation are made. The critical question, then, is whether the demand for coal will (1) continue to stagnate, (2) rise at a rate that maintains its present percentage of primary energy supply, or (3) rise sharply as oil availability declines.

A number of conditions will have to be met before coal can again fill a growing share of the world's energy needs. People must first recognize that alternatives to oil are needed and that coal is one of the most abundant and lowest-cost fuels. This implies an awareness of the longer-term energy outlook, policy decisions by governments to encourage both the production and use of coal, and public attitudes that make such policies feasible.

Widespread burning of coal will pollute the air, which could have severe environmental and climatic impacts. These issues must be faced and ultimately resolved. If use of coal is to increase greatly, new and improved techniques for the clean burning of coal are necessary. Government policies should support research, development, and demonstration in such areas as flue-gas desulfurization, fluidized-bed combustion, and chemical coal cleaning. Governments must also settle debates on clean-air standards and encourage study of the long-term impact of fossil fuel combustion on the earth's atmosphere.

Extensive facilities for handling and burning coal will have to be built. Construction of these facilities will require large investments and the development of better techniques for handling coal in a clean and convenient fashion. Better handling techniques are especially important to industrial, commercial, and domestic users, who would otherwise prefer oil or gas.

In addition to the need for realistic policies, new pollution control, and coal using facilities, consumers should be given an economic incentive to choose coal over other available fuels. The cost of coal should be competitive with that of oil and natural gas, after allowing for the costs to consumers of facilities to use coal and control pollution.

In the WAES unconstrained cases (assuming oil is not limited), the integrations of desired energy demand and potential energy supply yield little change in coal's share of its traditional markets. Its share of electricity generation declines, especially in the WAES "nuclear" cases, and its use in transportation becomes insignificant.

In global integrations limited by oil supply, however, a different pattern emerges. Total non-Communist world energy supplies are sufficient to meet demand only if massive fuel substitutions enable available fossil fuels to be used in the most efficient manner. This involves the elimination of fossil fuels, including coal, from electricity generation to avoid the 60-70 percent generating losses. Oil is reserved exclusively for transportation and petrochemical feedstocks, while coal is used directly in industrial, commercial, and domestic markets.

The degree to which energy consumers will be willing and able to convert to more extensive coal use is uncertain, as is the desirability of removing nearly all fuels from electricity generation. However, our analyses give a clear indication of the potential use of coal as a major alternative to oil. The magnitude of the fuel switching that may be required underscores the difficulties to be faced in making the transition from oil. Individual national governments will have to choose the extent to which they will encourage or discourage the development of expanded coal-consuming systems. Because long lead times are required in changing to an energy system based on coal, decisions must be made quickly if such a transition is to be made. The choices made in the next few years are likely to set the course for the rest of the century.

5.5 COAL SUPPLY POTENTIAL

5.5.1 WOCA Production Estimates

Potential coal production estimates for our scenarios were prepared by each WAES national team. A special analysis of potential coal production and exports from the rest of WOCA was also undertaken. These production estimates are summarized in Figure 5.4.

Throughout the period between 1975 and 2000, Western Europe and Japan will be net importers of coal, potentially on a large scale. Areas with coal export potential will be North America, parts of Latin America, South Africa, South Asia, Australia, and certain Communist countries. The eventual level of coal available for export from

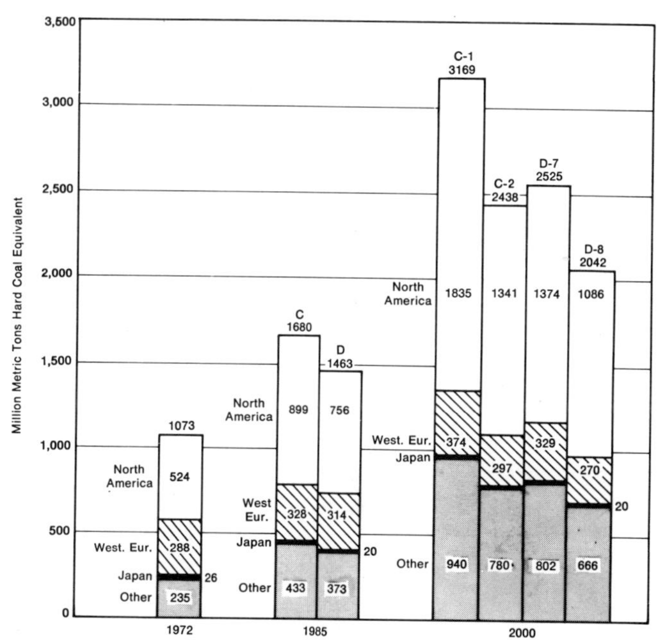

Figure 5.4
Potential coal production in WOCA.

each area is uncertain because it depends on estimates of both supply and demand for coal.

While the rest of WOCA region as a whole is likely to be a net exporter of coal, many countries in that region will import coal from sources within the region such as Australia. Our estimates of potential coal supply, demand, and net exports from the rest of WOCA are summarized in Figure 5.5.

Of the total coal exports from the Australia/New Zealand/South Africa group, approximately two-thirds are likely to come from Australia and the remainder from South Africa. Exports of coal from OPEC countries would be largely from Indonesia, with some from Venezuela. India will produce about half of the non-OPEC developing countries' total coal output, possibly in the range of 200–250 million metric tons in the year 2000.

The large coal reserves in the Communist areas indicate the potential for significant coal export. WAES has not analyzed this potential but assumes net export availability similar to the current levels of 40 million tons per year. These exports might be expanded if the Communist governments choose to make the coal available and non-Communist consumers choose to buy it.

WAES did not attempt to distinguish between steam coal and metallurgical coal. Coal supply data shown in the exhibits include both steam coal and coking coal, for thermal, metallurgical, or chemical purposes. The majority of incremental tonnages available for export will consist of steam coal, which is the major concern of this analysis. In the longer term, the distinction between coking coal and steam coal may become blurred, as new technology widens the types of coal usable for metallurgical purposes.

5.5.2 The United States as a Potential Coal Producer and Exporter

Since the United States possesses over half of the non-Communist world's coal reserves and is the largest producer and consumer of coal, a special review of potential U.S. coal production is appropriate. Estimates of maximum potential coal development were prepared as part of the U.S. supply study for WAES, as shown in Figure 5.6, and are independent of any estimates of the demand for U.S. coal. The results of this assessment give a range of maximum *potential* coal production of 1.1 to 2.0 billion metric tons annually by the year 2000. The low estimate assumes constant real energy prices and nuclear energy as the principal replacement fuel for oil. The high estimate assumes that the price of energy is rising and that coal is the principal replacement fuel.

While 70 percent of the U.S. coal reserves require underground mining, it is anticipated that most

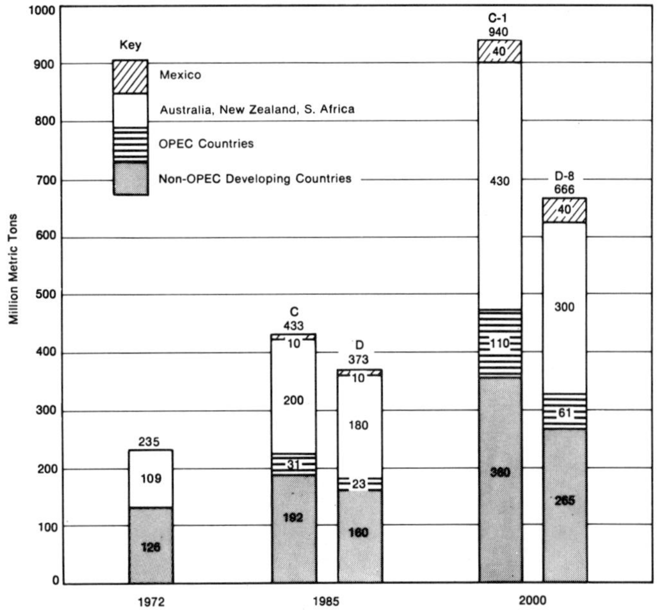

Figure 5.5
Coal production in the rest of WOCA.

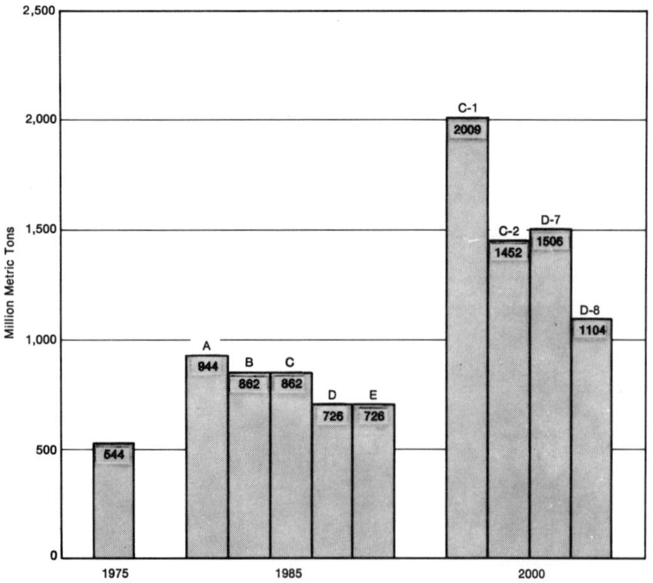

Figure 5.6
Potential U.S. coal production.

new coal production will come from surface mining in the West, particularly in the Northern Great Plains Region where coal seams are thick, near the surface, and located under relatively flat land. Much of this western coal also has a low sulfur content and can meet current emission standards for new facilities without stack gas scrubbing.

However, widespread surface mining in previously undisturbed areas is opposed by environmental groups and local agricultural interests. The two most important objections to western surface mining are (1) the potential despoiling of large tracts of land and (2) potentially undesirable social and economic impacts in a primarily rural, sparsely populated, agriculturally oriented society of small towns and isolated ranches. Ranchers are concerned that some of the ranch land currently used to graze livestock may be ruined for future agricultural use and that water requirements for mining and reclamation may conflict with the needs of farmers.

New mines are now being developed in the Powder River Basin of the Northern Great Plains. Soon some of these mines will be operating, and the mine developers will be in a position to demonstrate whether they can conduct mining operations in an acceptable manner and effectively rehabilitate disturbed lands. The mining companies are confident that this can be accomplished at a reasonable cost relative to the value of the coal produced. If the acceptability of these rehabilitation programs is demonstrated, as has been done in Europe, environmental issues are not likely to be a major constraint on future coal mine development.

Recent federal U.S. legislation has provided assistance to local governments in planning for and meeting the cost of expanded social and community services. Most of the states involved have also adopted strict surface mining regulations that will allow new mine development with concurrent reclamation to prevent the type of abuse that has occurred in the mountainous coal mining areas of the eastern United States.

On the other hand, Congress strongly supports federal legislation to regulate surface mining activities at the federal level and to limit mining in certain areas. This possible legislation and other federal/state conflicts—over regulatory jurisdiction, royalty sharing, land use planning, and local government assistance—may slow the expansion of western surface mining activities. The significance of these issues and the related policy choices is demonstrated by the difference of 550 million metric tons in estimated U.S. annual production potential between the rising energy price/coal scenario (C-1) and the rising energy price/nuclear scenario (C-2).

The scenarios with higher estimates also include substantial increases in eastern underground mining in the United States. Development of this high-cost coal is consistent with the rising energy price scenarios. The relative costs of underground coal may continue to rise because of increasingly stringent mine safety regulations, associated declines in underground mining productivity, and the gradual depletion of readily accessible reserves. But development of eastern underground coal would stimulate employment and economic activity in a region of the country that has had chronic economic development and employment problems for many years.

Would the United States accept the environmental consequences and the depletion of a nonrenewable national resource to provide energy for other countries? Although the United States is capable of producing great quantities of coal, whether it would be willing to undertake extensive coal mining and related transportation infrastructure expansion to provide coal for export is uncertain.

There are a number of reasons for adopting such a policy if the need for the coal is clearly established. Coal in the United States is not a scarce resource. Exporting it to meet the energy needs of others and to offset the cost of imported oil would be fully consistent with U.S. foreign policy goals. It would have a positive effect on economic development, employment, and the balance of payments. But it will not happen unless (1) demand for coal exports is assured by firm long-term contracts, based on buyers' expectations that the coal will be available for export; (2) the environmental acceptability of western surface mining has been demonstrated; and (3) there is sufficient advance planning to provide for infrastructure requirements, especially deep-water port facilities.

5.5.3 Infrastructure Requirements for Coal Development

Substantial increase in WOCA coal production will

require the construction and operation of a large mining and coal transportation network to produce and move coal to market. A special analysis assessing the material, equipment, and manpower requirements for coal development in the highest U.S. coal production case was prepared by Bechtel Corporation.

The Bechtel Energy Supply Planning Model was used to calculate the resources and infrastructure needed to expand U.S. coal production from current annual levels (about 500 million metric tons) to 2000 million metric tons by the year 2000 following the growth profile in Figure 5.7. The results are summarized in Tables 5.3 and 5.4.

In addition, WAES prepared a special analysis of transportation facilities required for rapid expansion of U.S. coal production to 1985. This study concluded that the transportation capacity required to handle a doubling of production can be provided by either rail or slurry pipeline or probably some combination of the two. Rail transport is most suitable for local movements and low-to-moderate quantities, or where maximum flexibility is required. Slurry pipelines are most economical for high and consistent volume and long-distance transport. At the present time, the development of slurry pipelines in the United States faces severe political obstacles that would have to be resolved. The study also concluded that barges would probably not play a significant role in the movement of western coal because the barge traffic on the Missouri/Mississippi River system is already heavy.

Although the United States is the world's leading coal exporter (56 million metric tons in 1974), port facilities are generally deficient for any substantial expansion of coal exports. The United States has limited capacity to accommodate the larger bulk carriers that would be most economical for large coal shipments. In addition, coal facilities are presently concentrated in East Coast and Great Lakes ports far from the abundant western reserves. Clearly, if significant quantities of coal are to be exported, seaport expansion would be required.

The infrastructure requirements are large but not insurmountable given the time period involved and the 6 percent maximum annual rate for U.S. coal production in the WAES scenarios. But any delays

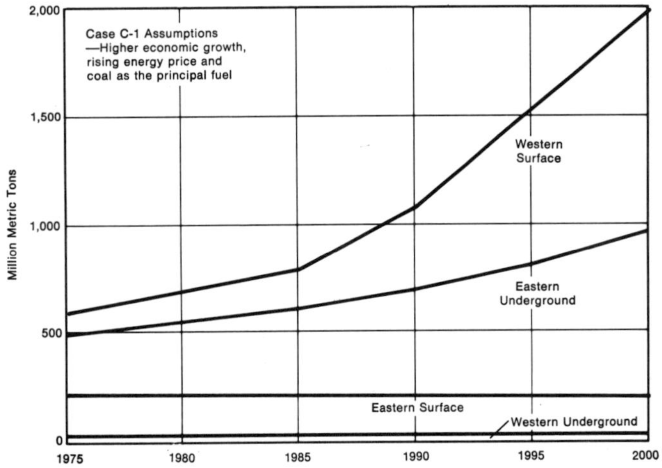

Figure 5.7
U.S. coal production capacity, Case C-1.

Table 5.3
Cumulative coal infrastructure requirements for U.S. coal production test case

New coal mine requirements to 2000
377 eastern underground mines producing 2×10^6 tons/year
75 eastern surface mines producing 4×10^6 tons/year
232 western surface mines producing 6×10^6 tons/year
8 western underground mines producing 2×10^6 tons/year
\$32 billion in capital for coal mining (constant 1975 \$)
New coal-related transportation facilities
1400 unit trains
3200 conventional trains
500 barges
9400 trucks
9 slurry pipelines carrying 25×10^6 tons/year
\$86 billion in capital for transportation (constant 1975 \$)

Table 5.4
Annual requirements of selected resources for U.S. coal production test case

	Units	1975	2000
Construction			
Total technical manpower	man-years/year	7,200	22,000
Total manual manpower	man-years/year	14,000	42,000
Total carbon steel	million tons	2.1	5.2
Draglines	cubic yards	1500	2800
Operations			
Total technical manpower	man-years/year	31,000	80,000
Underground miners	man-years/year	50,000	110,000
Total manual manpower	man-years/year	170,000	420,000

in the early stages of development will make it more difficult, if not impossible, to achieve the levels of coal production indicated in the WAES U.S. projections by the end of the century.

5.6 POLICY ISSUES IN THE DEVELOPMENT OF COAL PRODUCTION

Coal fueled the Industrial Revolution in Europe and North America and was the main industrial and domestic fuel for more than one hundred years. This long history of coal consumption has provided sophisticated coal production technology. However, underground mining is still labor intensive, especially in Western Europe, and deep-mined coal production costs are therefore sensitive to wage inflation. While constant or rising real energy prices will continue to make underground mining economically attractive in many areas, significant increases in coal production will also require sizable development of surface mining. Productivity in surface mining is much better than in deep mines; however, it has its own problems, most notably the environmental objections to disturbing large tracts of land during mining and before restoration.

These environmental, economic, and social issues are potential causes of political conflict in areas of the world that have not previously seen major coal mining. Whether environmental and social opposition will be a constraining factor in the expansion of coal production outside the United States is not clear. However, success or failure of U.S. efforts to reclaim surface-mined areas in the western states may have a direct impact on attitudes toward coal development in other countries.

Although the developing countries are unlikely to follow precisely the historic pattern of industrialization based on inexpensive and plentiful energy, coal could be a major source of energy for their economic growth and development. For those countries with indigenous reserves, it provides a domestic alternative to expensive imported oil and a potential source of export income.

5.7 SYNTHETIC OIL AND GAS FROM COAL

Coal's market penetration could be expanded by converting coal to synthetic liquid and gaseous fuels. Such conversion could be an attractive alternative to rising imports of oil or liquefied natural gas for countries with large coal reserves. It would allow consumers to use existing oil- and gas-fired equipment, thereby putting the burden of new investments on the energy producers and processors rather than on distributors and consumers. It could also simplify air pollution control by concentrating coal use in fewer locations. Conversely, converting coal to other forms of fuel adds to the consumers' cost of energy because of the relatively low efficiency of the conversion process (on the order of 60–70 percent) and the cost of the conversion facilities. The low efficiency of conversion necessitates mining significantly larger tonnages of coal and will require additional water for cooling and land reclamation.

The WAES national studies indicate limited potentials for the conversion of coal to synthetic fuels to the year 2000 within the scenario assumptions. Limited coal gasification is assumed under all scenarios, but significant development is expected only in the high growth/high price scenarios. Coal liquefaction is a factor only in the same scenarios as a response to both higher prices and policies to encourage coal and synthetic fuel development. The maximum contribution of coal-based synthetics in these scenarios is estimated at 2 MBDOE, with two-thirds of the synthetic gas and virtually all of the synthetic oil produced in North America. The North American development is expected to be based on conventional technology while Europe may use some high-temperature nuclear reactors to gasify coal. The global quantities of synthetic products would require roughly 600 million metric tons of coal per year.

Such a level of synthetic fuel development will require early action to support demonstration of commercial synthetic fuel technologies. Lead times are long, but activity could begin promptly. For example, several major U.S. companies have already commissioned the design of plants and have at least preliminary approval from some regulatory agencies. If favorable government decisions are made in 1977, production could begin before 1985 and could reach significant levels by 2000.

Because of high costs, synthetic fuels will most likely require vigorous policy support. Synthetic gas from coal is believed to cost near the $17.25/bbl assumed in the high oil price WAES scenarios. The technological and economic risk associated with new processes may discourage private investments.

Synthetic oil is currently estimated to cost more than $17.25/bbl, but some synthetic oil is projected in the WAES rising energy price coal scenarios, reflecting expected government assistance.

Given the potential for large-scale conversion of coal to synthetic oil and gas, we decided to explore the capital, manpower, equipment, and material resources needed to develop synthetic fuels at levels beyond that assumed in the WAES scenarios. A test case was developed for the United States calling for 3 MBDOE synthetic oil and gas from coal and 0.5 MBD of oil shale, half from conventional retorting and half from in situ recovery. We assumed that the facilities start from zero in 1985 and reach the target level in the year 2000 as shown in Figure 5.8. The results of the test case indicate the need for capital investments of about $90 billion (U.S.) over the period from 1981 to 2000. As shown in Figure 5.9, annual increments of new facilities increase rapidly. Although manpower, equipment, and material resources required are not likely to be constraints, the projected rate of growth beginning in the early 1980s would require commitments during the next few years to realize the levels of production assumed in this test case by the year 2000. The lead times for constructing specific facilities range from three to five years. The total lead time—including policy decisions, investment commitments, design, and engineering—stretches out from six to ten years. Policy decisions must therefore be made in the next few years for facilities to come on stream by 1985 and for production to be significant by 2000. Otherwise, synthetic fuels from coal are not likely to contribute significantly toward our total energy needs in this century.

5.8 IMPLICATIONS FOR ACTION

Coal is the one fossil fuel likely to remain abundant at relatively low cost for the remainder of this century and well into the next. It is one of the major alternative replacement fuels available to bridge the gap from the current era of abundant oil and gas to the future era of renewable energy resources. Coal will be even more important if energy-consuming nations are reluctant to move aggressively to implement nuclear energy. Conversely, decisions to avoid or limit the consumption of coal will increase pressure for nuclear development and energy conservation as alternatives to unsatisfied oil demand.

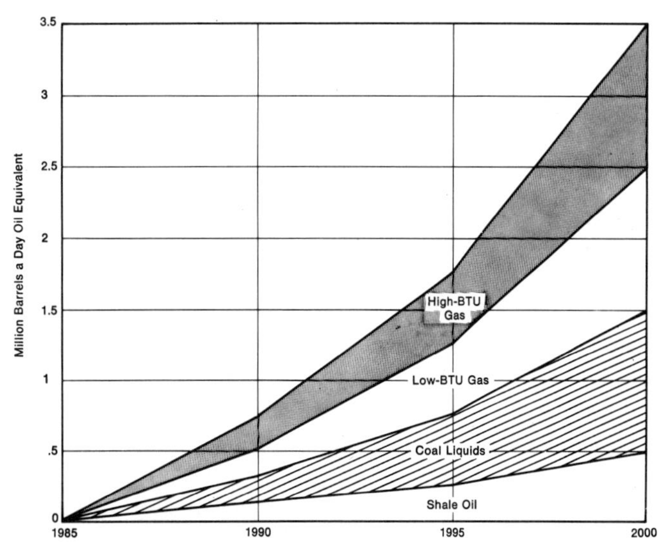

Source: Bechtel test case.

Figure 5.8
Synthetic fuel test case profile.

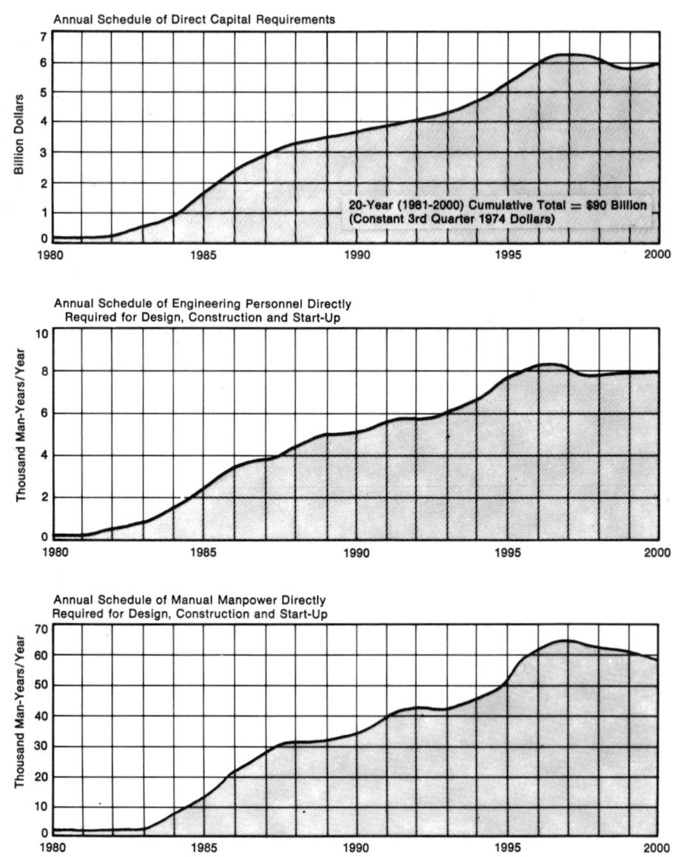

Source: Bechtel Corporation.

Figure 5.9
Synthetic fuel test case requirements.

Realizing coal's potential will, however, depend on recognizing the need to use that potential. Energy-consuming countries must adopt the necessary policies to permit this use. Realizing coal's potential also requires accommodative policies in coal-producing nations, the development of transportation systems (including deep-water port facilities), and satisfactory solutions to the environmental problems associated with mining and burning coal.

6 Nuclear Energy

6.1 THE WAES APPROACH[1]

Our projections of the future available supply of nuclear energy for 1985 and 2000 are based on estimates by WAES national teams for their own countries and on our joint estimates for non-WAES Europe, OPEC, and the rest of the world outside Communist areas (WOCA).

When we began our work in 1974, most of the WAES countries had embarked upon major programs to expand their electric power systems with large nuclear-fueled plants. Each plant of 1000 MWe capacity (1000 MWe = 1 GWe = 10^9 watts installed capacity) would be large enough to supply electricity for a large city in an industrialized country. The total capacity of such plants in WOCA was projected at that time to grow from about 70,000 MWe in 1974 to more than 2 million MWe by the year 2000—at an annual growth rate of about 15 percent—and at a cost of about $0.4 to $1.0 billion (1974 U.S.$) per 1000 MWe (excluding cost of transmission, distribution, and fuel cycle facilities, which together might be more than half of the power plant cost). Such massive building programs reflected confidence that (1) electricity use could and should grow more rapidly than other energy forms; (2) nuclear power plants were safe and could produce electricity at lower cost than oil, gas, or coal; and (3) use of nuclear energy would reduce national dependence on imports of traditional fuels.

Such confidence was also based on the outstanding safety record of operation of nuclear reactors and related facilities such as enrichment and reprocessing plants over more than twenty-five years. Early recognition of the hazards of nuclear radiation led to adoption of stringent and costly safeguards to protect workers and the public. Strict enforcement of such safeguards under close control by public authorities has given this industry up to now an impressive safety record.

Yet the unique character of nuclear energy as a source of radioactivity and its potential for destruction have led to resistance in many countries to the growth of the nuclear power industry. Such resistance has focused on various concerns including reactor safety, possible diversion of nuclear materials to military or terrorist use, the safe disposal of the highly radioactive waste, and the safety of fast breeder reactors.

Proponents of nuclear energy point to the fact that radioactivity is a phenomenon long known to science, that procedures for its control and containment are well established and demonstrated, and that other equally dangerous industrial activities have been accepted by society. Opponents of nuclear energy argue that it is unrealistic to expect the good safety record of the industry to be maintained in the wake of the growth and spread of nuclear capacity. The lack of a clear resolution of such differences has been a major factor in slowing down the growth of the industry from earlier expectations and is the main source of uncertainty as to its future growth.

6.2 WAES ESTIMATES OF FUTURE INSTALLED NUCLEAR CAPACITY

WAES did not attempt to seek agreement on how and when the nuclear debate would be resolved in various countries. Instead we decided to show the scale of potential contributions to the world's primary energy needs in 1985 and 2000 that could come from nuclear energy if an intensive nuclear power program were continued for twenty-five years. In this way we could show how the pace of nuclear development would affect the total supply of primary energy required to meet the world's future energy needs.

Because nuclear power might contribute substantially to the world's future energy supply, we selected it as a scenario variable for our projections over the period from 1985 to 2000. Our scenarios for this period specify a possible range of future installed nuclear capacity for electric power generation. The low nuclear estimate is based on the assumption that coal will be the major replacement fuel for oil; the high estimate assumes that nuclear power is the principal replacement fuel. Each WAES national team developed for their country estimates of nuclear power for 1985 and 2000.

1. Section 6.4 describes the fuel cycles for the principal types of nuclear power reactors and defines terms commonly used to describe nuclear systems. Some readers may wish to review this section before reading the rest of this chapter.

The maximum nuclear estimate (Case C-2) for year 2000 is defined as *maximum likely*, and the minimum nuclear estimate (Case D-7) is defined as *minimum likely*. These national projections reflect the national teams' best judgment as of October 1976 of the future installed nuclear capacity in their countries under the WAES scenario assumptions. The national estimates were generally based on or derived from the latest official forecasts. Nuclear projections appear to change fairly often, and therefore the date of October 1976 should be noted. For non-WAES countries, estimates were made by WAES Associates who were familiar with nuclear programs in these countries. In other cases we used OECD Nuclear Energy Agency/International Atomic Energy Agency estimates.

Our first nuclear estimates were made about the time of the December 1975 release of a joint report by the OECD Nuclear Energy Agency and the International Atomic Energy Agency, *Uranium—Resources, Production, and Demand, Including Other Nuclear Fuel Cycle Data.* This valuable study contains national estimates of future installed nuclear capacity to the year 2000 submitted by many countries early in 1975. It provided a framework for our technical assessment of the various parts of the nuclear fuel cycle (we shall summarize these briefly later in this chapter). This study covering the entire nuclear fuel cycle was made under the direction of thirty experts from twenty countries. Such a study carrying the authority of the two principal international bodies best informed about the nuclear industry was an invaluable aid in making our assessment. We have used their analyses and data, but we recognize that changes have occurred since early 1975 when the data were gathered and changes will continue to occur. Revised estimates of nuclear power growth appearing in the *IAEA Bulletin* of October 1975 (Vol. 18, No. 5/6, p. 6), for example, are comparable for the year 2000 in their high case with the WAES maximum likely Case C-2. The WAES minimum likely figures for 2000 (Case D-7) are substantially below their lower figures. This difference is due partly to the WAES scenario variable of low nuclear/high coal in Case D-7.

Reserves of uranium fuels, planned enrichment capacity, and planned reprocessing capacity were studied to determine whether they would be adequate to support the maximum likely and minimum likely levels of the WAES projections. A special review of the outlook for fast breeder reactors was made by a WAES team. The results of that review are reflected in our estimates in later sections of this chapter.

If the estimated maximum and minimum levels of installed nuclear capacity are achieved and if proven plants operate at 60 percent load factor to meet electricity demand, then the percentage of WOCA primary energy supplied by nuclear in the WAES cases would be as shown in Table 6.1.

The relationship of these nuclear energy projections to total energy needs as shown in Table 6.1 is very important. In the maximum case for the year 2000, the expected contribution of nuclear energy is 43 MBDOE—about as much as total 1975 oil production in WOCA. If public decisions delay or reduce the projected rapid growth of nuclear power (our estimates imply an 11-14 percent annual growth rate between 1975 and 2000), these energy needs will have to be borne by other sources or demand will have to be reduced with attendant consequences.

None of the maximum supply potentials that we have estimated for oil, natural gas, coal, and other sources seems likely to be capable of much expansion without higher energy prices or supervigorous policy actions. Therefore, shortfalls in nuclear power would most likely increase the prospective shortage between the demand for desired fuels and probable supply. Answers are needed to the following questions: How likely is a shortage in the WAES nuclear power projections? Would it occur in some countries but not in others? What kinds of uncertainties may affect the realization of the WAES nuclear projections?

6.3 SOME SPECIAL CHARACTERISTICS OF THE NUCLEAR SAFETY DEBATE

Development of nuclear power is being carried forward in an atmosphere of concern about radioactivity—the main difference between nuclear and other forms of energy. In spite of the positive safety record of reactor operation up to this time, the nuclear industry has been called upon to prove the safety of its reactors and the safety of every step in the fuel cycle. The question of safety has been debated publicly and often in terms similar to those that

Table 6.1
Summary of WAES estimates of installed WOCA nuclear power capacity

	1974	1985		2000	
	Installed capacity	Maximum likely	Minimum likely	Maximum likely	Minimum likely
Nuclear capacity (GWe)	66.9	412	291	1772	913
Share of total primary energy consumed in WOCA (%)	2	9	6	21	14
Oil equivalent (MBDOE)	1.7	10	7	43	22

Note: We have adopted the convention of expressing nuclear electricity production in terms of the fuel input that would be required to produce the equivalent amount of electricity output from fossil-fueled power stations. The assumed generating efficiency is 0.35 in 2000. The formula used to convert GWe to MBDOE, assuming 0.35 efficiency and a 0.6 load factor, is

$$\text{GWe} \times \frac{8700}{620{,}000} \times \frac{1}{0.35} \times 0.6 = \text{MBDOE primary energy input.}$$

have dominated discussions of atomic weapons and the threats they represent to human survival.

There are several serious problems around which debate has turned. Probably the most serious is that of containment of radioactivity. The safe transport, storage, and treatment of spent fuel elements and the resulting highly radioactive wastes that remain active for hundreds and thousands of years is a matter of great public concern. Further research and demonstration are required before the most acceptable processing methods and the most suitable sites for waste disposal can be selected.

Still another concern is the safety of nuclear power plants themselves, particularly the consequences of a failure that might lead to a release of radioactivity. Although it may become widely accepted that the probability of such a failure is very low, there is a fear of the unpredictable and possibly widespread effect of such an accident. The proposal to make prototype fast breeder reactors which produce heat in a small space has heightened this concern over reactor safety, despite the required introduction of special safety features in the design to provide for the possibilities of melting down of the core.

Many people are concerned about which materials are usable for nuclear weapons. The explanation given here may help the reader. Uranium for power reactor fuel at 2–4 percent enrichment in U-235 is useless for weapons. Similarly U-233 mixed with natural uranium to the level needed for reactor fuel (below 15 percent U-233) is useless for weapons. Such fuel-grade uranium can only be made into weapon-grade materials in an enrichment plant. Uranium enriched significantly above a U-235 content of 20 percent, U-233 made from thorium, and plutonium are all usable in nuclear weapons.

With plutonium a conventional chemical process can be used for obtaining materials from spent fuel elements for use in weapons, although it requires relatively complicated equipment. No means for rendering plutonium useless for weapons is known.

Nations that have made nuclear weapons first used nuclear reactors only to produce plutonium; any sizable country could, without insurmountable difficulties, obtain the same result in the near future, without resorting to nuclear power generating technology. Nevertheless, there is public concern about this part of the nuclear fuel cycle. Most nuclear power reactors can also partially use plutonium in the form of oxide as fuel, and large fast breeder reactors (FBRs) will require several tons of plutonium (or highly enriched U-235).

Another area of concern is the "front end" of the nuclear fuel cycle where the fissionable fraction (uranium 235) in natural uranium can be concentrated; nuclear power reactors require 2-4 percent U-235 concentration as fuel in the form of uranium oxide. Enrichment plants employing the same technology can produce the highly enriched U-235 for use in weapons, and they have been doing so for the past twenty-five years. This latter method for obtaining weapon-grade material is more difficult than the plutonium route.

The public debate on such issues is widespread and goes on in many countries at different levels of intensity. Some argue for the cessation of nuclear activities such as additions to fuel reprocessing facilities and the development of the fast breeder reactor. Others point out that in spite of the extensive experience and small-scale experiment to date, these issues can only be resolved by continuing development and demonstration. The attitudes of the general public on these questions are beginning to be reflected in the political process. In some countries political action has prevented or delayed the siting of nuclear plants. On the other hand, 1976 referenda in seven states in the United States all gave a large majority of support for a qualified continuation of nuclear development.

The increasing desire of many countries, including some in the developing world, for nuclear capacity to meet their energy needs has been accompanied by a rising international concern about the spread of the capability for making weapons based on plutonium and a move towards more effective controls including the possibility of multilateral operation and control of fuel reprocessing plants and stricter international control of waste disposal.

Later in this chapter we call attention to three stages of choice in the nuclear fuel cycle and point out the possibility of considering separately (1) reactor operation involving a single use of uranium without fuel reprocessing, (2) the processing of used fuel to extract and recycle plutonium and uranium, and (3) operation of fast breeder reactors.

We shall also point out some of the consequences of continued delay in decisions and the effects of such uncertainties on the WAES energy supply projections to the year 2000. Here we emphasize only that the nuclear industry needed to supply and fuel future nuclear power plants may not survive widespread moratoria or long delays on nuclear plant construction.

Although the debate on these issues continues, our projections necessarily assume the existence of a large and vigorous global nuclear industry capable of building one or two thousand nuclear power plants over the next twenty-five years, along with all of the associated parts of the nuclear fuel cycle. Such an industry would have to be capable of supplying power plants and related facilities whose total cost over the twenty-five years would range from \$1000 billion to \$3000 billion in constant 1975 U.S. dollars.

6.4 NUCLEAR FUEL CYCLES

6.4.1 Reactor Types

There are several different designs of nuclear power reactors. Each type has distinctive performance characteristics, economic considerations, fuel requirements, and so on. The OECD/IAEA study (reference 1), for example, projects seven kinds of reactors in their estimates of reactor types to be built by the year 2000. For analysis of fuel cycle requirements for WAES cases, we were guided generally by the OECD/IAEA projections of reactor mix because the mix of reactor types affects requirements in stages of the fuel cycle. For 1985 we used LWR 90 percent, HWR 5 percent, other 5 percent. For 2000 we used LWR 80 percent, HWR 7 percent, FBR 5 percent, other 8 percent. We confine our assessment to these three principal nuclear power reactor types: the light water reactor (LWR), of which the pressurized water reactor (PWR) and boiling water reactor (BWR) are subtypes; the heavy water reactor (HWR) (CANDU-type); and the fast breeder reactor (FBR).

6.4.2 Reactor Design Factors

Nuclear plants are similar to fossil fuel plants in that both systems generate steam to turn turbines connected to generators that produce electricity. The difference is that nuclear plants use nuclear fuels.

A nuclear energy plant—a reactor—is based on controlling a chain reaction in fissile material. The fissile material (U-235) is "diluted" with other material (U-238), but the fissile U-235 nuclei are the nuclear fuel. When a neutron is absorbed into a U-235 nucleus, the nucleus can split, releasing a tremendous amount of energy that is ultimately converted to heat and ejecting two to three neutrons. Three things may happen to these neutrons: (1) some may go out through the surface of the uranium mass and get lost; (2) some may be absorbed by a nucleus of U-238, which then becomes plutonium (Pu-239), a fissile material like U-235; and (3) some may be absorbed into other U-235 nuclei. At least one neutron must enter a U-235 nucleus to keep the chain reaction going. The U-235 nucleus splits, releases heat, and ejects some neutrons.

The "desired reproduction factor of 1" means that from each U-235 nucleus that splits, at least one neutron is absorbed into another U-235 nucleus to produce fission. When a reactor has a reproduction factor of 1, the system is "critical" and the chain reaction, which produces the heat, continues.

The system depends on a large enough quantity of uranium, the critical mass, to contain enough nuclei to maintain a chain reaction. Naturally occurring uranium (an element found in natural deposits) contains only 0.7 percent fissile U-235 nuclei. Enrichment increases this concentration to about 3 percent. This still relatively low concentration is adequate to sustain a chain reaction in a reactor provided the neutrons from fission are slowed down by a "moderator," a material with light nuclei, such as ordinary water, heavy water (D_2O), or carbon in the form of graphite.

A nuclear reactor, then, consists of fissile material, usually in the form of rods, in association with moderators. Adjustment of control rods, which are high neutron absorbers, and moderators in the reactor "core" where the reaction takes place keeps the system "critical" and allows shutdown when the control rods are pushed into the core to absorb neutrons, stopping the reaction.

6.4.3 The Light Water Reactor

There are two types of light water reactors—the pressurized water reactor and the boiling water reactor. They use the same fuel cycle. The fuel cycle for the LWR (see Figure 6.1) begins with uranium ore, a natural resource that must be discovered, mined, and processed to form a concentrate (U_3O_8) called yellowcake.

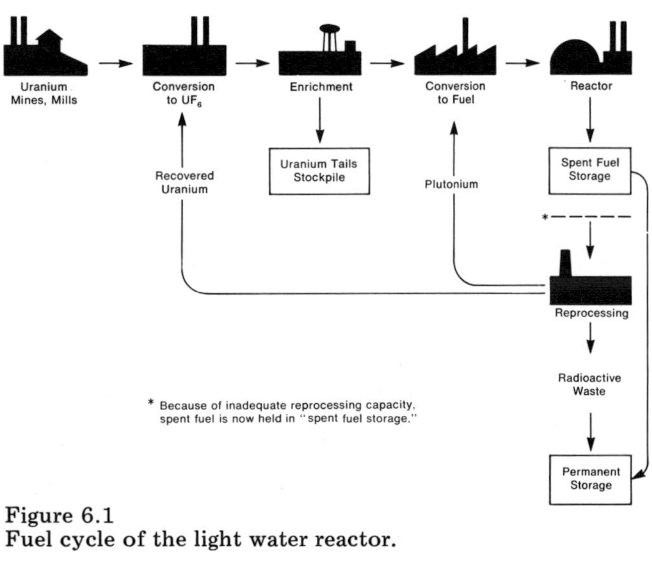

Figure 6.1
Fuel cycle of the light water reactor.

The uranium concentrate (U_3O_8) is then converted to a gas (uranium hexafluoride, UF_6), which goes to an enrichment plant. Uranium tails— uranium depleted in U-235 to 0.2-0.3 percent—are a by-product of the enrichment processes and are saved for use in fast breeder reactors. Fabrication follows enrichment and includes converting UF_6 to uranium dioxide (UO_2), which is the fuel, pelletizing, encapsulating in tubes, and assembling the uranium dioxide tubes into fuel elements. Finally, the fuel elements are loaded into reactors, which use the heat from nuclear fission to produce steam to drive the turbines and generators that produce electricity.

These steps are known as the "front end" of the nuclear fuel cycle. The "back end" of the cycle begins when the used fuel elements are removed from the reactor. According to a predetermined schedule, 20-30 percent of the fuel is removed each year. Used fuel is highly radioactive and produces some heat because of the slow decay of the radioactive fission products. The used fuel must be stored under water at the reactor site for many months to ensure removal of the heat and to shield against the radiation emitted by the fission products.

The used fuel elements can then be transported in specially shielded containers to a reprocessing plant. Here they are mechanically chopped up, dissolved in acid, and go through a chemical process to separate out three components: (1) the remaining uranium, (2) plutonium, and (3) the radioactive fission products. Such reprocessing plants are complex because the materials to be recovered (U and Pu) are mixed with highly radioactive fission products, although these eventually reach the point where they can be handled without shielding. Many operations must be performed by remote control, and operators are shielded from radioactive wastes by thick walls. Such plants have been operating for up to thirty years to recover Pu for weapons, so the technology for remote handling has been well developed.

At this stage there are a number of alternatives. The recovered uranium can be converted to UF_6 as feed for an enrichment plant, or it can be stored.

The uranium in reprocessed fuel from LWRs is normally richer in U-235 (0.8 percent) than is natural uranium. The plutonium can be recycled into the LWR for use with uranium as reactor fuel; it can be stored; or it can be used as fuel for a fast breeder reactor. The radioactive waste must be stored in permanent and "leakproof" facilities. In a single-use system the used fuel elements are not reprocessed but put directly into permanent storage repositories.

Plutonium, which does not occur in nature, is a by-product of all nuclear reactor operations where U-238 is present. The U-238 absorbs neutrons and is transmuted into plutonium. If recovered plutonium and uranium were recycled into reactors as fuel, they could replace about 20 percent of the fresh uranium required for fueling the LWR. Requirements for natural uranium could be reduced proportionately.

6.4.4 The Heavy Water Reactor (CANDU-type)

The heavy water reactor fuel cycle (see Figure 6.2) differs substantially from that of the LWR. HWR nuclear power plants using natural uranium as fuel and heavy water (D_2O) rather than light water (H_2O) as the moderator have been developed, and the four CANDU units at Toronto (2.0 GWe) have been in operation since 1972. Nuclear capacity in operation and under construction in Canada (up to and including Pickering B) now totals 8.8 GWe. In the HWR fuel cycle the uranium fuel is used only once. The used fuel is now stored without plans for reprocessing (single use).

The HWR cycle uses natural uranium as the fuel, thereby eliminating the conversion to UF_6, the enrichment step, and the creation of a stockpile of depleted uranium tails. In Canada today used fuel elements are stored until it is decided whether they will be reprocessed to recover uranium and plutonium or be placed in permanent storage repositories without reprocessing.

6.4.5 The Fast Breeder Reactor

The LWR and HWR are called thermal reactors because they operate with thermal neutron energies. Heat is generated by energy released during nuclear fission. This heat is carried away by a coolant and is used to raise steam.

The fast breeder reactor (FBR) is so named because it operates with fast neutrons (energies $10^7 \times$ thermal) and "breeds" more fuel than it consumes. It also operates as a power plant to generate steam. It provides a way to use most of the 99 percent of natural uranium that is the nonfissionable U-238. The breeder converts U-238 by absorbing neutrons to produce plutonium, which is fissionable. Thus, by extracting energy from most of the uranium that is unused by present reactors, fast breeders offer the possibility of multiplying the usability of natural uranium resources perhaps fifty times. The initial fuel loading of an FBR requires large amounts of plutonium, which must be obtained by reprocessing used fuel elements from thermal reactors. Figure 6.3 shows the fast breeder reactor fuel cycle.

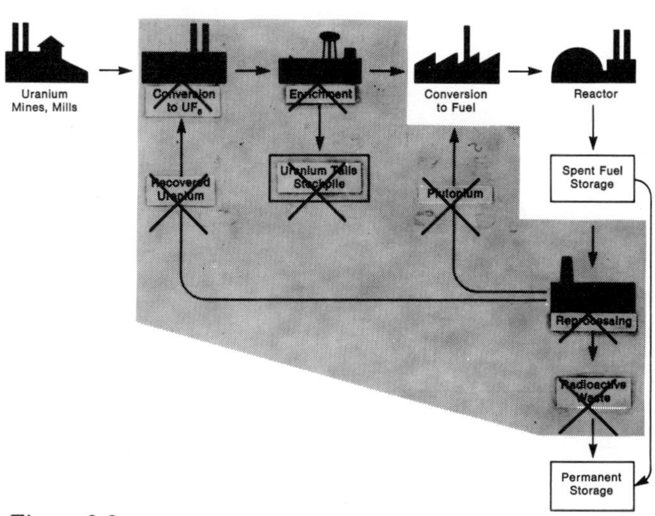

Figure 6.2
Fuel cycle of the heavy water reactor.

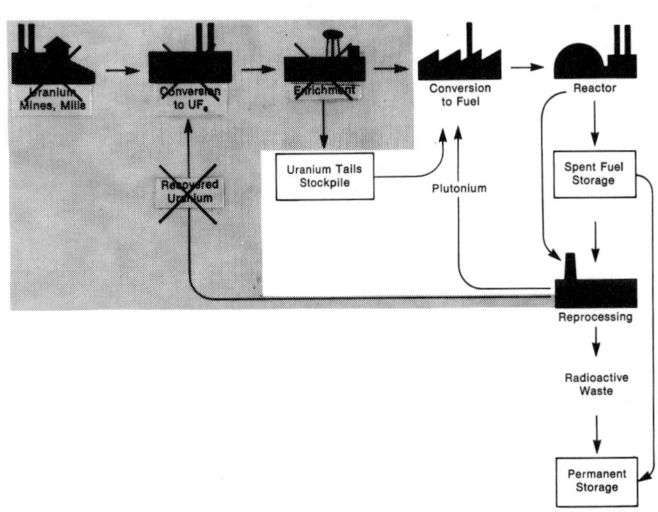

Figure 6.3
Fuel cycle of the fast breeder reactor.

6.4.6 The High Temperature Reactor

Successful development of the high temperature reactor (HTR) could create new uses for nuclear energy because such reactor systems are expected to deliver heat at high temperature for such industrial purposes as coal gasification and hydrogen production. Large-scale HTR prototypes have been built in Germany and in the United States. Problems encountered in commercial-scale units and the accompanying fuel cycle have not yet been resolved.

6.5 THE WAES NUCLEAR ASSESSMENT

6.5.1 WAES Scenario Estimates of Future Installed Nuclear Capacity

In the introduction to this chapter we explained how our estimates were developed and presented the nuclear energy share of total projected primary energy. Table 6.2 lists the actual installed nuclear capacity in 1974 for the countries and regions we examined and our estimates of future installed nuclear capacity for 1985 and year 2000. The table lists projections for Mexico, a member of WAES, separately from the "Rest of WOCA."

Figure 6.4 shows the projections of installed nuclear electric generating capacity based on estimates by WAES national teams of maximum likely and minimum likely levels for 1985 and 2000.

6.5.2 Uranium Resources

Uranium reserves are classified as "reasonably assured reserves" (RAR) and "estimated additional resources" (EAR) (see reference 1). The grades of such reserves are in the range 0.05–0.20 percent or 0.5 to 2.0 kg of U_3O_8 per ton of ore. In the OECD/IAEA Report (reference 1) the total of RAR was

Table 6.2
WAES scenario estimates of installed nuclear capacity (GWe) (as of October 1976)

	1974	1985		2000			
				Rising energy price High economic growth		Constant energy price Low economic growth	
Region or country	Installed capacity[a]	Minimum likely	Maximum likely	C-1 "coal"	C-2 "nuclear" (maximum likely)	D-7 "coal" (minimum likely)	D-8 "nuclear"
Denmark	0	1	1	2	5	2	5
Finland	0	2.2	3.2	6.6	16.2	4.2	8.2
France	2.9	35	45	90	140	70	100
Germany	4.0	21	33	50	120	50	80
Italy	0.6	5.4	9.4	20	60	20	60
The Netherlands	0.5	1.5	2.5	10	16	6.5	10
Norway	0	0	0	4	4	2	3
Sweden	2.6	3.8	7.4	10	20	10	20
United Kingdom	5.8	14	15	33	58	29	50
Non-WAES Europe[b]	2.5	23	31	72	94	72	94
Total Europe	18.9	106.7	147.5	297.6	533.2	265.7	430.2
Canada	2.5	8	12	44.5	74	44.5	74
United States	40.4	127	166	380	620	380	620
Total North America	42.9	135	178	424.5	694	424.5	694
Japan	3.9	25	40	75	120	75	100
Rest of WOCA	1.2	24	47	212	425	148	300
(including Mexico)	(0)	(2)	(3)	(40)	(80)	(20)	(30)
Total WOCA	66.9	290.9	412.5	1009.1	1772.2	913.2	1524.2

[a] From The Atlantic Council, "Nuclear Policy," as corrected by WAES countries.
[b] Non-WAES Europe includes Belgium, Luxembourg, Ireland, Austria, Greece, Yugoslavia, Portugal, Spain, and Switzerland.

estimated to be about 1 million tons of U_3O_8 at prices up to \$15/lb U_3O_8, and an additional 730,000 tons at prices up to \$30/lb U_3O_8. Estimated additional resources were 1 million tons up to \$15/lb and 680,000 tons at prices up to \$30/lb U_3O_8. Of the combined total of RAR and EAR, 72 percent lies in Australia, Canada, South Africa, and the United States.

In addition to these reserves and resources, estimates of resources may range higher for three reasons: (1) Until now, only a fraction of the earth's crust has been explored. (2) The price increases of the last two years to levels above \$30/lb U_3O_8 are not yet reflected in published reserve figures. (3) Uranium is known to be found in much lower concentrations that could be produced at much higher costs. Because fuel cost is a small fraction of the cost of the electricity produced in nuclear power plants, increases in uranium costs have much less effect than increases in fuel costs in fossil fuel plants.

Another potential fuel, with reserves of approximately the same order of magnitude as uranium, is thorium. When thorium is placed in a region of neutron flux, for example, within or surrounding a nuclear reactor core, thorium nuclei (Th-232) absorb a neutron and are transmuted into U-233, a fissile material that behaves like U-235 and Pu-239 in reactors. Thorium conversion to U-233 could become important in various types of thermal converter reactors, including heavy-water-moderated thermal reactors and high temperature reactors, if major research and development were undertaken.

Figure 6.5 shows estimated annual uranium requirements, and Figure 6.6 shows cumulative uranium requirements to the year 2000 to meet the needs of the WAES maximum and minimum likely projections. Estimates of planned and projected uranium production capacities are based on the OECD/IAEA Report and were made before the uranium price increases of the past two years. Much production capacity must be added to avoid uranium shortages in the early 1980s. From Figure 6.6 reasonably assured reserves appear to be sufficient to meet uranium needs of the WAES maximum case until the early 1990s. Therefore, to the extent that estimated additional resources are discovered and developed, uranium shortages would not appear until after the year 2000. The WAES low case could be met from reasonably assured reserves until the year 2000. If electric power companies make long-term contracts covering uranium supply for ten to twenty years of reactor operation, the total of RAR and EAR would be committed sooner at a lower level of installed nuclear capacity than in the WAES projections.

As in the rest of the nuclear industry, uncertainties about future demand discourage large exploration programs as well as the opening of new mines,

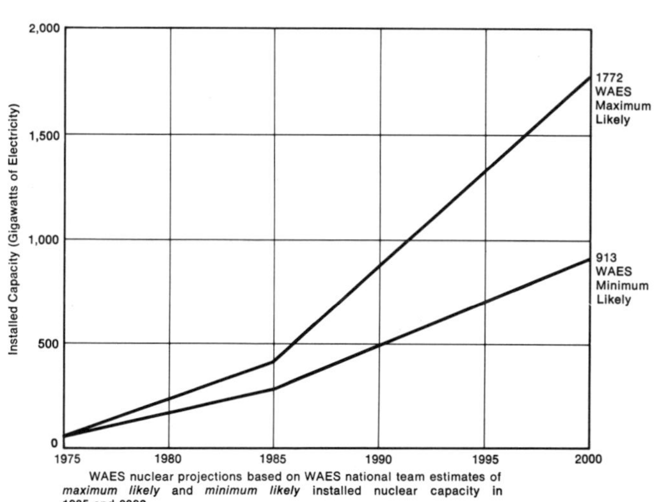

WAES nuclear projections based on WAES national team estimates of *maximum likely* and *minimum likely* installed nuclear capacity in 1985 and 2000.

Figure 6.4
Projected growth of nuclear power in WOCA.

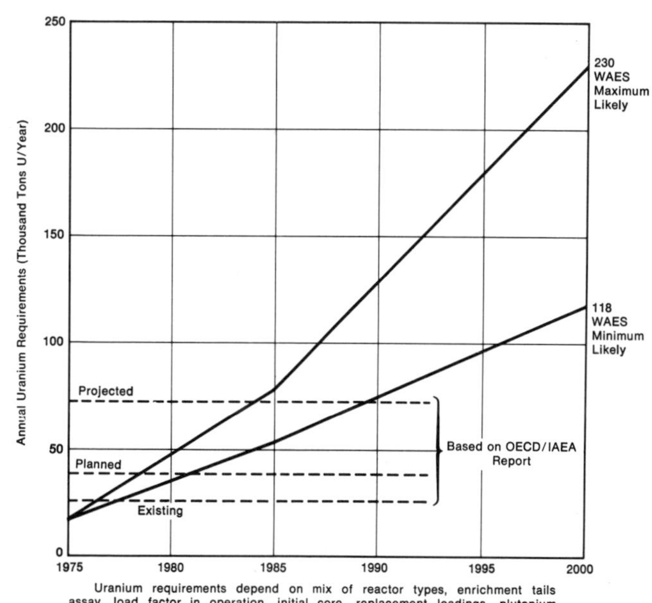

Uranium requirements depend on mix of reactor types, enrichment tails assay, load factor in operation, initial core, replacement loadings, plutonium recycle, delays in fuel cycle, etc. Required amounts shown here are computed on basis used in OECD/IAEA Report without plutonium recycle.

Figure 6.5
Projected annual uranium requirements in WOCA.

which may take six to seven years to start and can cost as much as $100 million before the first uranium is shipped. Moreover, the uranium mining industry remembers the lean years of low demand and low prices of the 1960s. The recently formed Uranium Institute may serve as a forum for producers and users of uranium and may lead to long-term forward commitments for uranium by the electric power industry. These commitments will allow the uranium mining industry to undertake exploration programs and to raise the capital for the needed large additions to production capacity.

Availability of uranium will also depend upon the policies of uranium-exporting countries. Canada, for example, earmarks for domestic use uranium reserves sufficient to fuel the nuclear reactors in Canada for thirty years. At this time it is not clear what conditions will be imposed on buyers of Australian uranium.

6.5.3 Enrichment Capacity

The unit of measure for enrichment capacity (a measure of the energy required for the enrichment process) is a separative work unit (SWU). The enrichment process is illustrated in Figure 6.7. Energy is applied to the natural uranium, and the products of the enrichment process are enriched uranium and depleted uranium called "tails." Separation or enrichment processes rely on the difference in mass between U-235 and U-238. In the centrifuge rotating at very high speed the lighter U-235 is separated from the heavier U-238.

Estimated annual separative work requirements are shown in Figure 6.8. An increase in tails assay from 0.25 to 0.30 percent would reduce separative work requirements by about 10 percent and increase uranium requirements by about the same percentage. Conversely, reducing tails assay from 0.25 to 0.20 percent would increase separative work requirements by about 10 percent and reduce uranium requirements by about the same percentage. Recycling recovered plutonium would further reduce separative work requirements by approximately 10 percent. The projected uranium enrichment capacity was developed from the OECD/IAEA projections.

From Figure 6.8 it appears that if projected enrichment capacity is built without delay, the needs for enriched fuel for LWRs in the WAES maximum case could be met until the late 1980s, when additional capacity would be needed. Currently projected capacity could meet the WAES minimum case needs until the year 2000. Enrichment capacity

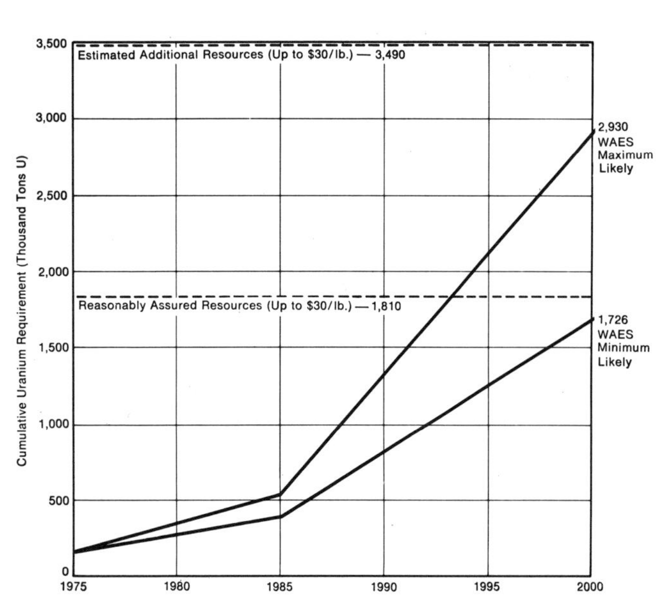

Figure 6.6
Projected cumulative uranium requirements in WOCA in relation to "reasonably assured reserves" and "estimated additional resources."

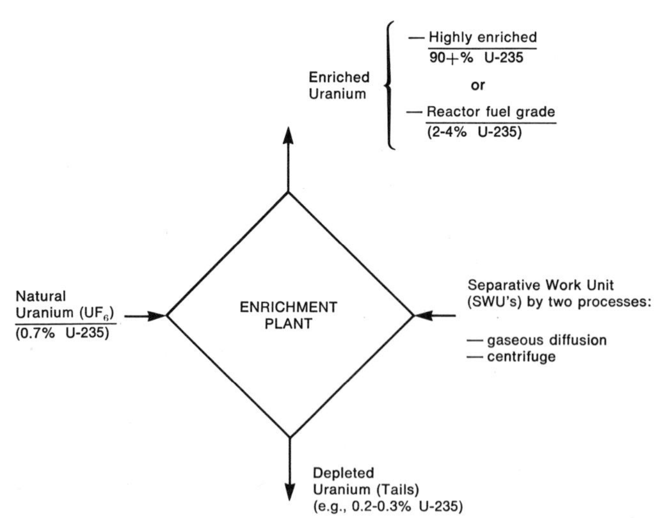

Figure 6.7
Schematic representation of the uranium enrichment process.

Table 6.3
Planned and projected additions to reprocessing capacity for LWR fuel (tonnes of heavy metal per year)

Countries and projects	1979	1980	1981	1982	1983	1984	1985	1986	1987	1988
Europe	600	860					2600	4100	5600	6600
Japan	210								1700	→
United States										
AGNS	1500									→
NFS				750						→
Other								600	2100	4200
Total capacity	2310	2570		3320			5060	7160	11,600	14,760

Source: OECD/IAEA Report, revised to reflect announced cancellation of an NFS facility and other reported delays projected as of September 1976.

Note: Assuming spent fuel output of 30 tons/year per GWe of LWR capacity at 0.6 load factor, the annual output of LWR oxide fuel for reprocessing is as shown below.

	1985		2000	
	C (max)	D (min)	C-2 (max)	D-7 (min)
LWR capacity (GWe)	371[a]	261[a]	1417[b]	728[b]
Spent fuel (tons U per year)	11,130	7830	42,510	21,840

[a]LWR = 90 percent of total.
[b]LWR = 80 percent of total.

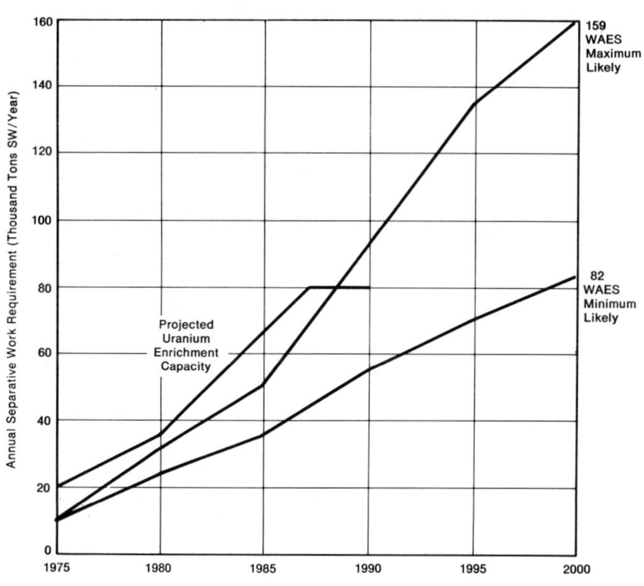

Figure 6.8
Projected annual separative work requirements in WOCA.

must be sufficient to meet fuel needs for new plants and to ensure a continuing supply of enriched fuel for existing nuclear plants.

6.5.4 Reprocessing Capacity

Table 6.3 shows projected additions to reprocessing capacity for LWR fuel as indicated in the OECD/IAEA Report, modified to reflect the announced cancellation of a 600 ton/year facility in the United States and delays on some other projects.

From such additions to existing capacity, we conclude that the uranium reprocessing capacity to recover U-235 and plutonium for LWR recycle or FBR initial fuel loadings will be insufficient until the late 1980s under the assumptions of the WAES high nuclear case, thus requiring storage of used fuel in the interim. The WAES high nuclear case could still be feasible despite this reprocessing undercapacity provided interim storage facilities for used fuel are expanded, thus allowing storage of spent fuel for several years or more before reprocessing.

6.5.5 Other Parts of the Fuel Cycle

While the basic technology for the LWR fuel cycle is well developed, some parts of the fuel cycle face complex problems that will have to be solved soon to provide nuclear power on the scale and rate projected in the WAES assessments. Methods of disposing of radioactive wastes, in their ultimate repository, are still debated. If demonstrated systems for handling such wastes become a prerequisite for reactor licensing, WAES projections could be delayed several years in countries with such policies.

6.5.6 Fast Breeder Reactors

Fast breeder reactors (FBRs) were projected by OECD/IAEA in their high case to amount to about 10 percent of total nuclear capacity by year 2000. We have reassessed the current position and probable future rate of development of the FBR in Europe. Our conclusion is that not more than 5 percent of the 533 GWe estimated for Western Europe in the WAES high nuclear case could be supplied by FBRs. By 1990, about 10 GWe of FBRs could be operating in France, Germany, and Great Britain, with another 10–20 GWe possible by the year 2000. On such a basis, it is unlikely that power from FBRs could provide more than 5 percent of nuclear energy by 2000. Estimated delays are due partly to schedules for construction and operation that allow time to gain experience with first-commissioned FBRs before additional units are ordered and built.

In the United States the situation is more uncertain, but recent official estimates (reference 2) indicate that FBRs might begin commercial operation in the United States in 1995. To achieve and sustain the growth of a nuclear program the size of the WAES high nuclear case by year 2000 would require the FBR to become commercial by the turn of the century or earlier. In addition, extra uranium supplies will be required to fuel the then-existing stock of thermal reactors during a transitional period until breeders begin to add to the fuel supply.

"Bred" plutonium from FBRs could add significantly to nuclear fuel supplies after year 2000, but schedules for such additions to fuel supply depend on the doubling time of commercial FBRs. (The *doubling time* is the time required for the FBR to breed double the amount of plutonium in the initial fuel loading.) Depleted uranium stocks from the enrichment of fuel for the LWRs would be sufficient to provide U-238 for FBRs for a long time.

It is difficult to make a detailed technical assessment of the FBR at this time; few reactors of this type are now actively operating in the world, and changes in designs will probably be made in the next generation of FBRs. Several countries are now planning to build FBRs because of their long-term fuel advantage. FBRs are needed to sustain a large-scale nuclear power program after the end of this century.

6.6 STAGES OF NUCLEAR CHOICE

The debate on nuclear energy has made it very difficult for the public to understand that there are choices between accepting the entire nuclear system and foregoing entirely the benefits of nuclear energy.

Early expectations for the rate of nuclear power growth implied such heavy demands on uranium that the projected program could be met only by

Table 6.4
Nuclear energy: stages of choice

Stage 1 Single use of uranium	Stage 2 Reprocessing and plutonium recycling	Stage 3 Fast breeder reactor
• Single use of uranium • Storage of used fuel without chemical reprocessing • Defer decision on U-235 and plutonium recovery Disposal of liquid waste	• Chemical reprocessing of used fuel to recover plutonium and uranium • Handle plutonium • Disposal of radioactive liquid waste	• Stage 2 plus fast breeder reactors (Breeders eventually multiply uranium fuel value 50 times) • Large-scale plutonium handling

introducing fast breeder reactors at a rapidly increasing rate by 1990. Such plans also required acceptable chemical processing of used fuel, plutonium handling on a large scale, and means for handling radioactive wastes. Conventional uranium reserves seemed unlikely to furnish enough fuel beyond the 1990s. Low-grade sources (below 100 parts per million U_3O_8) in shales (25–80 parts per million), in granite (10–20 parts per million), and in seawater (0.003 parts per million) were not expected to furnish any substantial part of the projected needs. Given time and a major research and development program for exploration, mining, and milling techniques for these low-grade uranium sources, some of them could develop into significant sources in the medium-to-long term, but WAES has not regarded them as commercial uranium sources in this century.

Lower estimates of projected nuclear buildup in most countries, with the accompanying reduction and stretchout of requirements for uranium, now allow a clearer view of the stages of choice. Some countries have already selected a course of action, others will have to choose. Table 6.4 describes the three stages of nuclear choice.

6.6.1 Stage 1: Single Use of Uranium Fuel

Three thermal reactor types have been developed and operated as full-scale commercial electric power plants: (1) the British Magnox and AGR, (2) the light water reactors (PWR and BWR), and (3) the heavy water reactor (CANDU). Stage 1 can be followed for all three reactor types by storing used fuel under water, where it cools thermally and the water acts as a shield against radiation. If at a later time it is decided to chemically reprocess the used fuel to recover plutonium and enriched uranium, such products will still be recoverable, and chemical reprocessing will be somewhat simplified because of reduced thermal and radioactive levels of the used fuel. In any case, at some future time the used fuel must either be reprocessed or permanently stored.

6.6.2 Stage 2: Reprocessing and Plutonium Recycling

This stage requires acceptable systems for transporting used fuel, handling, transporting, and fabricating plutonium into reactor fuel, decontaminating enriched uranium to levels acceptable in enrichment plants, and packaging and disposing of radioactive wastes. For LWRs, uranium recovery plus plutonium recycling might add fuel equivalent to roughly 20 percent more uranium.

6.6.3 Stage 3: The Fast Breeder

Stage 3 now seems essential if nuclear energy is to remain a substantial source of world energy beyond the end of this century. It implies acceptable resolution of all of the problems of stage 2 as well as the issues of safety of FBRs, which are different from those of thermal reactors. Implementing stage 3 requires that decisions on stage 2 be taken in time to provide the large quantities of plutonium needed for initial fuel loadings of FBRs.

6.7 PROJECTIONS TO THE YEAR 2000

6.7.1 Choice in WAES Scenarios

The role of nuclear energy in the WAES scenarios is based on an assumption that most countries choose stage 3, although FBRs are not expected to add more than 5 percent to the nuclear energy supply by the year 2000. If stage 3 is rejected or long postponed for reasons relating to plutonium handling or waste disposal, stage 2 might be similarly rejected or postponed. The effects of such a sequence of events are difficult to estimate. However, unless uranium resources are greatly increased, or unless thorium becomes an accepted supplement to uranium, a lengthy continuation of stage 1 will greatly reduce the future role of nuclear energy and lead to greater dependence on other energy sources.

6.7.2 Some Implications of Single Use

If single use of fuel becomes a standard mode of operation for nuclear power plants, uranium economy (the amount of useful energy obtained per ton of uranium) will assume critical importance. Special consideration needs to be given to a comparison of LWR and HWR types in this respect. At first the HWR appears to have a 4 to 1 advantage over the LWR because the HWR uses natural uranium (0.7 percent U-235). In the enrichment process, for every five parts of uranium fed into the process, four parts remain as tails (depleted to 0.25 percent U-235) and one part comes out as LWR fuel (enriched to 3.0 percent U-235). On the other hand, in the HWR cycle all five parts of the uranium are used as fuel.

At the next step, however, the burn-up of uranium fuel in the reactor (measured in MWd/T, megawatt days per ton of fuel) appears to give the LWR a two- to fourfold advantage over the HWR. If the burn-up is 15,000 MWd/T for LWRs and 7500 MWd/T for HWRs, the HWR retains a net twofold advantage over the LWR in its economy in the use of uranium. On the other hand, if LWR burn-up is 30,000 MWd/T, the LWR has a fourfold advantage over the HWR as measured by burn-up and there is no net difference in uranium economy between the LWR and the HWR.[2] These are problems deserving careful study if a single-use mode of operation is to be considered as a possible national policy.

Single use would also concentrate attention on the other potential reactor fuel—U-235 converted from thorium. Produced in thermal reactors, it might be brought forward far enough so that its utility and problems can be evaluated. If it can be used, thorium could then be added to the nuclear fuel resources. Thorium deposits are estimated to be of the same order of magnitude as high-grade uranium resources (RAR plus EAR), although exploration for thorium has been less intensive than for uranium.

6.7.2 Implications of Uncertainty on the Nuclear Industry

This discussion of choices does not take into account the implications for the nuclear industry of switches among choices or of delays in decisions. Unless the climate for a vigorous development of the nuclear industry becomes more favorable soon, there is doubt that the levels of capacity projected in WAES scenarios can be met. The multibillion dollar scale ($0.4–$1.0 billion for a 1 GWe plant), the long time (six to ten years) between start of a project and full-power operation, and the capital cost and lead times associated with other critical steps in the fuel cycle such as mines and enrichment plants, all make the dynamics of nuclear power supply relatively slow compared to changes in demand. The high capital cost requires investors to be confident that such investments are sound and will pay off.

6.8 NUCLEAR FUSION

In the WAES time frame to the year 2000, power from fusion reactors is not expected to contribute to global energy supply. It is uncertain whether a controlled nuclear fusion reaction sufficiently long to be a power source will be demonstrated even in the laboratory before the year 2000. In theory, controlled nuclear fusion of deuterium and tritium could provide an almost limitless amount of energy, but formidable engineering problems must be overcome before a heat source at 50 million degrees Celsius can be used to make steam that will drive turbines. Moreover, the intense neutron flux in the fusion reaction makes everything nearby intensely radioactive. If a large and growing research and development effort is continued for several decades, fusion power might be added to the global energy supply in the twenty-first century.[3]

2. Another method for estimating uranium utilization is by comparison of replacement fuel loadings; see *IAEA Bulletin*, Vol. 18, No. 5/6, p. 8.

3. For assessment of the problems and promise of fusion see David J. Rose and Michael Feirtag, "The Prospect of Fusion." *Technology Review*, December 1976.

6.9 THE TASK OF WAES

The reader will now appreciate some of the difficulties in estimating with confidence the contribution nuclear energy might make to the world's primary energy needs in the year 2000. A vigorous program of building nuclear power plants sustained over the next twenty-five years could permit nuclear energy to contribute as much as 21 percent of primary energy in the year 2000. This would be equal to the energy contained in total WOCA oil consumption in 1975. However, unless the many obstacles now facing such growth are quickly and favorably resolved, the nuclear energy contribution will be less, perhaps much less. Differences between countries may become much greater than in the past as the people in each country decide on the risks and benefits from nuclear energy. Our assessment of the amount of primary energy that *could* be obtained from nuclear energy should also make clear the size of any shortfall. This shortfall would have to be met by other fuels if projected primary energy demand in the WAES cases for the year 2000 is to be satisfied.

REFERENCES

1. The OECD Nuclear Energy Agency and the International Atomic Energy Agency, *Uranium: Resources, Production, and Demand, Including Other Nuclear Fuel Cycle Data* (Paris: Organization for Economic Cooperation and Development), 1976.
2. E.J. Hanrahan. "World Requirements and Supply of Uranium." Presented at the Atomic Industrial Forum, September 14, 1976, Geneva, Switzerland.

7 Other Fossil Fuels and Renewables

7.1 INTRODUCTION

Energy sources other than oil, natural gas, coal, and nuclear could make a growing contribution to energy supply before 2000. The supply sources described in this chapter include other fossil fuels—oil sands, heavy oil, and oil shale—as well as renewable energy sources—hydroelectricity, geothermal, and solar in its several forms. In 1972 hydroelectricity contributed approximately 5.7 MBDOE to the energy mix; other "renewables" and other fossil fuels provided only negligible amounts. By 2000, our scenario cases show a production total for these fuels of about 17 MBDOE in the C-1 case, of which hydroelectricity accounts for 12 MBDOE. In the D-8 case, the total is 12 MBDOE, of which hydroelectricity accounts for 9 MBDOE.

The first part of this chapter describes oil sands, heavy oil, and oil shales. This is followed by some observations on the potentials and constraints of the renewable energy sources.

7.2 OTHER FOSSIL FUELS

Known deposits of oil sands, heavy oil, and oil shale are much larger than the world's proven reserves of conventional oil. Such deposits represent a potential means for supply of petroleum long after conventional oil and gas fields are exhausted. To date, they have received very little attention compared with conventional oil, which has been abundant and inexpensive.

7.2.1 Oil Sands

In Canada oil or tar sands lie in beds 50 to 100 feet thick under an area of 12,000 square miles near the MacKenzie River in northern Alberta. Both in situ and surface mining methods can be used to extract the estimated 300 billion barrels of oil which are considered recoverable.

The amount minable from the surface depends on acceptable ratios of oil sand to overburden. At least 90 percent of the oil sands lie too deep for surface mining and must be recovered by in situ methods. Active development of in situ systems, including pilot plant operations, is aimed at recovery of near 270 billion barrels. Capital costs for in situ systems might be substantially lower than for surface mining, and environmental problems of land reclamation involving very large quantities of sand would be avoided. In situ techniques separate oil from sand in place, allowing the oil to be pumped to the surface. Water flooding, thermal means, and diluents are being investigated as possible methods. Technically feasible methods for separating the very viscous oil (called bitumen) from sand particles may present formidable problems of practical field operation. Hot water or steam must not leak into the surrounding sand or clay. Chemicals must not escape and can be reused if they are recovered. The amount of energy contained in the heat or chemicals used should be much less than the energy contained in the oil that is recovered.

Surface mining and processing of oil sands have been well demonstrated during several years of operation on a commercial scale at the Great Canadian Oil Sands (GCOS), where production is about 50,000 barrels per day (BD) of oil. A second 130,000 BD operation, scheduled to begin operation in 1978, will incorporate drag lines instead of bucket wheel excavators as at GCOS. Processing oil sands involves separating oil from sand with hot water, partial refining to obtain several petroleum fractions, and desulfurizing. Viscosity is reduced to facilitate pumping at the low winter temperatures in the region. The oil obtained is of good quality.

Problems of year-round operation in a hostile climate, excavating and transporting the sticky mass of oil sands, and disposing of the very large quantities of sand tailings make such operations difficult and costly. Reserves accessible through surface mining are currently estimated to be about 30 billion barrels, comparable with North Sea oil reserves. If fully exploited, such oil sand reserves would support about fifteen plants each producing 200,000 BD or a total of 3.0 MBD for twenty-five years.

Such operations require a large capital investment—about $20,000 per barrel of daily capacity (1975 Canadian $) resulting in a cost of about $4.0 billion for a 200,000 BD plant. A work force of 30,000 people would be needed to build one such plant per year. The cost of the oil produced is in the vicinity of present world oil prices if royalties and taxes are excluded.

The rate of development of production from oil sands will depend upon (1) trends in world oil prices, (2) prospects for in situ recovery that might involve lower capital costs, and (3) Canadian policy toward foreign investment. Given present and planned programs, it is unlikely that the technically possible oil sand production figures will be realized. Much more likely, based on present indications, is a Canadian production rate of some 800,000 BD by year 2000.

7.2.2 Heavy Oil

Significant deposits of heavy oil have been found in Canada and in Venezuela. It is estimated that there are 167 billion barrels of heavy oil in place in the Cold Lake deposits of Canada, with a possible recovery of 15 to 30 billion barrels. Estimates for Lloydminster-type crude oil are 10 to 15 billion barrels, with 2 to 4.5 billion barrels recoverable. Recovery requires in situ systems somewhat similar to those needed for recovery of oil from oil sands or for tertiary recovery from conventional oil fields.

A complex accumulation of hydrocarbons, mainly heavy and extra heavy oil, was discovered in the south of Venezuela's Orinoco basin as a result of geophysical exploration carried out between 1925 and 1935. The boundaries of this area, called the Orinoco Oil Belt, have not yet been accurately established. A preliminary estimate of in situ oil made in 1967 was 700×10^9 barrels. Information obtained after the first formal exploratory drilling program in 1973 demonstrated that the accumulations extended farther south than originally estimated.

Table 7.1
Estimated shale oil production cost

Technique	Investment cost ($/daily bbl)	Operating cost ($/bbl)	Production cost ($/bbl; 15% DCF, 100% equity)
Modified in situ	5000–7000	3.50–5.00	8.00–11.00
Surface retorting	14,000–23,000	4.00–5.00	16.00–25.00

Source: Reference 1.

Note: The oil produced by these methods may be prerefined to a quality comparable with that of Arabian crudes for an additional $1 to $4 per barrel.

In addition, the thickness of the oil sands is greater than first estimated, and the crude's gravity is not low enough to indicate the presence of bitumens.

The Venezuelan government is now considering the Orinoco Oil Belt Integral Study and Evaluation Program, whose objectives are to test and identify the most adequate methods for production and upgrading. The results of this program will be used to plan for the eventual and gradual commercial exploitation of this important oil-bearing accumulation.

7.2.3 Oil Shale

Oil shale has the potential to become a large energy resource. The largest known reserves are in the United States, with significant amounts in Brazil, Canada, Burma, the U.S.S.R., and China. Before the 1950s small oil shale industries existed in Sweden, the United Kingdom, and a few other countries, but low-cost imported oil halted these industries.

In situ recovery techniques and mining are two methods of extracting oil from oil shales. The present substantial efforts to develop in situ production make a combination of in situ and mining a likely pattern for the future. Underground mining by room and pillar methods in seams 100 feet thick leaves about half the shale in the pillars that support the roof. The oil in the shale, an organic material known as kerogen, is extracted by retorting the crushed shale at high temperature and treating it with hydrogen to produce a semirefined fuel oil. Part of the kerogen decomposes into gas, which can be burned on-site to furnish power or converted into pipeline quality gas. Consolidating the spent shale from which the oil has been extracted and restoring the vegetation takes about 1.4 barrels of water per barrel of oil recovered. Of this about a quarter barrel must be fresh water. This process may be largely avoided by in situ retorting in the mined cavern.

U.S. oil shale deposits are estimated to contain some 2000 billion barrels of oil. However, only about 6 percent of the deposits are accessible and are in seams 30 or more feet thick, which would contain more than 30 gallons of oil per ton of rock, and only such rich deposits are of commercial interest at this time. Although intermittent federal research and development and private efforts to develop oil from shale have been underway since 1948, there has been no commercial production to date. Table 7.1 gives current estimates of shale oil production costs in the United States.

Economic, technical, and environmental problems have hampered the development of oil shale. Inflation has increased the estimated cost of commercial plants beyond the range of usual industrial funding. Environmental factors including disposal of spent shale and water requirements present major obstacles to developing a large oil shale industry in the United States, and in situ technologies that might diminish the environmental impact have not yet been demonstrated to be commercially feasible. Uncertainty about future oil prices increases the investment risks.

Federal subsidies for at least the early stages of development appear necessary to demonstrate the potential of shale oil production. Such support must include vigorous development of shale oil to demonstrate not only technological and economic feasibility but also environmental acceptability in a region where water is scarce and already committed for other uses.

The WAES scenario cases assume a production level of oil from shale of 2 MBDOE in the world outside Communist areas (WOCA) in the rising energy price scenarios in 2000 (Cases C-1 and C-2). Nearly all of the production is assumed to be in the United States.

7.3 RENEWABLE ENERGY SOURCES

7.3.1 Hydroelectricity

The use of hydro power for electricity has increased steadily during the twentieth century and is likely to continue to grow at a moderate rate until the year 2000.

Hydroelectric energy is the most widely used of renewable energy sources and will be important in providing electricity in the future. Most hydroelectric expansion is likely to be in the developing countries of the world, where only 4 percent of the potential sites have been developed (reference 2). Although the potential for hydroelectric expansion is vast in the developing countries, growth is somewhat constrained by the location of potential sites and by the long lead times involved with construction. We have conservatively estimated that primary hydroelectric generation in developing countries will increase from about 1 MBDOE in 1972 to as much as 4.4 MBDOE by year 2000 (in the high economic growth/low nuclear scenario). This estimate could be higher if a number of developed countries, because of energy shortages or for environmental reasons, choose to transfer some energy-intensive industries (such as aluminum production) to developing countries that have abundant hydro resources, access to required raw materials, and the desire to develop such industries.

Growth of hydroelectricity in the developed world—which currently accounts for 80 percent of WOCA hydroelectric capacity—will be lower than in the developing countries because most favorable dam sites are already developed and because environmental concerns may limit development of some potential sites. North American hydroelectricity generation could increase from about 2.4 MBDOE in 1972 to as much as 4 MBDOE by 2000. Western European and Japanese growth rates will probably be lower.

The WAES projections show a total world primary energy growth in hydroelectricity from 6 MBDOE in 1972 to as much as 12 MBDOE in the year 2000.

7.3.2 Geothermal

Geothermal energy depends on using thermal energy produced by various processes in the crust of the earth. Such energy is more concentrated in certain parts of the earth's crust, with the greatest amounts of potentially available energy in volcanic regions where tectonic shells meet. Stored energy may also be obtained in areas with significant temperature concentrations, for example, those caused by radioactive decay. Geothermal energy is not a renewable source outside the volcanic zones, and in these nonvolcanic areas energy is drawn from a historically formed thermal store. Geothermal energy must be used locally in the region where it is found.

Geothermal energy from dry steam is obtained by drilling in suitable areas. A more abundant but less efficient source is hot water (usually very saline), which is also obtained from wells. Geothermal fields, such as the Geysers (U.S.) and Larderello (Italy) require a combination of three geologic components: (1) a natural underground source of water, (2) an impermeable layer above that traps the water and permits the formation of steam, and (3) a mass of hot rock near the natural water system. Only rarely do these three components occur in the same area. Total installed global geothermal capacity was 1400 MWe in 1974, less than the output of two nuclear power plants of 1 GWe each. It has been estimated that exploitation of energy from hot dry rocks, without the impermeable cap rock and the natural water system, could create a supply of geothermal energy several times greater than that of geothermal fields now exploited (reference 3).

Some experiments in the United States involve fracturing hot rocks (creating a large surface for heat exchange), injecting water into the lower part of the fractured area, and recovering steam or superheated water from the upper area; the steam is then used to generate electricity. Fractures have been successfully created by drilling into the rock and pumping water under pressure into a field zone of the well. Such "hydrofracturing" of the igneous rock is a method adapted from secondary recovery technology for petroleum (reference 4).

Numerous problems remain; for example, the fracture system may leak water, preventing pressure from building up. In addition, recovery of heat and drilling in hot igneous rocks are difficult and costly. These and other potential problems must be solved to make hot rock exploitation economically feasible. By the year 2000, better answers to these technological problems will be known.

If research, development, and demonstration achieve commercial success by 1990, electricity production from hot rocks by the year 2000 will still be in its initial stages because of the time involved in exploration, drilling, and development of geothermal fields and the subsequent integration of the developed fields into the electricity generation and distribution system. Geographic areas of greatest hot rock potential include those of conventional geothermal energy production (Italy, the United States, Mexico, New Zealand, the U.S.S.R., Iceland) and other regions of recent volcanism.

7.3.3 Solar

The energy crisis of 1973 has led to a growing belief by many people that the sun must become a major source of energy. Solar energy can be collected and used in many ways. Radiant energy from the sun can be used directly—in thermal applications such as heating water and heating buildings. Solar energy can also be converted into electricity by photovoltaic systems or by mirror systems to produce steam for power stations. Solar energy can also be used indirectly through the use of wind temperature differences between surface and deep waters in the oceans.

Each year the world's solar energy income at ground level is about 500,000 billion BOE—about 1000 times the energy of the known reserves of oil. Efficient extraction of this abundant resource is difficult. Solar energy is a very dilute source, and it is available only at intermittent intervals. In tropical latitudes sunshine is abundant, but demand for space heating is small. In more temperate latitudes winter days are short and overcast skies reduce the solar radiation reaching the collector, yet these are the regions that most need heat during winter.

The need is to overcome these obstacles by finding ways to reduce the collector area required to capture this dispersed energy and by developing efficient systems to store collected solar heat for use at night or during overcast periods.

WAES projections assume that solar will contribute 1 to 2 MBDOE for all industrialized countries in the year 2000. Most of this amount comes from solar space heating and hot water systems. Significant amounts of electricity generated from solar sources seem unlikely before 2000.

Our solar projections result largely from the WAES energy price assumptions. Higher prices than in our scenarios would probably bring about larger contributions from solar energy. WAES studies do not look beyond 2000, the time when solar technology might be expected to make a small but rapidly growing contribution to the world's primary energy needs.

The rate at which solar energy will become widely used is highly uncertain for technical, economic, and institutional reasons. Research and development have only recently begun, making it difficult to forecast how well and how fast modern technology will overcome these barriers. Questions of costs, materials, reliability, and acceptability need to be answered.

The financial constraints to widespread use of solar energy involve the capital needed for technological development and the generally high initial costs of the systems themselves. Life-cycle costs in which high initial costs are offset by minimal operating costs could give solar systems a growing advantage over systems based on fuels at rising prices.

These limitations vary greatly among solar energy's many different forms. Likewise, the extent to which modern technology has the potential for overcoming these barriers varies considerably among the different solar technologies. The growth of each type of solar energy can be expected to exhibit special dynamics.

Despite such problems and complexities, solar energy after the year 2000 can be expected to play an increasingly important role in the energy mix. The Energy Research and Development Administration (ERDA) has suggested that solar energy could contribute 25 percent of the total U.S. primary energy by 2020.

WAES has not prepared a thorough analysis of the global potential of solar energy comparable to our analysis for the major fossil fuels and nuclear. Limited time has forced us to restrict our study to WAES national teams' estimates of the contributions of solar energy in the WAES countries. We do believe that research, development, and demonstration of solar and other renewables should be a high priority. Our limited treatment of this subject is not a reflection of the importance we assign to solar and other renewable technologies. Indeed, we consider such energy forms to be of great importance in the long-term energy future.

Solar Uses for Hot Water and Space Heating and Cooling

In the climates of North America, Europe, and Japan enough solar energy falls on the roof of the average house to supply all of its energy needs. Capturing and converting small portions of this energy for useful purposes appears economically feasible over some of this region, especially for water heating applications.

The best-known application of solar energy is for heating hot water. Improved technology, lower manufacturing costs, and strong marketing efforts—enhanced by public policy—may lead to the substantial introduction of solar hot water systems in markets where sunshine is strong during much of the year.

Systems for the solar heating and cooling of buildings will probably develop more slowly than solar hot water systems. For efficient use of solar heating, buildings should be properly oriented, well-insulated, and equipped with supplementary heating and energy storage. Currently, small production levels for solar units usually result in high costs. When produced in greater volume and with more experience, capital costs should be substantially reduced, and maintenance and reliability will be better understood.

Programs for applying modern technology to the collection of solar energy and for storage systems that allow economical recovery and use are now being started. Such collectors may be *passive*, such as heat-absorbing walls or roofs of buildings, or *active*, as when a moving fluid collects the solar energy and transfers the heat to warm the building or stores the heat to be used at night or during overcast days. An active system involves collectors that should be efficient, have long life, need little maintenance, and be aesthetically acceptable because very large collector areas will be needed. Much progress is being made in developing collectors and other parts of a solar heating system including pumps, controls, and efficient means for storage.

The turnover time of a nation's housing stock is approximately one hundred years. Thus progress in the implementation of solar heating in the industrialized world may be slow if applications of solar technologies are confined to new construction. For example, to achieve a solar energy usage equivalent to 1 MBD in the United States by the year 2000 would require the building of solar hot water and heating systems in two out of every three homes built between 1980 and 2000. Such an effort would satisfy only about 2 percent of the total U.S. energy needs in 2000. A key problem, then, is how to devise more economical solar designs for existing buildings and to provide incentives for such investments with high first cost and very low operating costs.

Rising prices of heating fuels would make solar systems, which need no fuel other than for the backup system, increasingly attractive. Indeed, the lifetime heating cost of a solar system may be much less than the lifetime cost of a conventional heating system requiring fuel. But most of the world's economic and financial systems do not take account of lifetime costs—total costs during a twenty- to thirty-year period. The building owner must make a large investment; consequently lowest first cost is typically favored. Moreover, the capital cost to provide electricity, gas, or oil is now borne by the supplier, not the user. Revision of current financing and other arrangements may be necessary for solar heating and hot water systems to become widely adopted.

Electricity from the Sun

Many possible technologies are available for the conversion of solar energy to electricity. Schemes are under consideration for producing electricity both on a *decentralized* scale (various types of solar cells) and on a *centralized* scale (central station power plants). It is unlikely that any of these schemes will provide large amounts of electricity by 2000.

Solar energy can be converted directly into electricity in solar cells made from special materials such as silicon crystals. Such cells have provided electricity in spacecraft. Much research and development is being done, and breakthroughs on efficiencies and costs of cells in mass production appear possible. Again, storage systems and/or backup systems will be needed. Just as in the case of solar collector systems for heat, the cost of such systems for solar electricity will fall on the user. Since the capital cost of supplying electricity is now borne by the electric power company, new incentives may be needed to induce the user to make such an investment himself.

Two basic kinds of technology for central station conversion of solar energy to electricity are under consideration: thermal and photovoltaic (ground and space). Significant applications of these technologies before 2000 appear to be highly doubtful.

The thermal method involves concentrating sunlight on a boiler to superheat a fluid which then drives a turbine generator. Because of cost considerations, these designs may have to be relatively large (100 MWe and above) to reduce the cost of power.

Engineering feasibility studies indicate that "power tower" designs may be preferable. These use a field of individually guided mirrors (heliostats) to focus sunlight on a boiler at the top of a large tower. When generating electricity, these towers use only direct sunlight and therefore are restricted to areas of low cloud cover. High capital costs and intermittent operation dependent on daylight, coupled with uncertainties in maintenance and lifetime, are major obstacles.

Studies are now being made of large central station photovoltaic systems. Implementation will have to await breakthroughs in the cost of manufacturing solar cells. Another proposal calls for satellites, placed in synchronous earth orbit, to convert solar energy into microwave energy and transmit it to earth for conversion to electricity. This requires no new technologies and avoids the problems of the earth's atmosphere and cloud cover. However, the satellites would have to be enormous—with solar panels in space more than 6 miles square—to produce 5000 MW at the ground receiving station. The costs for transporting the required materials into orbit and for building the system are very large.

Solar and Gravitational Energy from the Seas

Three schemes proposed for the extraction of some form of "renewable" energy from the sea are briefly mentioned below. The capital requirements are uniformly large. The possibility of such sources making any significant contribution to meeting energy needs in this century appears to be very small, and we have not included any energy from such sources in the WAES cases.

Ocean thermal systems would use the small temperature differences between the sun-heated upper ocean layers and the colder, deeper waters to generate electricity. While there appear to be economic possibilities for such schemes, the required systems are large and expensive and must operate in an unfriendly and corrosive environment.

Waves are also a source of energy. Although the total amount of energy stored in waves is much smaller than that in some of the other forms of solar energy, they might provide modest amounts of energy in certain countries like Great Britain, where much of the research in this field is currently being done.

Tides, which result from earth-moon interactions, are another possible "renewable" source for power generation. Two modern tidal power systems—a 240 MW plant at La Rance in France and a smaller one at Kislaya Guba in the U.S.S.R.—have demonstrated that tidal power can be economically harnessed. However, since only a few major estuaries in the world offer prospects for moderately large tidal projects, power from tides is expected to be small and localized.

Bioconversion

Each year photosynthesis stores seventeen times more energy in plant matter than the world now consumes. The use of municipal, agricultural, and forestry wastes, growing plants for fuel, and other methods for extracting solar energy from organic matter constitute a potentially significant resource.

Growing plants for fuel is a novel but potentially promising approach, especially in developing countries. Tree plantations and floating kelp farms are variants being studied. Land-based energy plantations may compete with agriculture for limited arable land. Agricultural wastes in some cases may be sufficient in quantity and density to be used as a fuel, as bagasse from sugar cane has been used for many years.

Burning municipal wastes for energy or conversion into methane is a standard practice in many parts of Europe. The consideration of such processes is beginning in the United States, but some form of tax credit for waste disposal will probably be required before they can become economic.

Windpower

Windpower is much less common now than it was fifty or one hundred years ago when it provided a significant portion of the energy consumed in rural areas. Yet there are many ways of extracting useful energy from the wind. Substantial research and development is now devoted to many modern windmill technologies (in blade dynamics and engine control) with some emphasis on devices with capacities of at least 1 MWe. New concepts are focusing on the use of wind concentrators, diffusers, and vortex generators that increase ambient wind velocity and so decrease the size of the turbines, the most expensive element in the system. Even with presently visualized cost reductions, the materials required suggest that either electrical generators must be of large scale or the price of electricity must increase substantially for winds to become a significant source of power.

Even with large-scale wind-driven generators, electricity output is small. It has been estimated that about 50,000 windmills with propeller diameters of 56 meters and average power of 500 kW would be needed to produce energy equivalent to 1 MBDOE.

Although the prospects for large-scale winds for power are uncertain, local small-scale uses could be important in many countries in the years ahead.

Solar Summary

This brief summary of the potentials and challenges of several forms of solar energy has been intended only to highlight some of the issues. We have treated some potentially critical long-term energy sources in cursory fashion, but we have completely neglected other possible sources (such as wood, passive solar systems, and a wide variety of solar-electric schemes). Our purpose has been to place solar energy in context globally and within our time frame to 2000.

The institutional obstacles that now impede the rapid adoption of solar energy deserve more attention than they now receive. Research and development of solar systems are receiving increasing support and attention. It is now time to find better institutional arrangements for widespread adoption of solar systems.

7.4 CONCLUSIONS

Our studies indicated that other fossil fuels—oil shale, oil sands, and heavy oil—and renewable sources will not make a major contribution to energy supplies before the year 2000. We have reason to believe they could play an important role in the twenty-first century if research, development, and demonstration are pursued vigorously.

The need to investigate, evaluate, and develop many of these potential energy sources is clear. They all present problems, hard choices, and tradeoffs, but we believe that doing nothing will result in infinitely more difficult problems, more difficult choices, and more unpleasant tradeoffs.

REFERENCES

1. *ERDA Weekly Announcements,* December 24, 1976.

2. *Energy and Petroleum in Non-OPEC Developing Countries, 1974-1980,* World Bank Staff Working Paper No. 229.

3. Bernardo F. Grossling. "Geothermal Energy," in *U.S. Energy Outlook* (Washington, D.C.: National Petroleum Council, 1972), pp. 15-26.

4. Paul Kruger, ed. *Geothermal Energy,* Stanford University Press, 1975.

Part III
National Energy Supply Studies

Canada

8.1 INTRODUCTION

This report presents projections of Canadian energy supply prospects by fuel type for five WAES scenarios to 1985 and four scenarios to the year 2000.

The assessment of supplies of oil, natural gas, and coal are based on assumptions that reflect the criteria established by WAES. The supply estimates and production schedules were calculated in the light of world economic growth, energy price, and national policy response.

The supply estimates for nonfossil fuels and their conversion processes to other energy forms were derived from available published data as well as from contact with knowledgeable professionals.

In assessing this study, one must recognize the highly speculative nature of the forecasts of ultimate recoverable fossil fuels. Although it is reasonable to believe that large volumes of oil, natural gas, and coal do in fact exist, the fortuitous combination of geological factors required for the existence of major reserves is by no means guaranteed. Canada's frontiers in particular have simply not undergone sufficient development for conclusive evaluation of hydrocarbon potential. Thus the study as it relates to these fuels must itself be regarded as highly speculative.

The scenarios developed in this submission to WAES were derived solely to meet WAES requirements. They are in no sense official projections of the National Energy Board of the Government of Canada and are not intended to reflect Canadian government policy.

8.2 SUMMARY AND CONCLUSIONS

Canada is relatively well endowed with all five main energy resources: coal, oil, gas, hydro power, and uranium. The course of future development of these resources to meet growing requirements is unclear. Whether the estimated range of energy supply under the scenarios to 1985 and 2000 will in fact materialize and permit Canada to be self-sufficient in these resources will depend on development costs and the availability of capital, materials, technology, and labor. Increasing Canadian self-reliance will require the development of additional supplies of indigenous oil and of other energy sources that can be easily substituted for oil imports. This report does not deal with energy development in response to national requirements; it only analyzes and projects the maximum potential availability of each energy resource under the assumptions of each scenario.

The estimated annual production rates for the various energy sources in 1985 and 2000 under the scenario assumptions are summarized in Table 8.1. The estimated 1985 production rate of oil ranges from 0.575×10^9 barrels for Scenario E up to 1.167×10^9 barrels for Scenarios A and B. Similarly, the estimated 1985 gas production ranges from 2.2×10^{12} cubic feet for Scenario E up to 3.68×10^{12} cubic feet for Scenarios A and B. These estimates of the production rates for gas and oil vary primarily because of the high prices forecast for Scenarios A and B.

In the year 2000 the oil production rate is estimated to be 0.84×10^9 barrels for Scenario D-8 and as high as 1.12×10^9 barrels for Scenarios C-1 and C-2. The gas production is estimated to range from 1.49×10^{12} cubic feet in Scenarios D-7 and D-8 up to 1.78×10^{12} cubic feet in Scenarios C-1 and C-2. This variance in gas and oil production rates is due primarily to high energy prices forecast for Scenarios C-1 and C-2.

Coal production is estimated to increase from 33 million short tons in 1985 in Scenarios D and E to 120 million short tons in year 2000 for Scenarios C-1 and D-7. This increase should occur as coal replaces oil and gas.

For the 1985 scenarios two projections of installed nuclear generation capacity are provided. The projected electricity demand from the low price scenario E is compared with a high growth nuclear generation capacity estimated at 14.7×10^3 megawatts. Projected electricity demand from the mid-1975 and high price scenarios A, B, C, and D is compared with a low growth nuclear generation capacity estimated at 12.5×10^3 megawatts. In addition, two projections of installed hydroelectric capacity are provided for the 1985 scenarios. For the high growth case the hydroelectric capacity is estimated at 61.9×10^3 megawatts; for the low growth case capacity is estimated at 55.5×10^3 megawatts.

For the year 2000 scenarios two projections of installed nuclear and hydro generation capacities

This report was prepared by Marc LeClerc.

Table 8.1
Canadian energy supply prospects

Energy source	Units/year	1985 scenarios					2000 scenarios			
		A	B	C	D	E	C-1	C-2	D-7	D-8
Crude oil (including NGL)	10^9 bbl	1.021	1.021	0.891	0.753	0.553	0.792	0.792	0.640	0.640
Oil sands[a]	10^9 bbl	0.146	0.146	0.146	0.124	0.022	0.325	0.325	0.298	0.298
Natural gas	10^{12} ft^3	3.68	3.68	2.70	2.29	2.20	1.78	1.78	1.49	1.49
Coal[b]	10^9 short tons	0.056	0.056	0.045	0.033	0.033	0.120	0.100	0.120	0.100
Uranium and nuclear[c]	GWe	12.5	12.5	12.5	12.5	14.7	45	74.5	45	74.5
Hydroelectric	GWe	55.5	55.5	55.5	55.5	61.9	74	74	60	60

[a]Based on present mining techniques only. Successful development and application of in situ technology could drastically affect potential supply.
[b]As a result of more intensive delineation of the inventory of coal these figures are considered conservative.
[c]Maximum installed capacity.

are provided. A high nuclear generation capacity is estimated at 74.5 × 10^3 megawatts, and a low nuclear generation capacity is estimated at 45 × 10^3 megawatts. A high hydroelectric generation capacity of 74 × 10^3 megawatts and a low hydro generation capacity of 60 × 10^3 megawatts are estimated.

Table 8.2 provides a summary of available energy supplies under Scenarios A, B, C, D, E, C-1, C-2, D-7, and D-8.

This study shows that in 1985, the key factors affecting expected annual production rates are primarily energy prices. A lesser but still important factor is the national policy response.

National policy response will be governed by domestic energy demands and will require serious consideration and cooperation by various levels of government and energy resource industries. The implementation of "vigorous" national policy initiatives to obtain the relatively high levels of production must be timely to provide for proper lead times in the development of energy resources.

In the scenarios to the year 2000 energy prices are maintained at the medium and high levels in all scenarios beyond 1985. National policy response varies from restrained to vigorous in individual cases. An alternate policy variable is the choice of either coal or nuclear as the principal replacement for either oil or gas. In all cases the supplies of conventionally produced crude oil and natural gas decline. Only in the frontier regions are the supplies increased, because development is anticipated in these areas beyond 1985. Coal, uranium, and nuclear supplies show substantial increases by the year 2000.

Other energy sources such as solar and wind are not expected to contribute significantly to Canada's energy supplies as these would account for at most 3 percent of total energy requirements.

8.3 OIL AND NATURAL GAS SUPPLY ESTIMATES

8.3.1 Methodology

Ultimate Recoverable Reserves

The ultimate recoverable reserves of oil and natural gas were arrived at by assessing data submitted to the Oil Hearing and the Natural Gas Hearing held in 1975 as well as data published by the Geological Survey of Canada and the Canadian Society of Petroleum Geologists. The data were modified where necessary by utilizing information obtained from contact with explorationists in the petroleum industry as well as from personal knowledge of the areas concerned. Historic finding rates were the main criteria employed in estimating the recoverable reserves of the conventional producing areas of the Western Canada sedimentary basin. The estimates of ultimate recoverable reserves of the frontier regions were largely based on consideration of geological factors coupled with data from results of exploration to date. The "ultimate recoverable" reserves differ from the often-quoted term "ultimate potential." Ultimate recoverable reserves are those that can be exploited commercially under defined economic and technological conditions, while ultimate potential may or may not be known in a favorable geological formation and requires further exploration to prove.

Table 8.2

WAES NATIONAL ENERGY SUPPLY STUDIES, 1972-2000
ENERGY SUPPLY SUMMARY

Country: CANADA

Scenario Assumptions **		1972	1985					2000			
WAES CASE		Base Year	A	B	C	D	E	C-1	C-2	D-7	D-8
GNP 1976-85; 1985-2000			High	Low	High	Low	High	High	High	Low	Low
Oil/Energy Price*			11.50-17.25	11.50-17.25	11.50	11.50	11.50-7.66	11.50-17.25	11.50-17.25	11.50	11.50
National Policy Response			Vig	Vig	Vig	Res	Res	Vig	Vig	Res	Res
Principal Replacement Fuel								Coal	Nuclear	Coal	Nuclear
Energy Supply	**Units**										
Crude Oil	10^{15}Btu	3.797	6.769	6.769	6.015	5.087	3.335	6.479	6.479	5.440	5.440
Natural Gas	10^{15}Btu	2.324	3.809	3.809	2.794	2.370	2.277	1.842	1.842	1.542	1.542
Coal	10^{15}Btu	.437	1.411	1.411	1.134	.958	.832	3.024	2.520	3.024	2.520
Nuclear	10^{15}Btu	1.867	.820	.820	.820	.820	.970	2.950	4.890	2,950	4.890
Hydro-electric			2.915	2.915	2.915	2.915	3.250	3.900	3.900	3.150	3.150
Geothermal											
Solar Thermal											
Solar Electric											
Total Energy Supply		8.425	15,724	15,724	13,679	12.150	10.664	18.195	19.631	16.106	17.542

*1975 US dollars per barrel of Arabian light crude oil, fob Persian Gulf.
All Figures are primary production prior to any processing or conversion.

**Oil and gas supply estimates assume reserves will be found in lightly explored frontier areas.

Estimates of Expected Annual Production

National Energy Board forecasts of future production rates of both oil and gas from conventional onshore producing areas and oil sands have been utilized freely in this study and reflect to a large degree studies conducted by the Energy Resources Group. Production forecasts for the frontier areas were developed using available published and unpublished data. They are predicted on reserves developed and large volumes yet to be found, giving consideration to the appropriate timing of transportation facilities together with potential throughput. The production rates from the conventional producing areas, the oil sands, and the frontier regions were modified in this study to reflect the assumptions in each scenario.

The estimates of production rates of oil and natural gas are not restricted by demand.

8.3.2 Scenario Assumptions

Scenarios A and B: 1985

- World economic growth (real): 3.5 or 6 percent
- 1985 crude oil price: $11.50 rising to $17.25/bbl in constant 1975 U.S. dollars, FOB Persian Gulf
- National policy response: Vigorous

In this scenario a strong incentive for the development of energy resources is created by rising crude oil price. A vigorous national policy response would lead to development of fossil fuel resources, particularly oil, which would reduce or eliminate import dependence. Specific government policy initiatives could include

1. An increase in domestic oil prices to at least the mid-1975 price level with appropriate price increases in other sources of energy.
2. A national energy conservation program.
3. Implementation of measures to increase exploration and development activity in the frontier regions of Canada.
4. Increased research and development with financial support for the commercialization of new energy technologies.
5. Encouragement of interfuel substitution.
6. Fiscal incentives to promote secondary and tertiary recovery of crude oil and to maintain natural gas deliverability.

Scenario C: 1985; Scenarios C-1 and C-2: 2000

- World economic growth (real): 6 percent
- 1985 crude oil price: $11.50/bbl in constant 1975 U.S. dollars, FOB Persian Gulf
- National policy response: Vigorous

The constant crude oil price in these scenarios provides continued incentive for the development of Canada's conventional fossil fuel resources. The price level does not provide for further oil sands development or for development of natural gas reserves in some frontier regions. Government initiatives are adopted to encourage development of domestic energy supplies and reduce import dependence. Specific government policy initiatives could include

1. An increase in domestic oil prices with appropriate price increases in other sources of energy.
2. A national energy conservation program.
3. Implementation of measures to increase exploration and development activity in the frontier regions of Canada.
4. Increased research and development with financial support for the commercialization of new energy technologies.
5. Encouragement of interfuel substitution.
6. Fiscal incentives to promote secondary and tertiary recovery of crude oil and to maintain natural gas deliverability.

Scenario D: 1985; Scenarios D-7 and D-8: 2000

- World economic growth (real): 3.5 percent
- 1985 crude oil price: $11.50/bbl in constant 1975 U.S. dollars, FOB Persian Gulf
- National policy response: Restrained

Although the constant crude oil price provides for continued development of Canada's conventional fossil fuel resources, it does not provide for further oil sands development or for development of other fossil fuel resources in some frontier regions. Government initiatives are insufficient to provide incentives for enhanced recovery techniques. The restrained government policies could include

1. No increase in domestic oil prices.
2. Minimal national energy conservation program.
3. No increase in gas prices or other sources of energy.

4. No financial support for research and development of new energy technologies.

5. No incentives for enhanced crude oil recovery schemes or for the maintenance of natural gas deliverability.

Scenario E: 1985

- World economic growth (real): 6 percent
- 1985 crude oil price: $11.50 falling to $7.66/bbl in constant 1975 U.S. dollars, FOB Persian Gulf
- National policy response: Restrained

Under these assumptions the falling crude oil price provides only nominal incentives for the development of crude oil reserves. The price structure does not, however, provide for further oil sands development or for development of natural gas resources. There is insufficient incentive to develop alternate energy sources or to implement enhanced oil recovery techniques. The restrained government policies could include

1. No increase in domestic oil prices.

2. Minimal national energy conservation programs.

3. No increase in gas prices or other sources of energy.

4. No financial support for research and development of new energy technologies.

5. No incentives for enhanced crude oil recovery schemes or for the maintenance of natural gas deliverability.

For a summary of these scenarios see Figure 8.1. Note that for cases A, B, and E an attempt has been made to project beyond 1985.

8.3.3 Crude Oil and Natural Gas Liquids Supply Prospects

Scenarios A and B: 1985

Under these scenarios, where a vigorous national policy response is maintained and the world economic growth ranges from 3.5 percent to a high of 6 percent, the dominant effect on the development of oil reserves in the conventional onshore areas is the price escalation from $11.50 to $17.25 per barrel. Proven oil reserves would increase to 18.96×10^9 bbl of the 20.0×10^9 bbl ultimate recoverable. However, the price increase to $17.25/bbl would not result in increased producing rates over Scenario C because of physical constraints on the oil producing reservoirs. The estimates for these mature conventional onshore areas are considered realistic in view of the geological and historical data available.

In the eastern Canadian offshore, specifically the Scotian Shelf region, it is assumed that sufficient reserves of crude oil will be developed under the high price scenario to provide an annual production rate of 0.200×10^9 bbl by 1985.

In the Beaufort-Sverdrup basins of northern Canada no significant reserves of oil have yet been found; hence any estimates of production rates are purely speculative. Under the high price assumption it may be assumed that 7.42×10^9 bbl will have been found with a production capability of 0.380×10^9 bbl per year. Considering the level of exploration required to find the necessary threshold volumes and the lead time required for transportation facilities, 1985 is estimated to be the first year of production from this region.

Scenario C: 1985

Scenario C is similar to Scenario A except that in C the crude oil price remains static at $11.50/bbl. This price stability still provides some incentive for exploration and development of oil reserves in the conventional onshore areas. The realization of lower net backs to producers adversely affects to some degree the development of oil reserves compared with Scenarios A and B. Oil production is expected to be maintained at 0.44×10^9 bbl per year in 1985.

In the eastern Canadian offshore the Scotian Shelf area is expected to have sufficient reserves developed and be in production by 1985. An annual production rate of 0.150×10^9 bbl is forecast for the region. This is slightly lower than Scenarios A and B, reflecting a decrease in the development of oil reserves because of lower level price stability.

Under Scenario C the Beaufort-Sverdrup basins may have sufficient reserves developed and transportation facilities built to be in production by the 1980s. The stability of price at $11.50/bbl dampens the incentive for vigorous exploration and development in this hostile and costly environment. The yearly production rate is expected to be 0.08×10^9 bbl less than in Scenarios A and B.

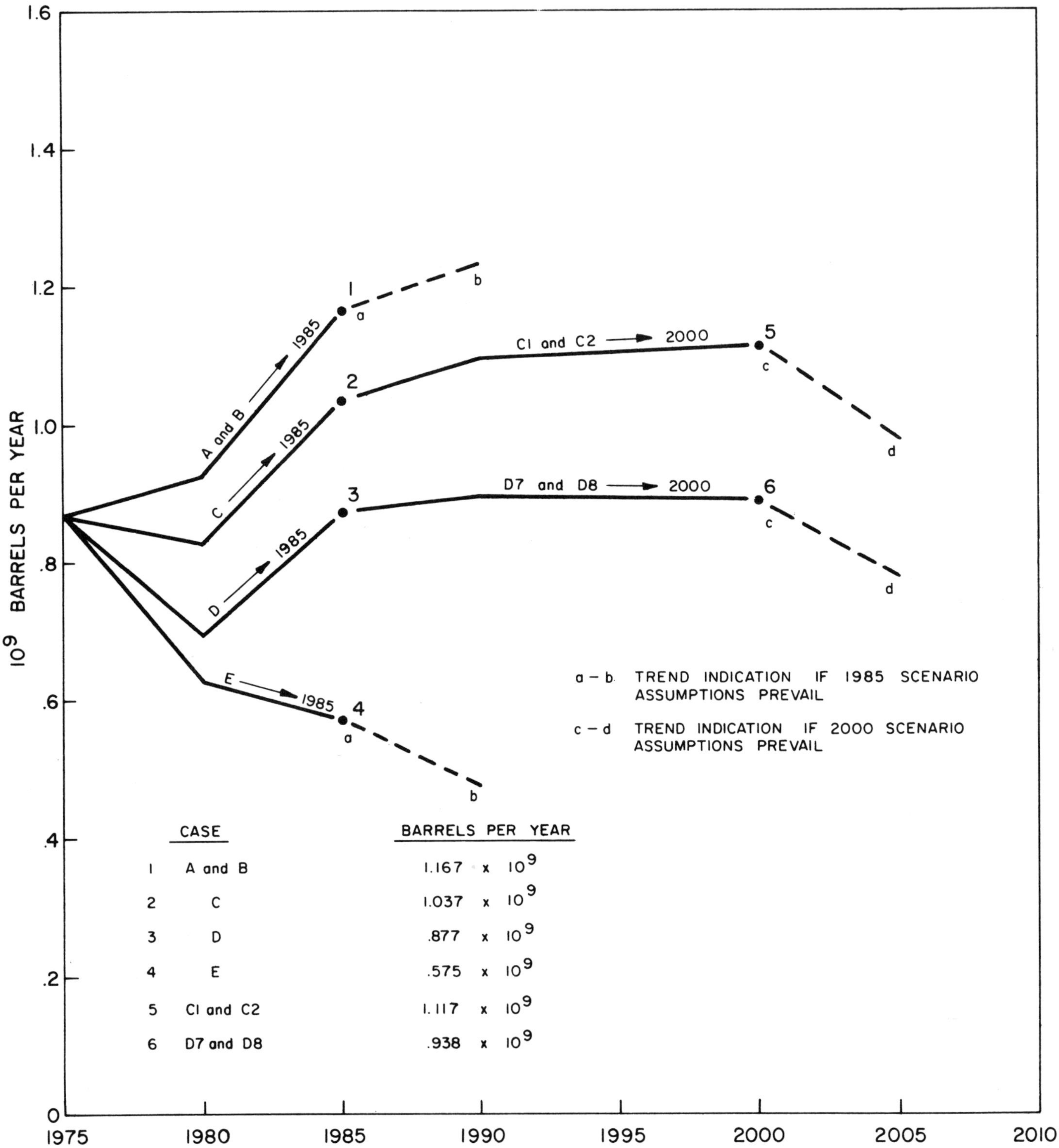

Figure 8.1
Canadian crude oil and NGL supply prospects (expected annual production).

Scenario D: 1985

Scenario D maintains the \$11.50/bbl price level with a restrained national policy response and a world economic growth rate of 3.5 percent. The level crude oil price still provides for some development of the conventional onshore reserves of crude oil. However, the restrained policy response does not provide incentives for enhanced crude oil recovery schemes. The effect of a restrained policy response is reflected in a decrease of 0.07×10^9 bbl in annual production compared with Scenario C.

In the eastern offshore area oil production of 0.127×10^9 bbl per year is anticipated. This will only materialize if exploration is reactivated on the Scotian Shelf, where nominal reserves have been found to date.

In the Beaufort and Sverdrup basins some exploration and development are expected, but they will be restricted to the onshore regions where exploratory costs are somewhat lower than in the offshore. A yearly production rate of 0.253×10^9 bbl may be attained by 1985.

Scenario E: 1985

In Scenario E the decreasing oil price from \$11.50 to \$7.66 per barrel and a restrained national policy response adversely affect the development of crude oil supplies. Although some development is expected, the declining price is the major factor preventing exploration for new reserves and implementation of enhanced recovery schemes. In the frontier regions—the Eastcoast offshore and the Beaufort and Sverdrup basins—development is expected to be restricted to onshore lower cost development areas. Annual production rates from the various regions are expected to be 0.369×10^9 bbl from conventional onshore, 0.061×10^9 bbl from the Eastcoast offshore, and 0.123×10^9 bbl from the Beaufort and Sverdrup basins. These rates are lower than in Scenario C, reflecting the effects of a restrained policy response in Scenario E.

Scenarios C-1 and C-2: 2000

A vigorous national policy response, high world economic growth, and high energy prices, together with the dominance of coal or nuclear replacement fuels, render the projection of anticipated supply of hydrocarbons to the year 2000 highly speculative. The uncertainty associated with any projection for the short term as well as the long term demands caution in using these estimates to formulate or recommend national policies.

In the conventional onshore producing areas of Canada the crude oil reserves that may be developed under this scenario would be the upper economically recoverable volumes, which would require the use of primary, secondary, and tertiary recovery techniques. It is anticipated that 98 percent of the ultimate recoverable reserves will have been developed by the year 2000. The expected production rate of 0.230×10^9 bbl per year will be 48 percent less than in 1985 under Scenario C.

In the eastern Canadian offshore 8.14×10^9 bbl are estimated to be developed from an ultimate recoverable of 10.0×10^9 bbl. Regardless of the assumptions, this estimate is considered optimistic, and perhaps unattainable. If these reserves are developed, the annual production rate is expected to be 0.187×10^9 bbl by the year 2000.

The Beaufort and Sverdrup basins are considered in the same category as the eastern Canadian offshore region. The reserves that may be proved by the year 2000 are 10.22×10^9 bbl, close to the upper limit of the ultimate recoverable of 13.6×10^9 bbl. If these reserves are developed, the annual producing rate may be expected to be 0.375×10^9 bbl, an increase of 20 percent over the 1985 production of Scenario C. This is the maximum production rate that can be expected beyond the year 2000.

Scenarios D-7 and D-8: 2000

With a restrained national policy response, low world economic growth, and medium energy prices, the oil reserves and corresponding producing rates will be lower than in Scenarios C-1 and C-2. The medium energy prices still provide for the implementation of secondary and tertiary recovery schemes. In the conventional onshore producing areas 19.02×10^9 bbl of the 20.0×10^9 bbl ultimate recoverable will have been developed. This development is the maximum attainable. Annual production is expected to be 0.175×10^9 bbl, 53 percent lower than in the 1985 Scenario D. The production decrease is indicative of the depleted state of the reservoirs and the maturity of the producing areas.

In the eastern offshore area oil production of 0.159×10^9 bbl per year is estimated. This is a maximum producing rate and is expected to decline beyond the year 2000. Production will only materialize if exploration is reactivated on the Scotian Shelf, where nominal reserves have been found to date.

In the Beaufort and Sverdrup basins 9.43×10^9 bbl are estimated to be developed by the year 2000, with an expected annual producing rate of 0.306×10^9 bbl, 21 percent greater than the 1985 Scenario D. This rate is the maximum that can be expected and will decline beyond the year 2000.

8.3.4 Oil Sands Supply Prospects

Canada is well endowed with nonconventional heavy oil (also called oil sands or tar sands). The total resource in place of the Athabaska and Cold Lake deposits and the Lloydminster deposits is estimated to be as high as 1000×10^9 bbl with a depth of burial ranging from 0 feet at the surface outcrop to 2500 feet in the subsurface. The ultimate recoverable reserves of up to 29×10^9 bbl are the estimates published by the Alberta Resources Conservation Board to be recovered by mining. Successful in situ recovery from all oil sands could exceed 300×10^9 bbl.

Because unlocking this vast resource poses considerable technical and economic problems, it is premature to rely on any forecast of future producibility regardless of scenario assumptions. The present great uncertainty regarding the rate at which developments will take place in the oil sands is largely attributable to the recently observed cost escalation in the Syncrude project and to some remaining doubts about the practicability of the yet unproved Syncrude technology. It is clear that barring major technological advances, future oil sands plants appear only marginally attractive at 1976 international oil prices, even with the special fiscal concessions made for the Syncrude project.

Conventional Heavy Oil

The conventional heavy oil deposits of Canada are located in the Lloydminster area of Alberta. The total resource in place is estimated to be 12.0×10^9 bbl; of this an estimated 4.0×10^9 bbl is considered proved. However, only some 5–6 percent of the oil can be produced using combined primary and waterflood methods. There is therefore a large potential for improved recovery using tertiary recovery techniques. Experimental work is being carried out, utilizing various methods of enhanced recovery. If successful, recovery is expected to increase by some 20 percent. Any speculation on projection of supply from these deposits is not practical at this time.

Nonconventional Heavy Oils

Because they require some form of enhanced recovery, the heavy oil deposits found principally in the Cold Lake area are generally referred to as "nonconventional," and they are generally included in the Board's definition of "tar sands." While the characteristics of the Cold Lake deposits are similar, it is important to differentiate between Lloydminster, Cold Lake, and the Athabasca deposits, since they differ widely in the recovery and processing techniques defined to produce commercial crude oil.

Estimates of total in-place Cold Lake deposits are 165×10^9 bbl with a possible recovery of 15 to 30 $\times 10^9$ bbl considered attainable using successful in situ thermal enhanced recovery.

The estimates of supply from the oil sands deposits under the various scenarios must be considered highly optimistic because of the inherent technical and economic problems associated with their development.

Table 8.3 is a summary of anticipated supply from the oil sands for each of the scenarios. These figures have also been included in Table 8.4 and in Figure 8.1.

Scenarios A, B, and C: 1985

The supplies that would be developed in Scenarios A and B are the same as in Scenario C. One processing plant is near operating and another is being built. The new plant is expected to be in operation in the late 1970s. Lead times preclude the

Table 8.3
Expected annual production from oil sands (10^9 bbl)

1985 scenarios					2000 scenarios			
A	B	C	D	E	C-1	C-2	D-7	D-8
0.146	0.146	0.146	0.124	0.022	0.325	0.325	0.298	0.298

completion of additional plants by 1985. An annual production rate of 0.146 × 10^9 bbl for the three cases is considered attainable by 1985. Some plant expansion can be anticipated.

Scenario D: 1985
A restrained policy response and low economic growth rate in Scenario D will constrain production. Any increase in plant capacity will be minimal. The two plants that will be in operation are expected to produce 0.124 × 10^9 bbl annually by 1985.

Scenario E: 1985
With the price of crude oil falling to $7.66/bbl development of oil sands production beyond the present one extraction plant is not possible.

Scenarios C-1 and C-2: 2000
Given a vigorous national policy response coupled with high energy prices and high world economic growth, one may anticipate fairly aggressive development of the oil sands. The decline of production in the conventional producing areas would create demand for this supply. The maximum that can be economically developed under these scenarios is 17.50 × 10^9 bbl of the 29.0 × 10^9 bbl ultimate recoverable at an annual production rate of 0.325 × 10^9 bbl by the year 2000.

Scenarios D-7 and D-8: 2000
In these scenarios the restrained national policy response and the medium energy price in particular preclude the development of additional resources beyond 1990. It is expected that 15.61 × 10^9 bbl of the 29.0 × 10^9 bbl of ultimate recoverable reserves will be developed at an annual production rate of 0.198 × 10^9 bbl by the year 2000.

8.3.6 Natural Gas Supply Prospects

Scenarios A and B: 1985
In these scenarios the dominant factor affecting development of natural gas reserves and production rates is the gas price at $2.88/$10^3$ ft^3. The effect on the conventional producing areas is significant, giving rise to escalated exploratory and development drilling. The main areas of development for natural gas are in the Rocky Mountain foothills and the deep basin areas of Alberta. The price structure also allows presently uneconomic reserves to be connected and programs to be implemented to increase deliverability from presently producing gas fields. Gas production from the conventional onshore areas is expected to be 3.23 × 10^{12} ft^3 annually, an increase of 25 percent over the 1975 production.

In the eastern Canadian offshore the ultimate potential is estimated to be 76.5 × 10^{12} ft^3. To 1975, perhaps 3.0 × 10^{12} ft^3 has been discovered on the Labrador and Scotian shelves. These areas pose many technological and environmental problems, and since the reserves have yet to be discovered, it is unlikely that gas will be produced from this area by 1985. Scenarios A and B do not provide for gas production from this area of Canada.

In the Beaufort-Sverdrup basins of northern Canada, it is reasonable to anticipate that under the $2.88/$10^3$ ft^3 price, natural gas will be produced

Table 8.4
Assessment of indigenous supply of crude oil and natural gas liquids (10^9 BOE)

	1985 scenarios					2000 scenarios			
	A	B	C	D	E	C-1	C-2	D-7	D-8
Total resource in place	1029.8	1029.8	1029.8	1029.8	1029.8	1029.8	1029.8	1029.8	1029.8
Ultimate recoverable	73.5	73.5	73.5	73.5	73.5	73.5	73.5	73.5	73.5
Reserves proved to date	47.25	47.25	44.17	39.77	23.50	55.55	55.55	51.11	51.11
Expected annual production	1.167	1.167	1.037	0.877	0.575	1.117	1.117	0.938	0.938
Cumulative production	17.13	17.13	16.53	15.35	14.27	31.45	31.45	28.65	28.65
Remaining proven reserves	30.12	30.12	27.64	24.42	9.23	24.10	24.10	22.46	22.46
Reserves/production ratio	25.8	25.8	39.0	27.8	16.0	21.6	21.6	23.9	23.9

by 1985. Estimates of reserves for the two areas as of 1975 are on the order of 15.0×10^{12} ft^3. Under Scenarios A and B it is expected that 0.45×10^{12} ft^3 will be produced annually from the Beaufort basin, which includes the MacKenzie Delta, where approximately 4.0×10^{12} ft^3 have been proved. It is conceivable that the Sverdrup Basin (Arctic Islands area) could also be in production by 1985.

Scenario C: 1985

Scenario C is similar to Scenario A except that in Scenario C the crude oil price remains static at \$11.50/bbl and gas at \$1.92/10^3 ft^3. This stability of price still provides for some exploration and development of natural gas reserves in the conventional onshore areas. The realization of lower net backs to producers to some degree adversely affects the development of natural gas reserves as compared with Scenarios A and B. A decrease of 0.93×10^{12} ft^3 in annual production from Scenarios A and B is expected. The stable lower price provides insufficient incentive for producers to maintain or increase deliverability rates.

In the eastern Canadian offshore gas production by 1985 is not anticipated. The threshold volumes yet to be attained, a hostile environment, and required lead times preclude the assumption of production from this region.

As in Scenarios A and B the Beaufort and Sverdrup basins are expected to be in production by the 1980s. Reserves estimates for the areas as of 1975 are on the order of 15.0×10^{12} ft^3. Under Scenario C it is reasonable to assume that 0.40×10^{12} ft^3 will be produced annually. This production is expected to be principally from the Mackenzie Delta, which is part of the Beaufort basin.

Scenario D: 1985

Scenario D maintains the \$1.92/10^3 ft^3 price level with a restrained national policy response and a world economic growth of 3.5 percent. The constant price still permits continued development of the conventional onshore reserves, but the restrained policy response does not provide incentives to maintain or increase gas deliverability. The effect of a restrained policy response is reflected in a decrease of approximately 16 percent in annual production compared with Scenario C.

In the eastern offshore, gas production is not anticipated by 1985. The hostile environment, long lead times for transportation facilities, yet-undeveloped technology, and undiscovered reserves make planning and policy in this area nearly impossible.

In the Beaufort and Sverdrup basins of northern Canada some exploration and development is expected at the \$1.92/10^3 ft^3 gas price, but only in the onshore regions where exploration costs are somewhat lower than in the offshore. Further reserves of natural gas beyond 1985 are not expected to be developed under this scenario. Yearly production rates of 0.34×10^{12} ft^3 are expected to be attained by 1985, principally from the Mackenzie Delta region of the Beaufort basin.

Scenario E: 1985

Scenario E, in which the gas price decreases from \$1.92/10^3 ft^3 to \$1.28/10^3 and the national policy response is restrained, adversely affects the development of natural gas supplies in all areas of Canada. The assumptions in this scenario preclude the development of new gas reserves by 1980. The price level of \$1.92/10^3 ft^3 in Scenario C is the minimum that will provide any incentive to continue development of the conventional and Beaufort-Sverdrup onshore gas supplies.

Scenarios C-1 and C-2: 2000

The vigorous national policy response, high world economic growth, high energy prices, and the dominance of coal or nuclear replacement fuels render estimates of supply of natural gas under these scenarios extremely difficult.

In the conventional onshore producing areas of Canada the natural gas reserves that may be developed would approach the upper economically recoverable volumes. The anticipated production rate of 1.20×10^{12} ft^3 per year would require the implementation of techniques to sustain deliverability in declining fields. Most of the gas fields will be in the production decline stage.

The assumptions in these scenarios do not provide for any supply prospects of natural gas from the eastern Canadian offshore regions.

In the Beaufort and Sverdrup basins the price assumption in Scenarios C-1 and C-2 does not provide sufficient incentive for exploration and development of gas reserves beyond 1985. Annual production rates are expected to be 0.58×10^{12} ft^3, greater than in Scenario C (1985). The production increase is a result of the production decline in the conventional producing areas.

Scenarios D-7 and D-8: 2000
With a restrained national policy response, low world economic growth, and medium energy prices, the gas reserves and corresponding producing rates will be lower than in Scenarios C-1 and C-2. The medium price of $\$1.92/10^3$ ft^3 still permits the implementation of techniques to maintain deliverability in the declining fields of the conventional onshore producing areas, but it does not provide sufficient incentive for additional exploration and development. An annual production rate of 1.01×10^{12} ft^3 is anticipated. This is 48 percent lower than the 1985 rate in Scenario C, indicating the maturity of the area.

In the eastern Canadian offshore region the scenario assumptions do not provide for any development of natural gas supplies.

In the Beaufort and Sverdrup basins, 16.1×10^{12} ft^3 are estimated to be developed by the year 2000. This level would be achieved by 1985 under Scenario D. The price assumption of $\$1.92/10^3$ ft^3 does not allow development of new reserves beyond 1985. An annual production rate of 0.48×10^{12} ft^3 is expected. This is 41 percent less than under Scenario D, reflecting production declines in the conventional producing areas.

These assessments of the indigenous natural gas supply are summarized in Table 8.5 and in Figure 8.2.

8.4 COAL SUPPLY ESTIMATES

Since the early 1970s the coal picture has changed dramatically in Canada. Production declined severely in the 1950s and 1960s because of the conversion of railways to diesel fuel and the expanded use of natural gas in the residential, commercial, and industrial sectors. However, the coal industry in Canada is presently enjoying a period of expansion that can be attributed to expanded use of thermal coal supplies and long-term export contracts of metallurgical coal.

In Canada each province has jurisdiction over its own resource developments. Thus supplies of coal will obviously depend on the favorable resolution of a number of issues and should be qualified accordingly.

Recent concern about the depletion of oil and gas reserves has caused western producing provinces to adopt a policy of restraint with regard to future coal development. British Columbia and Alberta, where the bulk of Canada's coal resources are located, have reassessed the environmental impact of coal operations to define clearly the areas where mining may be permitted and under what conditions. The Environmental Conservation Authority of Alberta has issued a report on land use and resource development on the eastern slopes of the Rocky Mountains. The government of Alberta is currently considering this report. Meanwhile no final approvals for coal development have been granted. Increases in royalties in the producing provinces have either been implemented or are

Table 8.5
Assessment of indigenous supply of natural gas (10^{12} cubic feet)

	1985 scenarios					2000 scenarios			
	A	B	C	D	E	C-1	C-2	D-7	D-8
Total resource in place	311.8	311.8	311.8	311.8	311.8	311.8	311.8	311.8	311.8
Ultimate recoverable	265.0	265.0	265.0	265.0	265.0	265.0	265.0	265.0	265.0
Reserves proved to date	153.0	153.0	121.9	109.9	103.4	126.0	126.0	115.9	115.9
Expected annual production	3.68	3.68	2.70	2.29	1.02	1.78	1.78	1.49	1.49
Cumulative production	61.5	61.5	56.5	52.0	68.2	87.8	87.8	77.6	77.6
Remaining proven reserves	91.5	91.5	65.4	57.9	35.2	38.2	38.2	38.3	38.3
Reserves/production ratio	24.9	24.9	19.2	25.3	34.5	21.2	21.2	37.9	37.9

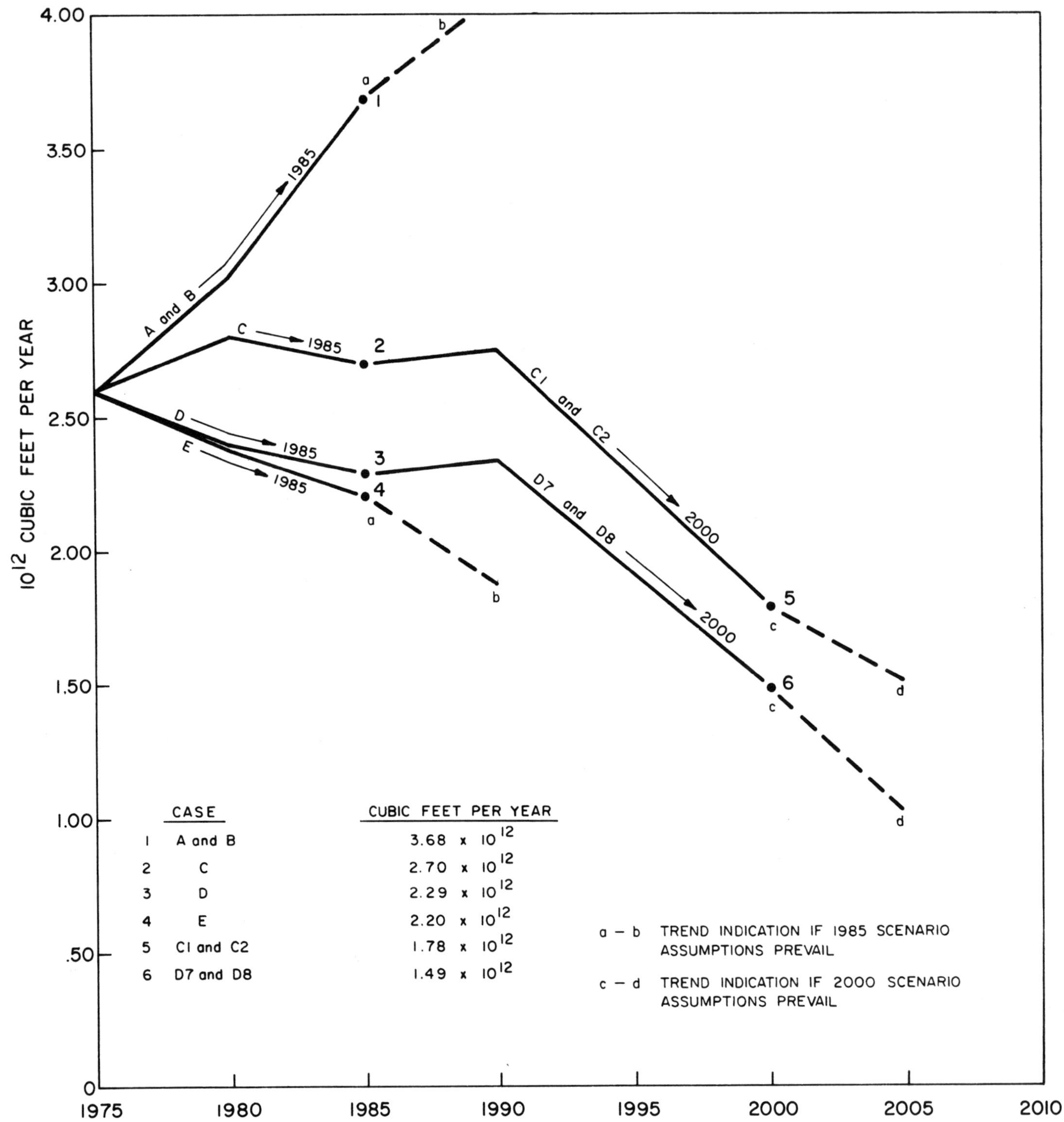

Figure 8.2
Canadian natural gas supply prospects (expected annual production).

being contemplated.

Although extensive exploratory and development drilling programs have been carried out in many areas, this period of reassessment has effectively resulted in cessation of new coal developments since 1974. New coal policies have just recently been formulated and published in Alberta and are expected in the near future from the government of British Columbia. The impacts of these new policies on coal development remain to be assessed.

National policy responses for coal are summarized in Table 8.6.

8.4.1 Methodology

Ultimate Recoverable Reserves

The ultimate recoverable Canadian coal reserves are based on estimates by the Geological Survey of Canada and the provincial resource departments. The estimates include all ranks—low and high thermal as well as metallurgical coal—that occur in Canada south of 60° north latitude. Coal deposits north of 60° north latitude are not included because data are insufficient to permit estimation of the resources. Further delineation is in progress. The following ultimate reserve estimates, therefore, are considered conservative.

Expected Annual Production

Expected annual production is estimated by projecting coal requirements from the actual production of 0.023×10^9 short tons in 1975. Estimates also consider the degree to which the demand for electric power increases since the bulk of the increase in the future domestic demand for coal is expected to be for electric power generation. Consideration has been given to interfuel substitution between coal and uranium/nuclear under the various scenarios.

Figure 8.3 shows the expected annual production under the assumptions of Scenarios A, B, C, D, E, C-1/D-7, and C-2/D-8.

8.4.2 Coal Supply Prospects

Scenarios A and B: 1985

In these scenarios the high oil price and vigorous government policy response tend to stimulate coal production. Annual production rates are expected to be 143 percent greater than in 1975, reflecting increases in demand for thermal coal for electric power generation and possible increases in exports of metallurgical coal.

Scenario C: 1985

In this scenario the high world economic growth rate, stable medium oil price, and a vigorous policy response encourage expansion of coal supplies by 1985 to 0.045×10^9 short tons or 95 percent over the 1975 production rate. This increase is somewhat less than in Scenarios A and B, reflecting the effects of the medium price level.

Table 8.6
National policy responses for coal

Vigorous Response	Restrained Response
1. Strip mining regulations are favorable for development in the eastern slopes of the Rocky Mountains.	1. Stringent land use regulations preventing development in the eastern slopes of the Rocky Mountains.
2. Land reclamation policies are favorable to stimulate production.	2. States of uncertainty in land reclamation policies.
3. Fair and equitable provincial royalty rates.	3. Escalating royalty rates.
4. Air quality standards permit the burning of coal without pollution control devices.	4. Air quality standards do not permit the burning of coal without pollution control devices.
5. Tax incentives created to stimulate development.	5. No tax incentives for the producers.
6. Government creates environment to attract labor to coal industry through wage rates and health and safety legislation.	6. Little or no government action to attract labor to coal industry.
7. Government conducts and supports research and development programs in mining techniques.	7. No government conducted or supported research and development programs.
8. Government assists with transportation facilities i.e., railways and slurry pipelines.	8. Government does not assist with transportation facilities.

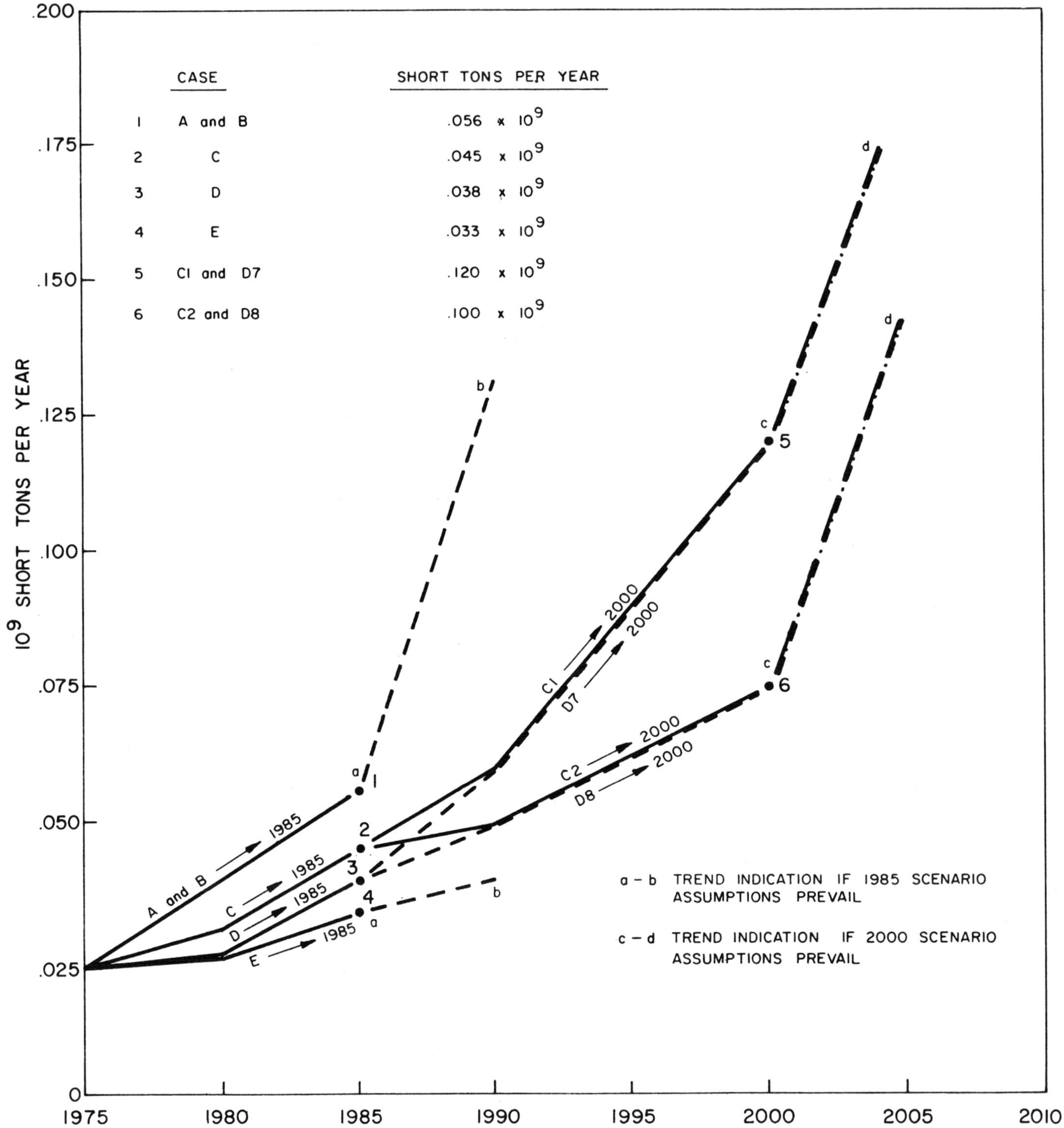

Figure 8.3
Canadian coal supply prospects (expected annual production).

Scenario D: 1985
The restrained national policy response together with a low world economic growth rate lowers the development of coal supplies by 15 percent from Scenario C. The effects of price level and stability outweigh the restrained policies in inducing thermal and metallurgical coal development.

Scenario E: 1985
The Scenario E assumptions of declining oil prices and a restrained policy response still allow for growth in coal production, but at a level 13 percent lower than in Scenario D.

Scenarios C-1 and D-7: 2000
In both these scenarios government policies emphasize coal as the principal replacement fuel. Energy prices range from high in C-1 to medium in D-7. Coal production is projected to increase to 0.120×10^9 short tons per year by the year 2000. The high energy price in C-1 may cause coal gasification to become economically viable. The decline in supplies of natural gas will increase the coal supply growth.

Scenarios C-2 and D-8: 2000
In these scenarios government policies emphasize uranium and nuclear as the principal replacement fuel. Energy prices range from high in C-2 to medium in D-8. Under C-2 and D-8 assumptions, with no increased governmental emphasis on coal supply, nuclear energy begins to increase its share of the electrical generation market. Even with the lack of emphasis on coal, the production of coal will still increase to 0.100×10^9 short tons per year in response to price and will result in the economic feasibility of coal gasification. Gasification may likely become prominent if frontier gas supplies are not developed in sufficient volumes to meet requirements.

Assessments of the coal supply are summarized in Table 8.7.

8.5 HYDROELECTRIC AND NUCLEAR

The expansion of electrical systems is primarily determined by considering how to meet anticipated regional market demands, with acceptable reserve margins, in the most efficient manner possible. The ultimate economic constraint on expanding electrical capacity arises from the price of electricity relative to the prices of alternative sources of energy. An electricity supply projection must therefore be based principally on a projection of future demand growth. Conceptually, electricity supply and demand should balance at a given price.

The demand for electricity, and hence the resulting supplies, depend on such factors as income, population, the availability and price of substitute fuels, capital costs of energy-using equipment, technology, convenience factors, and the price of electricity. Supply and price, in turn, will be affected by capital requirements for capacity expansion, financial characteristics of the utility concerned, and decisions of regulatory agencies.

For the 1985 scenarios, two projections of in-

Table 8.7
Assessment of indigenous supply of coal (10^9 short tons)

	1985 scenarios					2000 scenarios			
	A	B	C	D	E	C-1	C-2	D-7	D-8
Total resource in place	195.0	195.0	195.0	195.0	195.0	195.0	195.0	195.0	195.0
Ultimate recoverable	100.0	100.0	100.0	100.0	100.0	100.0	100.0	100.0	100.0
Reserves developed to date	23.6	23.6	21.5	18.1	17.9	32.3	27.0	32.3	27.0
Expected annual production	0.056	0.056	0.045	0.038	0.033	0.120	0.100	0.120	0.100
Cumulative production	1.6	1.6	1.4	1.3	1.3	3.3	3.0	3.3	3.0
Remaining developed reserves	22.0	22.0	20.1	16.8	16.6	29.0	24.0	29.0	24.0
Reserves/production ratio	392	392	446	442	520	242	240	242	240

stalled nuclear generation capacity are provided. The projected electricity demand from the low price scenario E is compared with a high growth nuclear generation capacity estimated at 14.7×10^3 megawatts. Projected electricity demands from the mid-1975 and high price scenarios A, B, C, and D are compared with a low growth nuclear generation capacity estimated at 12.5×10^3 megawatts. These estimates are based on forecasts provided by Energy, Mines and Resources Canada in its publication *1975 Assessment of Canada's Uranium Supply and Demand* (June 1976). The high estimate for the 1985 scenarios was obtained by assuming that the total in-service capacity forecast for 1986 would apply in 1985. The low estimate was obtained by assuming that the total projected in-service capacity for 1984 would apply in 1985.

Two projections of installed hydroelectric capacity are provided for the 1985 scenarios, based on estimates in *An Energy Strategy for Canada: Policies for Self-Reliance* (April 1976). For the high growth case the hydroelectric capacity is estimated at 61.9×10^3 megawatts; for the low growth case capacity is estimated at 55.5×10^3 megawatts.

WAES Associates agreed that the period from 1985 to 2000 may be characterized by a decline of oil and gas as the world's major sources of energy. It was noted that only coal or nuclear energy could reasonably be expected to be available widely enough and in sufficient quantities to fill the potential "gap" resulting from the relative decline of oil and gas.

The Canadian response to the proposed use of nuclear power as a principal replacement fuel is demonstrated in Scenarios C-2 and D-8. In these scenarios it was assumed that nuclear-generated electricity would represent an increasingly greater component of Canadian electricity requirements. To demonstrate a major role for coal as the principal replacement fuel (Scenarios C-1 and D-7), it was assumed that the substantial increase in demand for nuclear-generated electricity did not occur but was replaced by a demand for electricity generated by coal. This substitution took place to the extent that domestic coal production would allow.

These assumptions, particularly for coal, were made to meet the criteria established by WAES and should not be interpreted as a forecast of the likely course of events in Canada.

For the year 2000 scenarios, two projections of installed nuclear and hydro generation capacities are provided. A high nuclear generation capacity is estimated at 74.5×10^3 megawatts and a low nuclear generation capacity is estimated at 45×10^3 megawatts. A high hydroelectric generation capacity of 74×10^3 megawatts and a low hydro generation capacity of 60×10^3 megawatts are estimated.

Turning for a moment from the scenarios per se, several trends are evident as one follows the planning of the Electrical Utilities in Canada. First, increasing emphasis on the use of nuclear generation in Canada is expected, particularly toward the end of the century. Second, more emphasis on coal generation is expected, particularly in the Prairie region. Natural gas will continue to be used sparingly and will be reserved primarily for the domestic market and for petrochemical feedstock.

8.6 ENERGY SUPPLY WORKSHEETS

This section consists of the following worksheets (see Section 1.2.1 of Chapter 1 for the definitions of line items):

For Scenarios A and B:
Crude oil (6)
Natural gas (4)
Coal (1)
Uranium and nuclear (1)
Hydroelectric (1)

For Scenarios C, C-1, and C-2:
Crude oil (6)
Natural gas (4)
Coal (1 for Scenarios C, C-1, and D-7)
Uranium and nuclear (2)
Hydroelectric (1)

For Scenarios D, D-7, and D-8:
Crude oil (6)
Natural gas (4)
Coal (1 for Scenarios D, C-2, and D-8)
Uranium and nuclear (2)
Hydroelectric (1)

For Scenario E:
Crude oil (6)
Natural gas (4)
Coal (1)
Uranium and nuclear (1)
Hydroelectric (1)

INDIGENOUS SUPPLY ASSESSMENT WORKSHEET

CRUDE OIL

Country: CANADA

Crude Oil:
Offshore X
Onshore X
Total

WAES SCENARIO CASE (1985): A & B

WAES SCENARIO CASE (2000):

Total Resource in Place: 1029.8 x 10^9 bbls.

Domestic Oil Price: $17.25/bbl.

Technology Assumptions:

	(Units)	1972	1975	1980	1985	1990	2000	2010	2025
(1) Ultimate Recoverable Resource:	10^9 bbls.	73.5	-	-	-	-	-	-	73.5
(2) Discoveries:									
(3) Reserves Proved to Date:	10^9 bbls.	18.11	20.30	29.84	47.25	55.60	60.90	60.90	60.90
(4) Potential Production:									
(5) Expected Annual Production 10^9 bbls.	heavy) light)	.700	.865	.922	1.167	1.227	1.476	1.145	.710
(6) Cumulative Production:	10^9 bbls.	7.24	8.76	12.69	17.13	22.75	34.58	46.03	56.68
(7) Remaining Proved Reserves:	10^9 bbls.	10.87	11.54	17.15	30.12	32.85	26.32	14.87	4.22
(8) Reserve/Production Ratio:		15.5	13.3	18.6	25.8	26.8	17.8	12.8	5.9
(9) Host Government Take ($/unit of production):									
(10) Capital Cost ($/unit of daily capacity):									
(11) Operating Cost ($/unit of production):									

INDIGENOUS SUPPLY ASSESSMENT WORKSHEET

CRUDE OIL

Country: CANADA

Crude Oil:
Offshore ____ Oil Sands
Onshore X

WAES SCENARIO CASE (1985): A & B

WAES SCENARIO CASE (2000): ________

Total Resource in Place: 895 x 10^9 bbls.

Domestic Oil Price: $17.25/bbl.

Technology Assumptions: 1975 technology prevailing

	(Units)	1972	1975	1980	1985	1990	2000	2010	2025
(1) Ultimate Recoverable Resource:	10^9 bbls.	29.0	-	-	-	-	-	-	29.0
(2) Discoveries:									
(3) Reserves Proved (DEVELOPED) to Date:	10^9 bbls.	.70	1.84	4.12	15.10	17.50	17.50	17.50	17.50
(4) Potential Production:									
(5) Expected Annual Production 10^9 bbls.	heavy)								
	light)	.018	.022	.057	.146	.305	.370	.370	.370
(6) Cumulative Production: 10^9 bbls.		.08	.12	.29	.82	2.03	5.73	9.43	14.98
(7) Remaining Proved Reserves:	10^9 bbls.	.62	1.72	3.83	14.28	15.47	11.77	8.07	2.52
(8) Reserve/Production Ratio:		34.4	78.2	67.2	97.9	50.8	31.9	21.9	6.8
(9) Host Government Take ($/unit of production):									
(10) Capital Cost ($/unit of daily capacity):									
(11) Operating Cost ($/unit of production):									

INDIGENOUS SUPPLY ASSESSMENT WORKSHEET

CRUDE OIL

Country: CANADA

WAES SCENARIO CASE (1985): A & B

WAES SCENARIO CASE (2000):

Total Resource in Place: 134.8×10^9 bbls.

Domestic Oil Price: $17.25/bbl.

Technology Assumptions: 1975 technology prevailing

Crude Oil:
Offshore X
Onshore X
Total including N.G.L.

	(Units)	1972	1975	1980	1985	1990	2000	2010	2025
(1) Ultimate Recoverable Resource:	10^9 bbls.	44.5	-	-	-	-	-	-	44.5
(2) Discoveries:									
(3) Reserves Proved to Date:	10^9 bbls.	17.41	18.46	25.72	32.15	38.10	43.40	43.40	43.40
(4) Potential Production:									
(5) Expected Annual Production 10^9 bbls.	heavy)								
	light)	.682	.843	.865	1.021	.922	1.106	.775	.340
(6) Cumulative Production:	10^9 bbls.	7.16	8.64	12.40	16.31	20.72	28.85	36.60	41.70
(7) Remaining Proved Reserves:	10^9 bbls.	10.25	9.82	13.32	15.84	17.38	14.55	6.80	1.70
(8) Reserve/Production Ratio:		15.0	11.6	15.4	15.5	18.9	13.2	8.8	5.0
(9) Host Government Take ($/unit of production):									
(10) Capital Cost ($/unit of daily capacity):									
(11) Operating Cost ($/unit of production):									

INDIGENOUS SUPPLY ASSESSMENT WORKSHEET

CRUDE OIL

Country: CANADA

WAES SCENARIO CASE (1985): A & B

WAES SCENARIO CASE (2000): ________

Total Resource in Place: 60.6×10^9 bbls.

Domestic Oil Price: $17.25/bbl

Technology Assumptions: 1975 technology prevailing

Crude Oil:
Offshore ______ Conventional
Onshore __X__ including N.G.L.

	(Units)	1972	1975	1980	1985	1990	2000	2010	2025
(1) Ultimate Recoverable Resource:	10^9 bbls.	20.0	-	-	-	-	-	-	20.0
(2) Discoveries:									
(3) Reserves Proved to Date:	10^9 bbls.	17.01	17.23	18.34	19.20	19.52	19.90	19.90	19.90
(4) Potential Production:									
(5) Expected Annual Production 10^9 bbls.	heavy)								
	light)	.682	.843	.677	.441	.302	.230	.110	.100
(6) Cumulative Production:	10^9 bbls.	7.16	8.64	12.02	14.17	15.48	16.90	18.00	19.50
(7) Remaining Proved Reserves:	10^9 bbls.	9.85	8.59	6.32	5.03	4.04	3.00	1.90	.40
(8) Reserve/Production Ratio:		14.4	10.2	9.3	11.4	13.4	13.0	17.3	4.0
(9) Host Government Take ($/unit of production):									
(10) Capital Cost ($/unit of daily capacity):									
(11) Operating Cost ($/unit of production):									

INDIGENOUS SUPPLY ASSESSMENT WORKSHEET

CRUDE OIL

Country: CANADA

Crude Oil:
Offshore X Beaufort -
Onshore X Sverdrup

WAES SCENARIO CASE (1985): A & B

WAES SCENARIO CASE (2000):

Total Resource in Place: 41.30×10^9 bbls.

Domestic Oil Price: $17.25/bbl.

Technology Assumptions: 1975 technology prevailing

	(Units)	1972	1975	1980	1985	1990	2000	2010	2025
(1) Ultimate Recoverable Resource:	10^9 bbls.	13.6	-	-	-	-	-	-	13.6
(2) Discoveries:									
(3) Reserves Proved to Date:	10^9 bbls.	.30	.850	4.30	7.42	10.60	13.00	13.00	13.00
(4) Potential Production:									
(5) Expected Annual Production 10^9 bbls.	heavy)								
	light)				.380	.380	.466	.300	.130
(6) Cumulative Production:	10^9 bbls.				.76	2.66	7.32	10.32	12.27
(7) Remaining Proved Reserves:	10^9 bbls.	.30	.85	4.3	6.66	7.94	5.68	2.68	.73
(8) Reserve/Production Ratio:					17.5	20.9	12.2	8.9	5.6
(9) Host Government Take ($/unit of production):									
(10) Capital Cost ($/unit of daily capacity):									
(11) Operating Cost ($/unit of production):									

INDIGENOUS SUPPLY ASSESSMENT WORKSHEET

CRUDE OIL

Country: CANADA

Crude Oil:
Offshore X Eastern
Onshore ____ Canada

WAES SCENARIO CASE (1985): A & B

WAES SCENARIO CASE (2000): ____

Total Resource in Place: 32.90 x 10^9 bbls.

Domestic Oil Price: $17.25/bbl.

Technology Assumptions: Pipeline and production facilities can be built

	(Units)	1972	1975	1980	1985	1990	2000	2010	2025
(1) Ultimate Recoverable Resource:	10^9 bbls.	10.9	-	-	-	-	-	-	10.9
(2) Discoveries:									
(3) Reserves Proved to Date:	10^9 bbls.	.10	.38	3.08	5.53	7.98	10.50	10.50	10.50
(4) Potential Production:									
(5) Expected Annual Production 10^9 bbls.	heavy) light)			.188	.200	.240	.410	.365	.110
(6) Cumulative Production:	10^9 bbls.			.38	1.38	2.58	4.63	8.28	9.93
(7) Remaining Proved Reserves:	10^9 bbls.	.10	.38	2.70	4.15	5.40	5.87	2.22	.57
(8) Reserve/Production Ratio:				14.4	20.8	22.5	14.3	6.1	5.2
(9) Host Government Take ($/unit of production):									
(10) Capital Cost ($/unit of daily capacity):									
(11) Operating Cost ($/unit of production):									

INDIGENOUS SUPPLY ASSESSMENT WORKSHEET

NATURAL GAS

Country: CANADA

Natural Gas:
Offshore ______ Conventional
Onshore X Producing Area

WAES SCENARIO CASE (1985): A & B

WAES SCENARIO CASE (2000): ______

Total Resource in Place: 135.3×10^{12} c.f.

Domestic Gas Price: $2.88/$10^3$ c.f.(1)

Technology Assumptions: 1975 technology prevailing

	(Units)	1972	1975	1980	1985	1990	2000	2010	2025
(1) Ultimate Recoverable Resource:	10^{12} c.f.	115.0	-	-	-	-	-	-	115.0
(2) Discoveries:									
(3) Reserves Proved to Date:	10^{12} c.f.	79.0	85.4	104.1	109.0	110.6	113.4	113.4	113.4
(4) Potential Production:									
(5) Expected Annual Production:	10^{12} c.f.	2.30	2.59	3.03	3.23	2.37	1.29	1.20	.43
(6) Cumulative Production:	10^{12} c.f.	21.3	26.5	44.1	59.7	73.4	84.0	96.0	105.8
(7) Remaining Proved Reserves:	10^{12} c.f.	57.7	58.9	60.0	49.3	37.2	29.4	17.4	7.6
(8) Reserve/Production Ratio:		25.1	22.7	19.8	15.3	15.7	22.7	14.5	17.6
(9) Host Government Take ($/unit of production):									
(10) Capital Cost ($/unit of daily capacity):									
(11) Operating Cost ($/unit of production):									

(1) Based on BTU equivalent of 6000 cu. ft. gas per barrel of oil

INDIGENOUS SUPPLY ASSESSMENT WORKSHEET

NATURAL GAS

Country: CANADA

Natural Gas:
Offshore X Beaufort and
Onshore X Sverdrup

WAES SCENARIO CASE (1985): A & B

WAES SCENARIO CASE (2000): ________

Total Resource in Place: 100×10^{12} c.f.

Domestic Gas Price: $\$2.88/10^3$ c.f.(1)

Technology Assumptions: 1975 technology prevailing

	(Units)	1972	1975	1980	1985	1990	2000	2010	2025
(1) Ultimate Recoverable Resource:	10^{12} c.f.	85.0	-	-	-	-	-	-	85.0
(2) Discoveries:									
(3) Reserves Proved to Date:	10^{12} c.f.	12.0	15.0	21.0	27.0	31.0	31.0	31.0	31.0
(4) Potential Production:									
(5) Expected Annual Production:	10^{12} c.f.				.45	.74	.94	.68	.38
(6) Cumulative Production:	10^{12} c.f.				1.8	5.5	13.5	20.3	26.0
(7) Remaining Proved Reserves:	10^{12} c.f.	12.0	15.0	21.0	25.2	25.5	17.5	10.7	5.0
(8) Reserve/Production Ratio:					56.0	34.5	18.6	15.7	13.2
(9) Host Government Take ($/unit of production):									
(10) Capital Cost ($/unit of daily capacity):									
(11) Operating Cost ($/unit of production):									

(1) Based on BTU equivalent of 6000 cu. ft. gas per barrel of oil

INDIGENOUS SUPPLY ASSESSMENT WORKSHEET

NATURAL GAS

Country: CANADA

Natural Gas:
Offshore X
Onshore X
Total

WAES SCENARIO CASE (1985): A & B

WAES SCENARIO CASE (2000): ________

Total Resource in Place: 311.8×10^{12} c.f.

Domestic Gas Price: $\$2.88/10^3$ c.f.(1)

Technology Assumptions: ________

	(Units)	1972	1975	1980	1985	1990	2000	2010	2025
(1) Ultimate Recoverable Resource:	10^{12} c.f.	265.0	-	-	-	-	-	-	265.0
(2) Discoveries:									
(3) Reserves Proved to Date:	10^{12} c.f.	92.0	103.4	135.1	153.0	161.6	168.4	168.4	168.4
(4) Potential Production:									
(5) Expected Annual Production:	10^{12} c.f.	2.30	2.59	3.03	3.68	4.11	3.23	2.48	1.31
(6) Cumulative Production:	10^{12} c.f.	21.3	26.5	44.1	61.5	81.9	105.5	130.3	153.3
(7) Remaining Proved Reserves:	10^{12} c.f.	70.7	76.9	91.0	91.5	79.7	62.9	38.1	15.1
(8) Reserve/Production Ratio:		30.7	29.7	30.0	24.9	19.4	19.5	15.4	11.5
(9) Host Government Take ($/unit of production):									
(10) Capital Cost ($/unit of daily capacity):									
(11) Operating Cost ($/unit of production):									

(1) Based on BTU equivalent of 6000 cu. ft. gas per barrel of oil

INDIGENOUS SUPPLY ASSESSMENT WORKSHEET

NATURAL GAS

Country: CANADA

WAES SCENARIO CASE (1985): A & B

WAES SCENARIO CASE (2000): ________

Total Resource in Place: 76.5×10^{12} c.f.

Domestic Gas Price: $\$2.88/10^3$ c.f.[(1)]

Natural Gas:
Offshore X Eastern Canada
Onshore ________

Technology Assumptions: Production and pipeline facilities can be built

	(Units)	1972	1975	1980	1985	1990	2000	2010	2025
(1) Ultimate Recoverable Resource:	10^{12} c.f.	65.0	-	-	-	-	-	-	65.0
(2) Discoveries:									
(3) Reserves Proved to Date:	10^{12} c.f.	1.0	3.0	10.0	17.0	20.0	24.0	24.0	24.0
(4) Potential Production:									
(5) Expected Annual Production:	10^{12} c.f.					1.0	1.0	.6	.5
(6) Cumulative Production:	10^{12} c.f.					3.0	8.0	14.0	21.5
(7) Remaining Proved Reserves:	10^{12} c.f.	1.0	3.0	10.0	17.0	17.0	16.0	10.0	2.5
(8) Reserve/Production Ratio:						17.0	16.0	16.6	5.0
(9) Host Government Take ($/unit of production):									
(10) Capital Cost ($/unit of daily capacity):									
(11) Operating Cost ($/unit of production):									

(1) Based on BTU equivalent of 6000 cu. ft. gas per barrel of oil

INDIGENOUS SUPPLY ASSESSMENT WORKSHEET

COAL

Country: CANADA

WAES SCENARIO CASE (1985): A & B

WAES SCENARIO CASE (2000): ______

Total Resource in Place: 195×10^9 short tons

Domestic Coal Price: $57.44/short ton (1)

Technology Assumptions: 1975 technology prevailing

Coal:
Surface Mined X
Deep Mined X
Total

	1972	1975	1980	1985	1990	2000	2010	2025
(1) Ultimate Recoverable Resource: 10^9 sh. tons	100.0	-	-	-	-	-	-	100.0
(2) Reserves Proved to Date: 10^9 sh. tons	13.7	14.9	18.9	23.6	34.6	45.1	55.1	62.0
(3) Potential Production:								
(4) Expected Annual Production: 10^9 sh. tons	.021	.023	.038	.056	.132	.265	.412	.806
(5) Cumulative Production: 10^9 sh. tons	1.0	1.1	1.3	1.6	2.3	5.0	9.1	13.1
(6) Remaining Proved Reserves: 10^9 sh. tons	12.7	13.8	17.6	22.0	32.3	40.1	46.0	48.9
(7) Reserve/Production Ratio:	605	600	463	392	245	151	112	61
(8) Primary Transportation* —Capital Cost ($/unit of daily capacity): —Operating Cost ($/unit of daily capacity):								
(9) Capital Cost ($/unit of daily capacity):								
(10) Operating Cost ($/unit of production):								

*From mine mouth to first processing stage.

(1) Based on 1 ton of coal per 3.3 bbls. oil

NONFOSSIL SOURCES AND CONVERSION PROCESSES WORKSHEET

URANIUM AND NUCLEAR

Country: CANADA

WAES SCENARIO CASE (1985): A & B

WAES SCENARIO CASE (2000): ______

Total Resource in Place: 415,000 short tons U_3O_8

Domestic Uranium Price: Up to $20/lb U_3O_8

Technology Assumptions: ______

	(Units)	1972	1975	1980	1985	1990	1995	2000	2010
1. Ultimate Recoverable Uranium Resource:	(sh. tons)		189,000						
2. Reserve of Uranium Proved to Date:	(sh. tons)		82,000						
3. Potential Production of Uranium:	(sh. tons)			13,000	15,000				
4. Expected Production of Uranium:	(sh. tons)		4,600						
5. Reserve/Production Ratio:									
6. Maximum Installed Capacity (GWe)*:		1.75	2.5		12.5				
7. Capital Cost ($/KWe):			$500 to $700						
8. Operating Cost (¢/Kwhr)**:									

*Indicate assumed load factor and LWR or HWR.

**Excluding capital charges.

NONFOSSIL SOURCES AND CONVERSION PROCESS WORKSHEET

Hydroelectric X
Coal Gasification
Coal to Syncrude
Gas Liquifaction
Regasification
Other* (specify):

Country: CANADA

WAES SCENARIO CASE (1985): A & B

WAES SCENARIO CASE (2000):

Total Resource in Place:

Technology Assumptions:

	(Units)	1972	1975	1985	1990	2000	2010	2025
(1) Yield:								
(2) Maximum Potential Output	** Capacity (GWe)	31.5	36.8	55.5				
(3) Capital Cost ($/unit of daily capacity):								
(4) Cumulative Capital Investment:								
(5) Operating Cost ($/unit of production):								

*Specify other relevant non-fossil energy source or conversion process, i.e., geothermal, solar, wind, etc.

** "Maximum technological development" in GEMM

INDIGENOUS SUPPLY ASSESSMENT WORKSHEET

CRUDE OIL

Country: CANADA

Crude Oil:
Offshore X
Onshore X
Total

WAES SCENARIO CASE (1985): C

WAES SCENARIO CASE (2000): C1 & C2

Total Resource in Place: 1029.8×10^9 bbls.

Domestic Oil Price: $11.50/bbl.

Technology Assumptions: ______

	(Units)	1972	1975	1980	1985	1990	2000	2010	2025
(1) Ultimate Recoverable Resource:	10^9 bbls.	73.5	-	-	-	-	-	-	73.5
(2) Discoveries:									
(3) Reserves Proved to Date:	10^9 bbls.	18.11	20.29	27.93	44.17	51.27	55.55	55.55	55.55
(4) Potential Production:									
(5) Expected Annual Production 10^9 bbls.	heavy) light)	.700	.865	.828	1.037	1.094	1.117	.820	.716
(6) Cumulative Production:	10^9 bbls.	7.24	8.76	14.50	16.53	21.49	31.45	39.55	48.10
(7) Remaining Proved Reserves:	10^9 bbls.	10.87	11.53	13.43	27.64	29.78	24.10	6.00	7.45
(8) Reserve/Production Ratio:		15.5	13.3	17.2	39.0	27.2	21.6	19.5	10.4
(9) Host Government Take ($/unit of production):									
(10) Capital Cost ($/unit of daily capacity):									
(11) Operating Cost ($/unit of production):									

INDIGENOUS SUPPLY ASSESSMENT WORKSHEET

CRUDE OIL

Country: CANADA

WAES SCENARIO CASE (1985): C

WAES SCENARIO CASE (2000): C1 & C2

Crude Oil:
Offshore ____
Onshore X — Oil Sands

Total Resource in Place: 895×10^9 bbls.

Domestic Oil Price: $11.50/bbl.

Technology Assumptions: 1975 technology prevailing

	(Units)	1972	1975	1980	1985	1990	2000	2010	2025
(1) Ultimate Recoverable Resource:	10^9 bbls.	29.0	-	-	-	-	-	-	29.0
(2) Discoveries:									
(3) Reserves Proved (DEVELOPED) to Date:	10^9 bbls.	.70	1.84	4.12	15.10	17.50	17.50	17.50	17.50
(4) Potential Production:									
(5) Expected Annual Production 10^9 bbls.	heavy)								
	light)	.018	.022	.057	.146	.305	.325	.325	.325
(6) Cumulative Production:	10^9 bbls.	.08	.12	.29	.82	2.03	5.28	8.53	13.40
(7) Remaining Proved Reserves:	10^9 bbls.	.62	1.72	3.83	14.28	15.47	12.22	8.97	4.10
(8) Reserve/Production Ratio:		34.4	78.1	67.2	97.8	50.7	37.6	27.6	12.6
(9) Host Government Take ($/unit of production):									
(10) Capital Cost ($/unit of daily capacity):									
(11) Operating Cost ($/unit of production):									

INDIGENOUS SUPPLY ASSESSMENT WORKSHEET

CRUDE OIL

Country: CANADA

WAES SCENARIO CASE (1985): C

WAES SCENARIO CASE (2000): C1 & C2

Total Resource in Place: 134.8 x 10^9 bbls.

Domestic Oil Price: $11.50/bbl.

Technology Assumptions: 1975 technology prevailing

Crude Oil:
Offshore X
Onshore X
Total including N.G.L.

	(Units)	1972	1975	1980	1985	1990	2000	2010	2025
(1) Ultimate Recoverable Resource:	10^9 bbls.	44.5	-	-	-	-	-	-	44.5
(2) Discoveries:									
(3) Reserves Proved to Date:	10^9 bbls.	17.41	18.45	23.81	29.07	33.77	38.05	38.05	38.05
(4) Potential Production:									
(5) Expected Annual Production 10^9 bbls.	heavy)								
	light)	.682	.843	.771	.891	.789	.792	.495	.391
(6) Cumulative Production:	10^9 bbls.	7.16	8.64	14.21	15.71	19.46	26.17	31.02	34.70
(7) Remaining Proved Reserves:	10^9 bbls.	10.25	9.81	9.60	13.36	14.31	11.88	7.03	3.35
(8) Reserve/Production Ratio:		15.0	11.6	12.5	15.0	18.1	15.0	14.2	8.6
(9) Host Government Take ($/unit of production):									
(10) Capital Cost ($/unit of daily capacity):									
(11) Operating Cost ($/unit of production):									

INDIGENOUS SUPPLY ASSESSMENT WORKSHEET

CRUDE OIL

Country: CANADA

WAES SCENARIO CASE (1985): C

WAES SCENARIO CASE (2000): C1 & C2

Total Resource in Place: 60.6×10^9 bbls.

Domestic Oil Price: $11.50/bbl.

Technology Assumptions: 1975 technology prevailing

Crude Oil:
Offshore ____
Onshore X

Conventional including N.G.L.

	(Units)	1972	1975	1980	1985	1990	2000	2010	2025
(1) Ultimate Recoverable Resource:	10^9 bbls.	20.0	-	-	-	-	-	-	20.0
(2) Discoveries:									
(3) Reserves Proved to Date:	10^9 bbls.	17.01	17.23	18.08	18.96	19.28	19.69	19.69	19.69
(4) Potential Production:									
(5) Expected Annual Production 10^9 bbls.	heavy)								
	light)	.682	.843	.677	.441	.302	.230	.109	.109
(6) Cumulative Production	10^9 bbls.	7.16	8.64	12.02	14.17	15.48	16.95	18.04	18.45
(7) Remaining Proved Reserves:	10^9 bbls.	9.85	8.59	6.06	4.79	3.80	2.74	1.65	1.24
(8) Reserve/Production Ratio:		14.5	10.2	8.9	10.9	12.6	11.9	15.1	11.4
(9) Host Government Take ($/unit of production):									
(10) Capital Cost ($/unit of daily capacity):									
(11) Operating Cost ($/unit of production):									

INDIGENOUS SUPPLY ASSESSMENT WORKSHEET

CRUDE OIL

Country: CANADA

WAES SCENARIO CASE (1985): C

WAES SCENARIO CASE (2000): C1 & C2

Total Resource in Place: 41.30×10^9 bbls.

Domestic Oil Price: $11.50/bbl.

Technology Assumptions: 1975 technology prevailing

Crude Oil:
Offshore X Beaufort -
Onshore X Sverdrup

	(Units)	1972	1975	1980	1985	1990	2000	2010	2025
(1) Ultimate Recoverable Resource:	10^9 bbls.	13.6	-	-	-	-	-	-	13.6
(2) Discoveries:									
(3) Reserves Proved to Date:	10^9 bbls.	.30	.85	3.35	5.85	8.35	10.22	10.22	10.22
(4) Potential Production:									
(5) Expected Annual Production 10^9 bbls.	heavy)								
	light)				.300	.300	.375	.223	.150
(6) Cumulative Production	10^9 bbls.				.60	2.10	5.47	7.70	9.22
(7) Remaining Proved Reserves:	10^9 bbls.	.30	.85	3.35	5.25	6.25	4.75	2.52	1.00
(8) Reserve/Production Ratio:					17.5	20.8	12.7	11.3	6.6
(9) Host Government Take ($/unit of production):									
(10) Capital Cost ($/unit of daily capacity):									
(11) Operating Cost ($/unit of production):									

INDIGENOUS SUPPLY ASSESSMENT WORKSHEET

CRUDE OIL

Country: CANADA

Crude Oil:
Offshore X Eastern Canada
Onshore

WAES SCENARIO CASE (1985): C

WAES SCENARIO CASE (2000): C1 & C2

Total Resource in Place: 32.90×10^9 bbls.

Domestic Oil Price: $11.50/bbl.

Technology Assumptions: Pipeline and production facilities can be built

	(Units)	1972	1975	1980	1985	1990	2000	2010	2025
(1) Ultimate Recoverable Resource:	10^9 bbls.	10.9	-	-	-	-	-	-	10.9
(2) Discoveries:									
(3) Reserves Proved to Date:	10^9 bbls.	.10	.38	2.38	4.26	6.13	8.14	8.14	8.14
(4) Potential Production:									
(5) Expected Annual Production	heavy)								
10^9 bbls.	light)			.093	.150	.187	.187	.153	.131
(6) Cumulative Production:	10^9 bbls.			.19	.94	1.87	3.75	5.28	7.03
(7) Remaining Proved Reserves:	10^9 bbls.	.10	.38	2.19	3.32	4.26	4.39	2.86	1.11
(8) Reserve/Production Ratio:				23.4	22.1	22.7	23.4	18.6	8.4
(9) Host Government Take ($/unit of production):									
(10) Capital Cost ($/unit of daily capacity):									
(11) Operating Cost ($/unit of production):									

INDIGENOUS SUPPLY ASSESSMENT WORKSHEET

NATURAL GAS

Country: CANADA

WAES SCENARIO CASE (1985): C

WAES SCENARIO CASE (2000): C1 & C2

Total Resource in Place: 311.8×10^{12} c.f.

Domestic Gas Price: $\$1.92/10^{3}$ c.f.[(1)]

Technology Assumptions: ______

Natural Gas:
Offshore X
Onshore X
Total

	(Units)	1972	1975	1980	1985	1990	2000	2010	2025
(1) Ultimate Recoverable Resource:	10^{12} c.f.	265.0	-	-	-	-	-	-	265.0
(2) Discoveries:									
(3) Reserves Proved to Date:	10^{12} c.f.	92	103.4	116.4	121.9	123.4	126.0	126.0	126.0
(4) Potential Production:									
(5) Expected Annual Production:	10^{12} c.f.	2.30	2.59	2.81	2.70	2.75	1.78	1.23	.58
(6) Cumulative Production:	10^{12} c.f.	21.3	26.5	40.8	56.5	71.8	87.8	105.1	113.8
(7) Remaining Proved Reserves:	10^{12} c.f.	70.7	76.9	75.6	65.4	51.6	38.2	20.9	12.2
(8) Reserve/Production Ratio:		30.7	29.6	27.0	19.2	19.1	21.2	17.0	20.7
(9) Host Government Take ($/unit of production):									
(10) Capital Cost ($/unit of daily capacity):									
(11) Operating Cost ($/unit of production):									

(1) Based on BTU equivalent of 6000 cu. ft. of gas per barrel of oil

INDIGENOUS SUPPLY ASSESSMENT WORKSHEET

NATURAL GAS

Country: CANADA

WAES SCENARIO CASE (1985): C

WAES SCENARIO CASE (2000): C1 & C2

Total Resource in Place: 135.3×10^{12} c.f.

Domestic Gas Price: $1.92/$10^3$ c.f.(1)

Technology Assumptions: 1975 technology prevailing

Natural Gas:
Offshore ____
Onshore X
Conventional Producing Area

	(Units)	1972	1975	1980	1985	1990	2000	2010	2025
(1) Ultimate Recoverable Resource:	10^{12} c.f.	115.0	-	-	-	-	-	-	115.0
(2) Discoveries:									
(3) Reserves Proved to Date:	10^{12} c.f.	79.0	85.4	96.4	100.9	102.4	105.0	105.0	105.0
(4) Potential Production:									
(5) Expected Annual Production:	10^{12} c.f.	2.30	2.59	2.81	2.30	2.20	1.20	.92	.40
(6) Cumulative Production:	10^{12} c.f.	21.3	26.5	40.8	55.3	68.0	77.8	92.0	98.0
(7) Remaining Proved Reserves:	10^{12} c.f.	57.7	58.9	55.6	45.6	34.4	27.2	13.0	7.0
(8) Reserve/Production Ratio:		25.1	22.7	19.9	15.2	16.5	22.7	14.1	17.5
(9) Host Government Take ($/unit of production):									
(10) Capital Cost ($/unit of daily capacity):									
(11) Operating Cost ($/unit of production):									

(1) Based on BTU equivalent of 6000 cu. ft. gas per barrel of oil

INDIGENOUS SUPPLY ASSESSMENT WORKSHEET

NATURAL GAS

Country: CANADA

WAES SCENARIO CASE (1985): C

WAES SCENARIO CASE (2000): C1 & C2

Total Resource in Place: 100 x 10^{12} c.f.

Domestic Gas Price: \$1.92/10^3 c.f.[(1)]

Technology Assumptions: 1975 technology prevailing

Natural Gas:
Offshore X Beaufort and
Onshore X Sverdrup

	(Units)	1972	1975	1980	1985	1990	2000	2010	2025
(1) Ultimate Recoverable Resource:	10^{12} c.f.	85.0	-	-	-	-	-	-	85.0
(2) Discoveries:									
(3) Reserves Proved to Date:	10^{12} c.f.	12.0	15.0	17.0	18.0	-	-	-	18.0
(4) Potential Production:									
(5) Expected Annual Production:	10^{12} c.f.				.40	.55	.58	.31	.18
(6) Cumulative Production:	10^{12} c.f.				1.2	3.8	10.0	13.1	15.8
(7) Remaining Proved Reserves:	10^{12} c.f.	12.0	15.0	17.0	16.8	14.2	8.0	4.9	2.2
(8) Reserve/Production Ratio:					41.8	25.9	13.7	15.8	12.0
(9) Host Government Take ($/unit of production):									
(10) Capital Cost ($/unit of daily capacity):									
(11) Operating Cost ($/unit of production):									

(1) Based on BTU equivalent of 6000 cu. ft. gas per barrel of oil

INDIGENOUS SUPPLY ASSESSMENT WORKSHEET

NATURAL GAS

Country: CANADA

WAES SCENARIO CASE (1985): C

WAES SCENARIO CASE (2000): C1 & C2

Total Resource in Place: 76.5×10^{12} c.f.

Domestic Gas Price: $\$1.92/10^3$ c.f.(1)

Natural Gas:
Offshore X Eastern Canada
Onshore ____

Technology Assumptions: Production and pipeline facilities can be built

	(Units)	1972	1975	1980	1985	1990	2000	2010	2025
(1) Ultimate Recoverable Resource:	10^{12} c.f.	65.0	–	–	–	–	–	–	65.0
(2) Discoveries:									
(3) Reserves Proved to Date:	10^{12} c.f.	1.0	3.0	–	–	–	–	–	3.0
(4) Potential Production:									
(5) Expected Annual Production:									
(6) Cumulative Production:									
(7) Remaining Proved Reserves:	10^{12} c.f.	1.0	3.0	–	–	–	–	–	3.0
(8) Reserve/Production Ratio:									
(9) Host Government Take ($/unit of production):									
(10) Capital Cost ($/unit of daily capacity):									
(11) Operating Cost ($/unit of production):									

(1) Based on BTU equivalent of 6000 cu. ft. gas per barrel of oil

INDIGENOUS SUPPLY ASSESSMENT WORKSHEET

COAL

Country: CANADA

Coal:
Surface Mined X
Deep Mined X
Total

WAES SCENARIO CASE (1985): C

WAES SCENARIO CASE (2000): C1 & D7

Total Resource in Place: 195 x 10^9 short tons (1)

Domestic Coal Price: $38.30/short ton (1)

Technology Assumptions: 1975 technology prevailing

	1972	1975	1980	1985	1990	2000	2010	2025
(1) Ultimate Recoverable Resource: 10^9 sh. tons	100.0	-	-	-	-	-	-	100.0
(2) Reserves Proved (Developed) to Date: 10^9 sh. tons	13.7	14.9	17.7	21.5	25.5	32.3	35.4	40.0
(3) Potential Production:								
(4) Expected Annual Production: 10^9 sh. tons	.021	.023	.031	.045	.059	.120	.260	.400
(5) Cumulative Production: 10^9 sh. tons	1.0	1.1	1.2	1.4	1.7	3.3	5.9	9.3
(6) Remaining Proved Reserves: 10^9 sh. tons	12.7	13.8	16.5	20.1	23.8	29.0	29.5	30.7
(7) Reserve/Production Ratio:	605	584	515	446	397	242	113	77
(8) Primary Transportation* —Capital Cost ($/unit of daily capacity): —Operating Cost ($/unit of daily capacity):								
(9) Capital Cost ($/unit of daily capacity):								
(10) Operating Cost ($/unit of production):								

*From mine mouth to first processing stage.

(1) Based on 1 ton of coal per 3.3 bbls. oil

NONFOSSIL SOURCES AND CONVERSION PROCESSES WORKSHEET

URANIUM AND NUCLEAR

Country: CANADA

WAES SCENARIO CASE (1985): C

WAES SCENARIO CASE (2000): C-1

Total Resource in Place: 415,000 short tons U_3O_8

Domestic Uranium Price: Up to $20/lb U_3O_8

Technology Assumptions:

	(Units)	1972	1975	1980	1985	1990	1995	2000	2010
1. Ultimate Recoverable Uranium Resource:	(sh. tons)		189,000						
2. Reserve of Uranium Proved to Date:	(sh. tons)		82,000						
3. Potential Production of Uranium:	(sh. tons)			13,000	15,000				
4. Expected Production of Uranium:	(sh. tons)		4,600						
5. Reserve/Production Ratio:									
6. Maximum Installed Capacity (GWe)*:		1.75	2.5		12.5			45	
7. Capital Cost ($/KWe):			$500 to $700						
8. Operating Cost (¢/Kwhr)**:									

*Indicate assumed load factor and LWR or HWR.

**Excluding capital charges.

NONFOSSIL SOURCES AND CONVERSION PROCESSES WORKSHEET

URANIUM AND NUCLEAR

Country: CANADA

WAES SCENARIO CASE (1985): C

WAES SCENARIO CASE (2000): C-2

Total Resource in Place: 415,000 short tons U_3O_8

Domestic Uranium Price: Up to $20/lb U_3O_8

Technology Assumptions:

	(Units)	1972	1975	1980	1985	1990	1995	2000	2010
1. Ultimate Recoverable Uranium Resource:	(sh. tons)		189,000						
2. Reserve of Uranium Proved to Date:	(sh. tons)		82,000						
3. Potential Production of Uranium:	(sh. tons)			13,000	15,000				
4. Expected Production of Uranium:	(sh. tons)		4,600						
5. Reserve/Production Ratio:									
6. Maximum Installed Capacity (GWe)*:		1.75	2.5		12.5			74.5	
7. Capital Cost ($/KWe):			$500 to $700						
8. Operating Cost (¢/Kwhr)**:									

*Indicate assumed load factor and LWR or HWR.

**Excluding capital charges.

NONFOSSIL SOURCES AND CONVERSION PROCESS WORKSHEET

Hydroelectric X
Coal Gasification
Coal to Syncrude
Gas Liquifaction
Regasification
Other* (specify):

Country: CANADA

WAES SCENARIO CASE (1985): C

WAES SCENARIO CASE (2000): C1 & C2

Total Resource in Place:

Technology Assumptions:

	(Units)	1972	1975	1985	1990	2000	2010	2025
(1) Yield:								
(2) Maximum Potential Output**	Capacity (GWe)	31.5	36.8	55.5		74		
(3) Capital Cost ($/unit of daily capacity):								
(4) Cumulative Capital Investment:								
(5) Operating Cost ($/unit of production):								

*Specify other relevant non-fossil energy source or conversion process, i.e., geothermal, solar, wind, etc.

** "Maximum technological development" in GEMM

INDIGENOUS SUPPLY ASSESSMENT WORKSHEET

CRUDE OIL

Country: CANADA

Crude Oil: Total
Offshore X
Onshore X

WAES SCENARIO CASE (1985): D

WAES SCENARIO CASE (2000): D7 & D8

Total Resource in Place: 1029.8×10^9 bbls.

Domestic Oil Price: $11.50/bbl.

Technology Assumptions:

	(Units)	1972	1975	1980	1985	1990	2000	2010	2025
(1) Ultimate Recoverable Resource:	10^9 bbls.	73.5	-	-	-	-	-	-	73.5
(2) Discoveries:									
(3) Reserves Proved to Date:	10^9 bbls.	18.11	20.30	26.33	39.77	47.20	51.11	51.11	51.11
(4) Potential Production:									
(5) Expected Annual Production 10^9 bbls.	heavy)								
	light)	.700	.865	.694	.877	.952	.938	.710	.629
(6) Cumulative Production	10^9 bbls.	7.24	8.76	11.91	15.35	19.73	28.65	36.88	43.82
(7) Remaining Proved Reserves:		10.87	11.54	14.42	24.42	27.47	22.46	14.23	7.29
(8) Reserve/Production Ratio:		15.5	13.3	20.8	27.8	28.8	23.9	20.0	11.6
(9) Host Government Take ($/unit of production):									
(10) Capital Cost ($/unit of daily capacity):									
(11) Operating Cost ($/unit of production):									

INDIGENOUS SUPPLY ASSESSMENT WORKSHEET

CRUDE OIL

Country: CANADA

WAES SCENARIO CASE (1985): D

WAES SCENARIO CASE (2000): D7 & D8

Total Resource in Place: 895×10^9 bbls

Domestic Oil Price: $11.50/bbl.

Technology Assumptions: 1975 technology prevailing

Crude Oil:
Offshore ____
Onshore X Oil Sands

	(Units)	1972	1975	1980	1985	1990	2000	2010	2025
(1) Ultimate Recoverable Resource:	10^9 bbls.	29.0	-	-	-	-	-	-	29.0
(2) Discoveries:									
(3) Reserves Proved (DEVELOPED) to Date:	10^9 bbls.	.70	1.84	3.71	13.29	15.61	-	-	15.61
(4) Potential Production:									
(5) Expected Annual Production 10^9 bbls.	heavy)								
	light)	.018	.022	.048	.124	.283	.298	.298	.298
(6) Cumulative Production	10^9 bbls.	.08	.12	.24	.71	1.92	5.14	8.41	12.22
(7) Remaining Proved Reserves:	10^9 bbls.	.62	1.72	3.47	12.58	13.69	10.47	7.20	3.39
(8) Reserve/Production Ratio:		34.4	78.1	72.3	101.5	48.4	35.1	24.2	11.4
(9) Host Government Take ($/unit of production):									
(10) Capital Cost ($/unit of daily capacity):									
(11) Operating Cost ($/unit of production):									

INDIGENOUS SUPPLY ASSESSMENT WORKSHEET

CRUDE OIL

Country: CANADA

WAES SCENARIO CASE (1985): D

WAES SCENARIO CASE (2000): D7 & D8

Total Resource in Place: 134.8 x 10^9 bbls.

Domestic Oil Price: $11.50/bbl.

Technology Assumptions: 1975 technology prevailing

Crude Oil:
Offshore X
Onshore X
Total including N.G.L.

	(Units)	1972	1975	1980	1985	1990	2000	2010	2025
(1) Ultimate Recoverable Resource:	10^9 bbls.	44.5	-	-	-	-	-	-	44.5
(2) Discoveries:									
(3) Reserves Proved to Date:	10^9 bbls.	17.41	18.46	22.62	26.48	31.53	35.50	35.50	35.50
(4) Potential Production:									
(5) Expected Annual Production 10^9 bbls.	heavy)								
	light)	.682	.843	.646	.753	.669	.640	.412	.331
(6) Cumulative Production:	10^9 bbls.	7.16	8.64	11.67	14.64	17.81	23.51	28.47	31.60
(7) Remaining Proved Reserves:	10^9 bbls.	10.25	9.82	10.95	11.84	13.72	11.99	7.03	3.90
(8) Reserve/Production Ratio:		15.0	11.6	16.9	15.7	20.5	18.7	17.1	11.8
(9) Host Government Take ($/unit of production):									
(10) Capital Cost ($/unit of daily capacity):									
(11) Operating Cost ($/unit of production):									

INDIGENOUS SUPPLY ASSESSMENT WORKSHEET

CRUDE OIL

Country: CANADA

WAES SCENARIO CASE (1985): D

WAES SCENARIO CASE (2000): D7 & D8

Total Resource in Place: 60.6 x 10^9 bbls.

Domestic Oil Price: $11.50/bbls.

Technology Assumptions: 1975 technology prevailing

Crude Oil:
Offshore ____
Onshore X
Conventional including N.G.L.

	(Units)	1972	1975	1980	1985	1990	2000	2010	2025
(1) Ultimate Recoverable Resource:	10^9 bbls.	20.0	-	-	-	-	-	-	20.0
(2) Discoveries:									
(3) Reserves Proved to Date:	10^9 bbls.	17.01	17.23	17.63	17.82	18.62	19.02	19.02	19.02
(4) Potential Production:									
(5) Expected Annual Production 10^9 bbls.	heavy)								
	light)	.682	.843	.567	.373	.257	.175	.093	.093
(6) Cumulative Production:	10^9 bbls.	7.16	8.64	11.51	13.33	14.44	15.69	17.46	17.81
(7) Remaining Proved Reserves:	10^9 bbls.	9.85	8.59	6.12	4.49	4.18	3.33	1.56	1.21
(8) Reserve/Production Ratio:		14.5	10.2	10.8	12.0	16.3	19.0	16.7	13.0
(9) Host Government Take ($/unit of production):									
(10) Capital Cost ($/unit of daily capacity):									
(11) Operating Cost ($/unit of production):									

INDIGENOUS SUPPLY ASSESSMENT WORKSHEET

CRUDE OIL

Country: CANADA

Crude Oil:
Offshore X Beaufort -
Onshore X Sverdrup

WAES SCENARIO CASE (1985): D

WAES SCENARIO CASE (2000): D7 & D8

Total Resource in Place: 41.30×10^9 bbls

Domestic Oil Price: $11.50/bbls.

Technology Assumptions: 1975 technology prevailing

	(Units)	1972	1975	1980	1985	1990	2000	2010	2025
(1) Ultimate Recoverable Resource:	10^9 bbls.	13.6	-	-	-	-	-	-	13.6
(2) Discoveries:									
(3) Reserves Proved to Date:	10^9 bbls.	.30	.85	2.85	5.20	7.70	9.43	9.43	9.43
(4) Potential Production:									
(5) Expected Annual Production 10^9 bbls.	heavy)								
	light)				.253	.253	.306	.189	.127
(6) Cumulative Production:	10^9 bbls.				.51	1.78	4.64	6.53	7.83
(7) Remaining Proved Reserves:	10^9 bbls.	.30	.85	2.85	4.69	5.92	4.79	2.90	1.60
(8) Reserve/Production Ratio:					18.5	23.4	15.6	15.3	12.6
(9) Host Government Take ($/unit of production):									
(10) Capital Cost ($/unit of daily capacity):									
(11) Operating Cost ($/unit of production):									

INDIGENOUS SUPPLY ASSESSMENT WORKSHEET

CRUDE OIL

Country: CANADA

WAES SCENARIO CASE (1985): D

WAES SCENARIO CASE (2000): D7 & D8

Total Resource in Place: 32.90 x 10^9 bbls.

Domestic Oil Price: $11.50/bbls

Crude Oil:
Offshore X Eastern Canada
Onshore

Technology Assumptions: Pipeline production facilities can be built

	(Units)	1972	1975	1980	1985	1990	2000	2010	2025
(1) Ultimate Recoverable Resource:	10^9 bbls.	10.9	-	-	-	-	-	-	10.9
(2) Discoveries:									
(3) Reserves Proved to Date:	10^9 bbls.	.10	.38	2.14	3.46	5.21	7.05	7.05	7.05
(4) Potential Production:									
(5) Expected Annual Production 10^9 bbls.	heavy)								
	light)			.079	.127	.159	.159	.130	.111
(6) Cumulative Production:	10^9 bbls.			.16	.80	1.59	3.18	4.48	5.96
(7) Remaining Proved Reserves:	10^9 bbls.	.10	.38	1.98	2.66	3.62	3.87	2.57	1.09
(8) Reserve/Production Ratio:				25.1	20.9	22.7	24.3	19.7	9.8
(9) Host Government Take ($/unit of production):									
(10) Capital Cost ($/unit of daily capacity):									
(11) Operating Cost ($/unit of production):									

INDIGENOUS SUPPLY ASSESSMENT WORKSHEET

NATURAL GAS

Country: CANADA

Natural Gas:
Offshore X
Onshore X
Total

WAES SCENARIO CASE (1985): D

WAES SCENARIO CASE (2000): D7 & D8

Total Resource in Place: 311.8×10^{12} c.f.

Domestic Gas Price: $\$1.92/10^3$ c.f.(1)

Technology Assumptions:

	(Units)	1972	1975	1980	1985	1990	2000	2010	2025
(1) Ultimate Recoverable Resource:	10^{12} c.f.	265.0	-	-	-	-	-	-	265.0
(2) Discoveries:									
(3) Reserves Proved to Date:	10^{12} c.f.	92	103.4	105.0	109.9	111.3	115.9	115.9	115.9
(4) Potential Production:									
(5) Expected Annual Production:	10^{12} c.f.	2.30	2.59	2.39	2.29	2.34	1.49	1.04	.49
(6) Cumulative Production:	10^{12} c.f.	21.3	26.5	38.7	52.0	64.0	77.6	93.1	100.5
(7) Remaining Proved Reserves:	10^{12} c.f.	70.7	76.9	66.3	57.9	47.3	38.3	22.8	15.4
(8) Reserve/Production Ratio:	10^{12} c.f.	30.7	29.6	27.7	25.3	20.2	37.9	21.9	31.4
(9) Host Government Take ($/unit of production):									
(10) Capital Cost ($/unit of daily capacity):									
(11) Operating Cost ($/unit of production):									

(1) Based on BTU equivalent of 6000 cu. ft. of gas per barrel of oil

INDIGENOUS SUPPLY ASSESSMENT WORKSHEET

NATURAL GAS

Country: CANADA

WAES SCENARIO CASE (1985): D

WAES SCENARIO CASE (2000): D7 & D8

Total Resource in Place: 135.3×10^{12} c.f.

Domestic Gas Price: $1.92/10^{3}$ c.f.[1]

Technology Assumptions: 1975 technology prevailing

Natural Gas:
Offshore ____
Onshore X — Conventional producing area

	(Units)	1972	1975	1980	1985	1990	2000	2010	2025
(1) Ultimate Recoverable Resource:	10^{12} c.f.	115.0	-	-	-	-	-	-	115.0
(2) Discoveries:									
(3) Reserves Proved to Date:	10^{12} c.f.	79.0	85.4	86.7	90.8	92.2	96.8	96.8	96.8
(4) Potential Production:									
(5) Expected Annual Production:	10^{12} c.f.	2.30	2.59	2.39	1.95	1.87	1.01	.78	.34
(6) Cumulative Production:	10^{12} c.f.	21.3	26.5	38.7	51.0	61.77	70.1	83.0	88.1
(7) Remaining Proved Reserves:	10^{12} c.f.	57.70	58.90	48.00	39.80	30.43	26.70	13.80	8.70
(8) Reserve/Production Ratio:		25.1	22.7	20.1	20.4	16.3	26.4	17.7	25.6
(9) Host Government Take ($/unit of production):									
(10) Capital Cost ($/unit of daily capacity):									
(11) Operating Cost ($/unit of production):									

(1) Based on BTU equivalent of 6000 cu. ft. of gas per barrel of oil

INDIGENOUS SUPPLY ASSESSMENT WORKSHEET

NATURAL GAS

Country: CANADA

WAES SCENARIO CASE (1985): D

WAES SCENARIO CASE (2000): D7 & D8

Total Resource in Place: 100×10^{12} c.f.

Domestic Gas Price: $1.92/$10^3$ c.f.(1)

Technology Assumptions: 1975 technology prevailing

Natural Gas:
Offshore X Beaufort -
Onshore X Sverdrup

	(Units)	1972	1975	1980	1985	1990	2000	2010	2025
(1) Ultimate Recoverable Resource:	10^{12} c.f.	85.0	-	-	-	-	-	-	85.0
(2) Discoveries:									
(3) Reserves Proved to Date:	10^{12} c.f.	12.0	15.0	15.3	16.1	-	-	-	16.1
(4) Potential Production:									
(5) Expected Annual Production:	10^{12} c.f.				.34	.47	.48	.26	.15
(6) Cumulative Production:	10^{12} c.f.				1.0	2.2	7.5	10.1	12.4
(7) Remaining Proved Reserves:	10^{12} c.f.	12.0	15.0	15.3	15.1	13.9	8.6	6.0	3.7
(8) Reserve/Production Ratio:					44.4	29.6	17.9	23.1	24.6
(9) Host Government Take ($/unit of production):									
(10) Capital Cost ($/unit of daily capacity):									
(11) Operating Cost ($/unit of production):									

(1) Based on BTU equivalent of 6000 cu. ft. of gas per barrel of oil

INDIGENOUS SUPPLY ASSESSMENT WORKSHEET

NATURAL GAS

Country: CANADA

Natural Gas:
Offshore X Eastern
Onshore Canada

WAES SCENARIO CASE (1985): D

WAES SCENARIO CASE (2000): D7 & D8

Total Resource in Place: 76.5×10^{12} c.f.

Domestic Gas Price: $1.92/$10^3$ c.f.(1)

Technology Assumptions: Production and pipeline facilities can be built

	(Units)	1972	1975	1980	1985	1990	2000	2010	2025
(1) Ultimate Recoverable Resource:	10^{12} c.f.	65.0	-	-	-	-	-	-	65.0
(2) Discoveries:									
(3) Reserves Proved to Date:	10^{12} c.f.	1.0	3.0	-	-	-	-	-	3.0
(4) Potential Production:									
(5) Expected Annual Production:									
(6) Cumulative Production:									
(7) Remaining Proved Reserves:	10^{12} c.f.	1.0	3.0	-	-	-	-	-	3.0
(8) Reserve/Production Ratio:									
(9) Host Government Take ($/unit of production):									
(10) Capital Cost ($/unit of daily capacity):									
(11) Operating Cost ($/unit of production):									

(1) Based on BTU equivalent of 6000 cu. ft. of gas per barrel of oil

INDIGENOUS SUPPLY ASSESSMENT WORKSHEET

COAL

Country: CANADA

WAES SCENARIO CASE (1985): D

WAES SCENARIO CASE (2000): C2 & D8

Total Resource in Place: 195×10^9/short tons

Domestic Coal Price: \$38.30/short ton (1)

Technology Assumptions: 1975 technology prevailing

Coal:
Surface Mined X
Deep Mined X
Total

	1972	1975	1980	1985	1990	2000	2010	2025
(1) Ultimate Recoverable Resource: 10^9 sh. tons	100.00	-	-	-	-	-	-	100.00
(2) Reserves Proved to Date: 10^9 sh. tons	13.7	14.9	16.2	18.1	21.4	27.0	29.7	33.6
(3) Potential Production:								
(4) Expected Annual Production:	.021	.023	.026	.038	.049	.100	.218	.336
(5) Cumulative Production: 10^9 sh. tons	1.0	1.1	1.2	1.3	1.6	3.0	5.4	8.7
(6) Remaining Proved Reserves: 10^9 sh. tons	12.7	13.8	15.0	16.8	18.8	24.0	24.3	24.9
(7) Reserve/Production Ratio:	605	600	576	442	384	240	111	74
(8) Primary Transportation* —Capital Cost ($/unit of daily capacity):								
—Operating Cost ($/unit of daily capacity):								
(9) Capital Cost ($/unit of daily capacity):								
(10) Operating Cost ($/unit of production):								

*From mine mouth to first processing stage.

(1) Based on 1 ton of coal per 3.3 bbls. oil

NONFOSSIL SOURCES AND CONVERSION PROCESSES WORKSHEET

URANIUM AND NUCLEAR

Country: CANADA

WAES SCENARIO CASE (1985): D

WAES SCENARIO CASE (2000): D-7

Total Resource in Place: 415,000 short tons U_3O_8.

Domestic Uranium Price: Up to $20/lb U_3O_8

Technology Assumptions: ________

	(Units)	1972	1975	1980	1985	1990	1995	2000	2010
1. Ultimate Recoverable Uranium Resource:	(sh. tons)		189,000						
2. Reserve of Uranium Proved to Date:	(sh. tons)		82,000						
3. Potential Production of Uranium:	(sh. tons)			13,000	15,000				
4. Expected Production of Uranium:	(sh. tons)		4,600						
5. Reserve/Production Ratio:									
6. Maximum Installed Capacity (GWe)*:		1.75	2.5		12.5			45	
7. Capital Cost ($/KWe):			$500 to $700						
8. Operating Cost (¢/Kwhr)**:									

*Indicate assumed load factor and LWR or HWR.

**Excluding capital charges.

NONFOSSIL SOURCES AND CONVERSION PROCESSES WORKSHEET

URANIUM AND NUCLEAR

Country: CANADA

WAES SCENARIO CASE (1985): D

WAES SCENARIO CASE (2000): D-8

Total Resource in Place: 415,000 short tons U_3O_8

Domestic Uranium Price: Up to $20/lb U_3O_8

Technology Assumptions:

	(Units)	1972	1975	1980	1985	1990	1995	2000	2010
1. Ultimate Recoverable Uranium Resource:	(sh. tons)		189,000						
2. Reserve of Uranium Proved to Date:	(sh. tons)		82,000						
3. Potential Production of Uranium:	(sh. tons)			13,000	15,000				
4. Expected Production of Uranium:	(sh. tons)		4,600						
5. Reserve/Production Ratio:									
6. Maximum Installed Capacity (GWe)*:		1.75	2.5		12.5			74.5	
7. Capital Cost ($/KWe):			$500 to $700						
8. Operating Cost (¢/Kwhr)**:									

*Indicate assumed load factor and LWR or HWR.

**Excluding capital charges.

NONFOSSIL SOURCES AND CONVERSION PROCESS WORKSHEET

Hydroelectric X
Coal Gasification ____
Coal to Syncrude ____
Gas Liquifaction ____
Regasification ____
Other* (specify): ____

Country: CANADA

WAES SCENARIO CASE (1985): D

WAES SCENARIO CASE (2000): D7 & D8

Total Resource in Place: ____

Technology Assumptions: ____

	(Units)	1972	1975	1985	1990	2000	2010	2025
(1) Yield:								
(2) Maximum Potential Output**	Capacity (GWe)	31.5	36.8	55.5		60		
(3) Capital Cost ($/unit of daily capacity):								
(4) Cumulative Capital Investment:								
(5) Operating Cost ($/unit of production):								

*Specify other relevant non-fossil energy source or conversion process, i.e., geothermal, solar, wind, etc.

** "Maximum technological development" in GEMM

INDIGENOUS SUPPLY ASSESSMENT WORKSHEET

CRUDE OIL

Country: CANADA

WAES SCENARIO CASE (1985): E

WAES SCENARIO CASE (2000): ______

Total Resource in Place: 1029.8×10^9 bbls.

Domestic Oil Price: $7.66/bbl.

Technology Assumptions: ______

Crude Oil:
Offshore X
Onshore X
Total

	(Units)	1972	1975	1980	1985	1990	2000	2010	2025
(1) Ultimate Recoverable Resource:	10^9 bbls.	73.5	–	–	–	–	–	–	73.5
(2) Discoveries:									
(3) Reserves Proved to Date:	10^9 bbls.	18.11	18.50	20.95	23.50	26.27	28.55	28.55	28.55
(4) Potential Production:									
(5) Expected Annual Production 10^9 bbls.	heavy)								
	light)	.700	.865	.628	.575	.476	.447	.245	.208
(6) Cumulative Production:	10^9 bbls.	7.24	8.76	11.79	14.27	16.49	20.06	21.79	23.31
(7) Remaining Proved Reserves:	10^9 bbls.	10.87	9.74	9.16	9.23	9.78	8.49	6.76	5.24
(8) Reserve/Production Ratio:		15.5	11.3	14.6	16.0	20.5	18.9	27.6	25.2
(9) Host Government Take ($/unit of production):									
(10) Capital Cost ($/unit of daily capacity):									
(11) Operating Cost ($/unit of production):									

INDIGENOUS SUPPLY ASSESSMENT WORKSHEET

CRUDE OIL

Country: CANADA

WAES SCENARIO CASE (1985): E

WAES SCENARIO CASE (2000): ________

Total Resource in Place: 895 x 10^9 bbls.

Domestic Oil Price: $7.66/bbl.

Technology Assumptions: 1975 technology prevailing

Crude Oil:
Offshore ________
Onshore X Oil Sands

	(Units)	1972	1975	1980	1985	1990	2000	2010	2025
(1) Ultimate Recoverable Resource:	10^9 bbls.	29.0	-	-	-	-	-	-	29.0
(2) Discoveries:									
(3) Reserves Proved (DEVELOPED) to Date:	10^9 bbls.	.70	-	-	-	-	-	-	.70
(4) Potential Production:									
(5) Expected Annual Production 10^9 bbls.	heavy)								
	light)	.018	.022	.022	.022	.022	.022		
(6) Cumulative Production	10^9 bbls.	.08	.12	.23	.34	.45	.67	.69	.69
(7) Remaining Proved Reserves:	10^9 bbls.	.62	.58	.47	.36	.25	.03	.01	
(8) Reserve/Production Ratio:		34.4	26.3	21.3	16.3	11.3	1.3		
(9) Host Government Take ($/unit of production):									
(10) Capital Cost ($/unit of daily capacity):									
(11) Operating Cost ($/unit of production):									

INDIGENOUS SUPPLY ASSESSMENT WORKSHEET

CRUDE OIL

Country: CANADA

Crude Oil:
Offshore X Total
Onshore X including N.G.L.

WAES SCENARIO CASE (1985): E

WAES SCENARIO CASE (2000): ________

Total Resource in Place: 134.8×10^9 bbls.

Domestic Oil Price: $7.66/bbl.

Technology Assumptions: 1975 technology prevailing

	(Units)	1972	1975	1980	1985	1990	2000	2010	2025
(1) Ultimate Recoverable Resource:	10^9 bbls.	44.5	-	-	-	-	-	-	44.5
(2) Discoveries:									
(3) Reserves Proved to Date:	10^9 bbls.	17.41	17.80	20.25	22.80	25.57	27.85	27.85	27.85
(4) Potential Production:									
(5) Expected Annual Production 10^9 bbls.	heavy)								
	light)	.682	.843	.606	.553	.454	.425	.245	.208
(6) Cumulative Production:	10^9 bbls.	7.16	8.64	11.56	13.93	16.04	19.39	21.10	22.62
(7) Remaining Proved Reserves:	10^9 bbls.	10.25	9.16	8.69	8.87	9.53	8.46	6.75	5.23
(8) Reserve/Production Ratio:		15.0	10.9	14.3	16.0	20.9	19.9	27.5	25.1
(9) Host Government Take ($/unit of production):									
(10) Capital Cost ($/unit of daily capacity):									
(11) Operating Cost ($/unit of production):									

INDIGENOUS SUPPLY ASSESSMENT WORKSHEET

CRUDE OIL

Country: CANADA

WAES SCENARIO CASE (1985): E

WAES SCENARIO CASE (2000): ________

Total Resource in Place: 60.6×10^9 bbls.

Domestic Oil Price: $7.66/bbl.

Technology Assumptions: 1975 technology prevailing

Crude Oil:
Offshore ______
Onshore X

Conventional including N.G.L.

	(Units)	1972	1975	1980	1985	1990	2000	2010	2025
(1) Ultimate Recoverable Resource:	10^9 bbls.	20.0	-	-	-	-	-	-	20.0
(2) Discoveries:									
(3) Reserves Proved to Date:	10^9 bbls.	17.01	17.20	17.52	18.04	18.39	18.77	18.77	18.77
(4) Potential Production:									
(5) Expected Annual Production 10^9 bbls.	heavy)								
	light)	.682	.843	.568	.369	.253	.193	.091	.091
(6) Cumulative Production:	10^9 bbls.	7.16	8.64	11.48	13.29	14.39	15.58	15.75	15.91
(7) Remaining Proved Reserves:	10^9 bbls.	9.85	8.56	6.04	4.75	4.00	3.19	3.02	2.86
(8) Reserve/Production Ratio:		14.5	10.2	10.6	12.8	15.8	16.5	33.2	31.4
(9) Host Government Take ($/unit of production):									
(10) Capital Cost ($/unit of daily capacity):									
(11) Operating Cost ($/unit of production):									

INDIGENOUS SUPPLY ASSESSMENT WORKSHEET

CRUDE OIL

Country: CANADA

WAES SCENARIO CASE (1985): E

WAES SCENARIO CASE (2000): ________

Total Resource in Place: 41.30×10^9 bbls.

Domestic Oil Price: \$7.66/bbl.

Technology Assumptions: 1975 technology prevailing

Crude Oil:
Offshore X Beaufort
Onshore X Sverdrup

	(Units)	1972	1975	1980	1985	1990	2000	2010	2025
(1) Ultimate Recoverable Resource:	10^9 bbls.	13.6	-	-	-	-	-	-	13.6
(2) Discoveries:									
(3) Reserves Proved to Date:	10^9 bbls.	.30	.42	1.60	2.75	4.25	5.19	5.19	5.19
(4) Potential Production:									
(5) Expected Annual Production 10^9 bbls.	heavy)								
	light)				.123	.124	.155	.093	.062
(6) Cumulative Production	10^9 bbls.				.25	.87	2.26	3.19	3.81
(7) Remaining Proved Reserves:	10^9 bbls.	.30	.42	1.60	2.50	3.38	2.93	2.00	1.38
(8) Reserve/Production Ratio:					20.2	27.3	18.9	21.5	22.3
(9) Host Government Take (\$/unit of production):									
(10) Capital Cost (\$/unit of daily capacity):									
(11) Operating Cost (\$/unit of production):									

INDIGENOUS SUPPLY ASSESSMENT WORKSHEET

CRUDE OIL

Country: CANADA

Crude Oil:
Offshore X Eastern Canada
Onshore

WAES SCENARIO CASE (1985): E

WAES SCENARIO CASE (2000):

Total Resource in Place: 32.90 x 10^9 bbls.

Domestic Oil Price: $7.66/bbl.

Technology Assumptions: Pipeline and production facilities can be built

	(Units)	1972	1975	1980	1985	1990	2000	2010	2025
(1) Ultimate Recoverable Resource:	10^9 bbls.	10.9	-	-	-	-	-	-	10.9
(2) Discoveries:									
(3) Reserves Proved to Date:	10^9 bbls.	.10	.19	1.13	2.01	2.93	3.89	3.89	3.89
(4) Potential Production:									
(5) Expected Annual Production 10^9 bbls.	heavy)								
	light)			.038	.061	.077	.077	.061	.055
(6) Cumulative Production:	10^9 bbls.			.08	.39	.78	1.55	2.16	2.90
(7) Remaining Proved Reserves:	10^9 bbls.	.10	.19	1.05	1.62	2.15	2.34	1.73	.99
(8) Reserve/Production Ratio:				27.6	26.1	27.9	30.3	28.3	18
(9) Host Government Take ($/unit of production):									
(10) Capital Cost ($/unit of daily capacity):									
(11) Operating Cost ($/unit of production):									

INDIGENOUS SUPPLY ASSESSMENT WORKSHEET

NATURAL GAS

Country: CANADA

Natural Gas:
Offshore X
Onshore X
Total

WAES SCENARIO CASE (1985): E

WAES SCENARIO CASE (2000): ________

Total Resource in Place: 311.8×10^{12} c.f.

Domestic Gas Price: $\$1.28/10^3$ c.f.(1)

Technology Assumptions: ________

	(Units)	1972	1975	1980	1985	1990	2000	2010	2025
(1) Ultimate Recoverable Resource:	10^{12} c.f.	265.0	-	-	-	-	-	-	265.0
(2) Discoveries:									
(3) Reserves Proved to Date:	10^{12} c.f.	92.0	103.4	-	-	-	-	-	103.4
(4) Potential Production:									
(5) Expected Annual Production:	10^{12} c.f.	2.30	2.59	2.38	2.20	1.86	1.02	.87	
(6) Cumulative Production	10^{12} c.f.	21.3	26.5	36.8	49.1	59.9	68.2	76.9	
(7) Remaining Proved Reserves:	10^{12} c.f.	70.7	76.9	66.6	54.3	43.5	35.2	26.5	
(8) Reserve/Production Ratio:		30.7	29.6	27.9	24.7	23.4	34.5	30.4	
(9) Host Government Take ($/unit of production):									
(10) Capital Cost ($/unit of daily capacity):									
(11) Operating Cost ($/unit of production):									

(1) Based on BTU equivalent of 6000 cu. ft. of gas per barrel of oil

INDIGENOUS SUPPLY ASSESSMENT WORKSHEET

NATURAL GAS

Country: CANADA

WAES SCENARIO CASE (1985): E

WAES SCENARIO CASE (2000): ________

Total Resource in Place: 135.3×10^{12} c.f.

Domestic Gas Price: $1.28/$10^3$ c.f.[1]

Technology Assumptions: 1975 technology prevailing

Natural Gas:
Offshore ________
Onshore X Conventional producing areas

	(Units)	1972	1975	1980	1985	1990	2000	2010	2025
(1) Ultimate Recoverable Resource:	10^{12} c.f.	115.0	-	-	-	-	-	-	115.0
(2) Discoveries:									
(3) Reserves Proved to Date:	10^{12} c.f.	79.0	85.4	-	-	-	-	-	85.4
(4) Potential Production:									
(5) Expected Annual Production:	10^{12} c.f.	2.30	2.59	2.38	2.20	1.86	1.02	.87	
(6) Cumulative Production:	10^{12} c.f.	21.3	26.5	36.8	49.1	59.9	68.2	76.9	
(7) Remaining Proved Reserves:	10^{12} c.f.	57.7	58.9	48.6	36.3	25.5	17.3	8.5	
(8) Reserve/Production Ratio:		25.1	22.7	20.4	16.5	13.7	16.9	9.7	
(9) Host Government Take ($/unit of production):									
(10) Capital Cost ($/unit of daily capacity):									
(11) Operating Cost ($/unit of production):									

(1) Based on BTU equivalent of 6000 cu. ft. of gas per barrel of oil

INDIGENOUS SUPPLY ASSESSMENT WORKSHEET

NATURAL GAS

Country: CANADA

Natural Gas:
Offshore X
Onshore X
Beaufort - Sverdrup

WAES SCENARIO CASE (1985): E

WAES SCENARIO CASE (2000):

Total Resource in Place: 100×10^{12} c.f.

Domestic Gas Price: $\$1.28/10^3$ c.f.(1)

Technology Assumptions: 1975 technology prevailing

	(Units)	1972	1975	1980	1985	1990	2000	2010	2025
(1) Ultimate Recoverable Resource:	10^{12} c.f.	85.0	-	-	-	-	-	-	85.0
(2) Discoveries:									
(3) Reserves Proved to Date:	10^{12} c.f.	12.0	15.0	-	-	-	-	-	15.0
(4) Potential Production:									
(5) Expected Annual Production:									
(6) Cumulative Production:									
(7) Remaining Proved Reserves:	10^{12} c.f.	12.0	15.0	-	-	-	-	-	15.0
(8) Reserve/Production Ratio:									
(9) Host Government Take ($/unit of production):									
(10) Capital Cost ($/unit of daily capacity):									
(11) Operating Cost ($/unit of production):									

(1) Based on BTU equivalent of 6000 cu. ft. of gas per barrel of oil

INDIGENOUS SUPPLY ASSESSMENT WORKSHEET

NATURAL GAS

Country: CANADA

Natural Gas:
Offshore X Eastern Canada
Onshore

WAES SCENARIO CASE (1985): E

WAES SCENARIO CASE (2000):

Total Resource in Place: 76.5×10^{12} c.f.

Domestic Gas Price: $1.28/$10^3$ c.f.(1)

Technology Assumptions: Production and pipeline facilities can be built

	(Units)	1972	1975	1980	1985	1990	2000	2010	2025
(1) Ultimate Recoverable Resource:	10^{12} c.f.	65.0	-	-	-	-	-	-	65.0
(2) Discoveries:									
(3) Reserves Proved to Date:	10^{12} c.f.	1.0	3.0	-	-	-	-	-	3.0
(4) Potential Production:									
(5) Expected Annual Production:									
(6) Cumulative Production:									
(7) Remaining Proved Reserves:	10^{12} c.f.	1.0	3.0	-	-	-	-	-	3.0
(8) Reserve/Production Ratio:									
(9) Host Government Take ($/unit of production):									
(10) Capital Cost ($/unit of daily capacity):									
(11) Operating Cost ($/unit of production):									

(1) Based on BTU equivalent of 6000 cu. ft. gas per barrel of oil

INDIGENOUS SUPPLY ASSESSMENT WORKSHEET

COAL

Country: CANADA

WAES SCENARIO CASE (1985): E

WAES SCENARIO CASE (2000): ________

Total Resource in Place: 195 x 10^9 short tons

Domestic Coal Price: $25.50/short ton (1)

Technology Assumptions: 1975 technology prevailing

Coal:
Surface Mined X Total
Deep Mined X

	1972	1975	1980	1985	1990	2000	2010	2025
(1) Ultimate Recoverable Resource: 10^9 sh. tons	100.0	-	-	-	-	-	-	100.0
(2) Reserves Proved to Date: 10^9 sh. tons	13.7	14.9	16.0	17.9	19.9	23.3	23.9	25.0
(3) Potential Production:								
(4) Expected Annual Production: 10^9 sh. tons	.021	.023	.025	.033	.039	.079	.146	.263
(5) Cumulative Production: 10^9 sh. tons	1.0	1.1	1.2	1.3	1.4	2.2	3.7	8.8
(6) Remaining Proved Reserves: 10^9 sh. tons	12.7	13.8	14.8	16.6	18.5	21.1	20.2	16.2
(7) Reserve/Production Ratio:	605	601	592	520	473	266	134	61
(8) Primary Transportation* —Capital Cost ($/unit of daily capacity): —Operating Cost ($/unit of daily capacity):								
(9) Capital Cost ($/unit of daily capacity):								
(10) Operating Cost ($/unit of production):								

*From mine mouth to first processing stage.

(1) Based on 1 ton of coal per 3.3 bbls. oil

NONFOSSIL SOURCES AND CONVERSION PROCESSES WORKSHEET

URANIUM AND NUCLEAR

Country: CANADA

WAES SCENARIO CASE (1985): E

WAES SCENARIO CASE (2000): ______

Total Resource in Place: 415,000 short tons U_3O_8

Domestic Uranium Price: Up to $20/lb U_3O_8

Technology Assumptions: ______

	(Units)	1972	1975	1980	1985	1990	1995	2000	2010
1. Ultimate Recoverable Uranium Resource:	(sh. tons)		189,000						
2. Reserve of Uranium Proved to Date:	(sh. tons)		82,000						
3. Potential Production of Uranium:	(sh. tons)			13,000	15,000				
4. Expected Production of Uranium:	(sh. tons)		4,600						
5. Reserve/Production Ratio:									
6. Maximum Installed Capacity (GWe)*:		1.75	2.5		14.7				
7. Capital Cost ($/KWe):			$500 to $700						
8. Operating Cost (¢/Kwhr)**:									

*Indicate assumed load factor and LWR or HWR.

**Excluding capital charges.

NONFOSSIL SOURCES AND CONVERSION PROCESS WORKSHEET

Hydroelectric X
Coal Gasification
Coal to Syncrude
Gas Liquifaction
Regasification
Other* (specify):

Country: CANADA

WAES SCENARIO CASE (1985): E

WAES SCENARIO CASE (2000):

Total Resource in Place:

Technology Assumptions:

	(Units)	1972	1975	1985	1990	2000	2010	2025
(1) Yield:								
(2) Maximum Potential Output**	Capacity (GWe)	31.5	36.8	61.9				
(3) Capital Cost ($/unit of daily capacity):								
(4) Cumulative Capital Investment:								
(5) Operating Cost ($/unit of production):								

*Specify other relevant non-fossil energy source or conversion process, i.e., geothermal, solar, wind, etc.

** "Maximum technological development" in GEMM

9 Denmark

9.1 SUMMARY

Denmark has very little in the way of fossil fuel resources. Only the uranium ores and possibly some yet-undiscovered offshore hydrocarbons in Greenland might be of some international interest.

From a national point of view small discoveries of oil and gas in the Danish part of the North Sea are of some significance, although there is nothing to indicate that they might ever supply more than a small fraction of the total energy needed in Denmark. There are no official forecasts for the Danish oil and gas production in the North Sea. This paper is based on the assumption that most of the presently known reserves (reference 1) will in fact be produced by year 2000. The rationale for this assumption is the high price of energy and the extreme shortage of oil and gas projected for the end of the century.

These assumptions lead to an estimated Danish production of the order of 1.5 million metric tons of oil per year (30,000 BD) and of 2.5×10^9 m^3 of natural gas per year (40,000 BDOE). The proven reserves are sufficient to maintain this production for ten to twenty years. Unless new reserves are discovered, production will decline toward the end of the century. Some new reserves will probably be discovered, but it is unlikely that they will be major. Therefore new discoveries would serve mainly to lengthen the production plateau.

This analysis assumes that the Danish North Sea gas will be marketed in Denmark, although this would require the construction of a distribution network that does not exist today. If an analysis, which is presently being undertaken, should show that the potential gas production is too small to justify investment in a network in Denmark, the gas may be fed into the pipelines from Norwegian North Sea fields to the European continent.

Uranium ores have been located in southern Greenland at Kvanefjeld. The ores are of low grade (0.03 percent), but the location is easily accessible and might yield of the order of 1000 metric tons of uranium per year—the expected Danish demand around year 2000. Additional discoveries of uranium in Greenland are possible, and considerable quantities of thorium have also been discovered.

Nothing can be said about possible hydrocarbon resources in Greenland except that offshore exploration has been started in geologically favorable areas.

9.2 COMMENTS ON INDIVIDUAL RESOURCES

The estimates given in the following sections are based on published information (references 1, 2). The interpretation is, however, entirely the responsibility of the Danish WAES group.

9.2.1 North Sea Oil

Exploration for oil has been going on in Denmark since 1935. Several onshore drillings have given a good geological picture of the underground, but hydrocarbons have not been found in any of the onshore wells. Offshore drilling started in 1966 and resulted in the first discovery of oil in the North Sea. Unfortunately, discoveries there have all been small. Oil production was initiated in the Dan-field in 1972. Presently production is about 200,000 metric tons per year; efforts to increase production from this field continue, and a production of 500,000 metric tons per year is expected in 1977.

The proven plus probable reserves for the North Sea are estimated to be between 22 and 62 million metric tons of oil (reference 1). In this report an average value of 40 million metric tons is assumed. The oil in place is considerably larger, around 300 million metric tons. However, the limestone formations containing the oil have very low permeability, and only about 10 percent of the oil in place appears producible by present techniques.

Only scattered information is available about the future production from the Danish fields. Including condensate from the gas fields, the estimates range from 0.5 to 1.5 million metric tons per year for ten to twenty years. This analysis is based on the assumption that most of the proven reserves will be produced in this century and that the production will be of the order of 1.5 million metric tons per year. With these assumptions a production profile within the boundaries shown in Figure 9.1 is plausible.

Production must be expected to decline somewhat toward the end of the century unless new discoveries are made, but exploration continues and "promising" results have been announced. It therefore seems reasonable to assume a fairly flat plateau in the production curve because of the addition of new reserves.

The general philosophy behind these estimates is the same as that for the WAES evaluation of regional and global oil production. The estimates do in a crude manner allow for future discoveries. We stress, however, that for an area as small as the Danish North Sea, estimates are more uncertain than the average for a larger region. The production estimates in Figure 9.1 should therefore be regarded with caution, and actual production could be either higher or lower and could show a different profile.

9.2.2 North Sea Gas

A number of small to medium gas fields have been discovered in the Danish North Sea, and there is associated gas in the North Sea oil fields. Production from these fields is being considered, but no decision can be expected until 1978. As in the case of oil, this analysis is based on the assumption that the gas will be produced and the known reserves more or less depleted during this century.

Estimates by the contractors for proven plus probable reserves range from 27 to 50×10^9 m^3. Slightly larger reserves of 62×10^9 m^3 have been independently estimated by a consultant.

The estimated maximum production volume ranges from 1.3 to 2.5×10^9 m^3 per year. The maximum, which depends on the development of several minor fields, is here taken as indicative of the future production level. If this maximum cannot be achieved, it is doubtful that the construction of a distribution network in Denmark can be justified. In this case the gas might be fed into pipelines from the Norwegian fields to the continent. Such lines pass close to the prospective Danish fields.

Figure 9.2 shows a suggested production profile for gas in the period until 2000. Like the oil profile, the gas profile is highly uncertain in the later period where continued production depends on new discoveries.

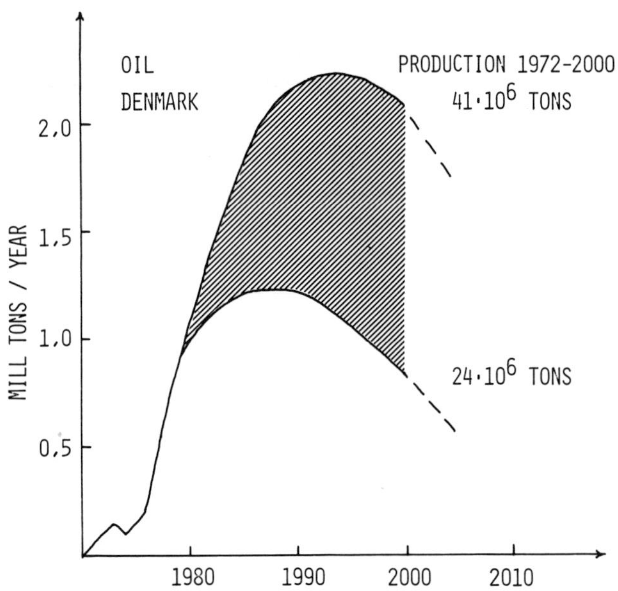

Figure 9.1
Estimated production profiles for oil and condensate from Danish North Sea fields. The hatched area indicates a probable range but must be regarded with caution, especially for the later period in which development will depend on discoveries not yet made.

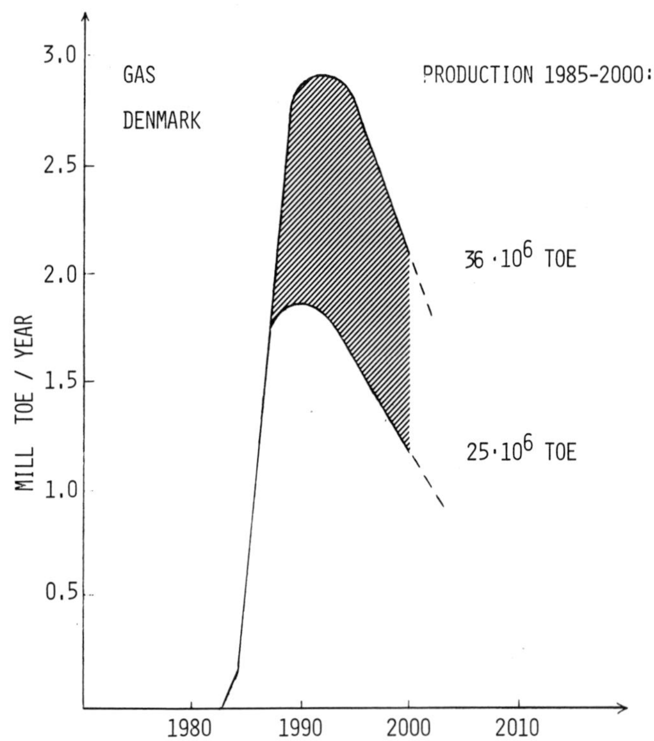

Figure 9.2
Estimated production profile for natural gas. The indicated range must be regarded with caution. Production from the fields has not yet been decided upon.

The development of the Danish North Sea gas depends on a number of other factors such as the price of energy. We have not attempted to relate the production levels to the energy price for either oil or gas. However, the upper estimates in Figures 9.1 and 9.2 are, of course, more likely to be realized in the high price cases (C cases) than in the low price cases (D cases).

Gas development might also be affected by the degree to which government participates in the construction of a gas distribution system. A direct government subsidy might be a condition for introducing gas on the Danish market. In addition, the outlook for future deliveries from Norway might affect decisions concerning the exploitation of the Danish fields. In view of the gas shortage in Europe expected after 1985, it is perhaps prudent to view the introduction of gas on the Danish market with some reservation.

9.2.3 Greenland Uranium

Deposits of low-grade uranium ore have been discovered at Kvanefjeld in the southern part of Greenland. The possible exploitation of these deposits is under the jurisdiction of the Danish government and will probably remain so if there is no formal change in the international status of Greenland. Decisions concerning exploitation of minerals in Greenland might, however, be strongly influenced by local authorities.

The proven plus estimated additional reserves of uranium at Kvanefjeld amount to 16,000 metric tons. The ores contain on the average 0.03 percent uranium. At the same location are large quantities of ore containing more than 0.02 percent uranium, and in general the resource base is believed to be much larger than the known reserves.

Uranium ores of this low grade have hitherto been considered too poor for commercial development. However, the Greenland ores are easily accessible, and although special techniques are required for extracting the uranium from these ores, it has been estimated that the uranium can be produced at prices well below recent world market prices (\$90/kg U_3O_8).

Given the assumptions concerning oil and natural gas in the North Sea and the prospective shortage of uranium, this analysis assumes that the Greenland uranium will in fact be produced starting around 1985.

The estimated production level is 500–1500 metric tons uranium per year, limited mostly by environmental concerns. This quantity of uranium corresponds to the operation of 4–12 GWe installed nuclear capacity at 60 percent load factor. Since the WAES estimate of installed nuclear capacity in Denmark is 2 to 5 GWe by 2000, the estimated uranium production would permit Denmark to supply its own uranium needs and have a small additional amount for export.

9.2.4 Renewable Energy Sources

Apart from some heat from waste combustion the Danish supply-demand integration does not include contributions from renewable energy sources because such contributions are estimated to provide fuel substitutions at prices well above the WAES scenario assumptions.

Expressed as primary fossil energy substituted, the government energy plan for 1995 (reference 3) includes 20 PJ/year of solar heating and 10 PJ/year of wind-generated electricity, a total of about 3 percent of the projected energy demand of 1050 PJ/year. An alternative energy plan (reference 4), which rejects the use of nuclear energy, assumes substitution of 44 PJ/year by solar heating and 40 PJ/year by wind-generated electricity, a total of about 9 percent of an energy demand reduced to 970 PJ/year by strong conservation measures.

Solar heating of hot water and generation of electricity from the wind under Danish climatic conditions are presented as modules that outline the technology assumed, the investment, the energy prices, and the maximum supply potential. The potential is expressed as the *maximum* amount of fossil fuel that can be substituted. If wind produces electricity, the potential given is the amount of boiler fuel saved in conventional power plants.

Solar water heating is one of the more promising possibilities for utilizing solar energy, even under Danish weather conditions. It can be accomplished by simple technical equipment and provide a useful *supplement* to the hot water supply. It is especially attractive in the summer months when many oil furnaces operate at low efficiency just for the provision of hot water.

The scenarios for year 2000 estimate total energy demand at about 60 PJ/year for hot water heating. This corresponds to approximately 4 percent of the country's total primary energy demand. Some 50 percent of this demand could be covered by simple flat plate solar collectors without additional hot water storage. However, since much of the hot water supply (approximately 40 percent) in the scenarios is obtained from district heating systems that utilize coal or uranium, the amount of oil or gas saved would be less than the full potential given in the module.

Wind generation of electricity was once practiced in Denmark on a small scale. The prospective energy shortage has renewed the interest in wind energy, and the government and the utilities are now engaged in an experimental program that includes restoring the windmill at Gedser (200 kW) and constructing one or two new experimental wind generators in the 1 MW range. The total cost of this program is 14 million D.kr. ($2.3 million).

The wind module data are based largely on earlier Danish experiences with electricity-producing wind generators but with allowance for technical improvements and benefits from mass production (reference 5). No storage facilities are considered in the module. This limits the potential to 5–10 percent of the public grid load, which in the scenarios is approximately 40 TWh/year by year 2000. Expressed as primary energy saved, the wind generators could provide

Module 9.1
Wind energy

Description
The module requires 1000 large wind energy generators placed in coastal regions. The generators produce electricity that is fed directly into the public grid. In the absence of energy storage the potential is limited to about 10 percent of the grid load because of the fluctuating input. Other limitations are available land and aesthetics.

Wind generator characteristics
Conventional propeller

Tower: 60 m

Propeller diameter: 56 m

Efficiency: 0.33

Area swept: 2500 m^2

Average wind power: 500 W/m^2

Rated power: 2000 kW

Electric production: 3.6 GWh/year

Generator cost: $820,000

Operating cost: $18,000/year

Capital charge (15%): $106,500/year

Electricity price (busbar): 4.0 ¢/kWh

National impact
Energy supply (fuel substituted in fossil plants): 37 PJ/year

Maximum potential (10% of grid load): 45 PJ/year

Start: 1985

Lead time: 10 years

Investment: $820 million

Operating cost: $18 million per year

Energy cost

Electricity at busbar: 4.0 ¢/kWh

Fuel substitution in power plants at 0.35 efficiency: $3.9/GJ = $24/bl

Fuel substitution for heating oil at 0.8 efficiency: $8.8/GJ = $54/bl

Module 9.2
Solar water heating

Description
All dwellings are supposed to be equipped with simple flat plate solar collectors supplying hot water for household use. Without extra storage such systems can supply around 50 percent of the hot water demanded, mostly during the summer months.

Water heater characteristics
Flat plate collectors with plastic or glass covering

Area: 8 m^2 per household

Insolation in Denmark: (56° north) 3.62 GJ/m^2/year

Efficiency of collector: 0.30

Hot water supplied (8 m^2 collector): 8.8 GJ/year

Normal hot water demand at 0.65 fossil efficiency: 25 GJ/dwelling/year

Fraction supplied: $8.8/25 \times 0.65 = 0.54$

Price of collector and plumbing (mass production): $150/$m^2$

Capital charges and depreciation: 15% p.a.

Energy cost: $13.3/GJ = $83/BOE = 7.4 ¢/kWh

National impact (2.2×10^6 dwellings)
Fossil fuel saved: 30 PJ/year (max) (2.2% of primary energy in scenario C-1)

Maximum potential: 30 PJ/year

Start: 1980

Lead time: 10–15 years

Investment: $2.64 billion

Operating cost: $15 million per year

Energy cost

Fuel substitution in furnaces at 0.65 efficiency: $13.3/GJ = $83/bl

Electricity substitution in water heaters at 1.0 efficiency: $20.5/GJ = 7.4¢/kWh

Table 9.1
Renewable energy sources in Denmark

	Maximum potential[a] (PJ/year)	Energy cost[b] ($/GJ)
Solar water heating	30	13
Wind generators	40	4
Straw	30	1
Biogas	20	?

[a]Estimated energy demand is 1200–1400 PJ/year.
[b]An oil price of $17.25/bbl corresponds to $2.80/GJ.

up to 40 PJ/year, or 3 percent of total primary energy. Only a small fraction of the energy saved (less than 20 percent) would be oil or gas.

Another renewable energy source is the considerable surplus straw in Danish agriculture. Approximately 2 million tons of straw is burned each year in the fields. The heating value of this straw is around 30 PJ/year or 2 percent of the projected primary energy demand for year 2000. Straw-burning furnaces presently being installed on many farms directly substitute for heating oil.

Other farm wastes (manure) have been considered for the production of biogas (methane). Total manure is estimated to yield a maximum potential for biogas production of 20 PJ/year or 1.5 percent of projected demand. The total contribution of the two modules and other renewable sources discussed is summarized in Table 9.1. These contributions refer to the *maximum* potentials; larger contributions are either unlikely or would imply higher energy costs. Most likely it would be quite difficult to utilize more than a small fraction of the maximum potentials.

9.3 REFERENCES

1. *Rapport om eneretsbevilling m.v.* Handelsministeriet (1975).

2. Henning Sørensen et al. *The Uranium Deposit at Kvanefield.* The Geological Survey of Greenland, no. 60, Copenhagen (1974).

3. *Dansk Energipolitik 1976.* Handelsministeriet (1976).

4. *Skitse til Alternativ Energiplan.* OOA og OVE, Copenhagen, ISBN 87-87625-00-8 (1976).

5. *Vindkraft.* Akademiet for de Tekniske Videnskaber (ATV), Copenhagen (1975).

9.4 ENERGY SUPPLY WORKSHEETS

This section includes worksheets in the following order (see Section 1.2.1 of Chapter 1 for the definitions of line items):

Energy supply summary

Uranium and nuclear

Crude oil

Natural gas

WAES NATIONAL ENERGY SUPPLY STUDIES, 1972-2000
ENERGY SUPPLY SUMMARY

Country: Denmark

Scenario Assumptions	1972	1985					2000			
WAES CASE	Base Year	A	B	C	D	E	C-1	C-2	D-7	D-8
GNP 1976-85; 1985-2000		High	Low	High	Low	High	High	High	Low	Low
Oil/Energy Price*		11.50-17.25	11.50-17.25	11.50	11.50	11.50-7.66	11.50-17.25	11.50-17.25	11.50	11.50
National Policy Response		Vig	Vig	Vig	Res	Res	Vig	Vig	Vig	Vig
Principal Replacement Fuel							Coal	Nuclear	Coal	Nuclear
Energy Supply **Units**	PJ/y									
Crude Oil	4	63	63	63	63	63	59	59	59	59
Natural Gas	0	82	82	82	82	82	67	67	67	67
Coal										
Nuclear	0	225	225	225	225	225	450	450	450	450
Hydro-electric										
Geothermal										
Solar Thermal										
Solar Electric										
Refuse	2	10	10	10	10	10	10	10	10	10
Total Energy Supply	6	380	380	380	380	380	586	586	586	586

*1975 US dollars per barrel of Arabian light crude oil, fob Persian Gulf.
All Figures are primary production prior to any processing or conversion.

Crude oil: An oil production of 10^6 tons/year equals 42 PJ/y.

Natural gas: A gas production of 10^9 m^3/y equals 41 PJ/y. By 1985 the gas production is expected to rise sharply. Consequently the production figures become very time dependent.

Nuclear: An uranium production of 10^3 tons/year corresponds to a fossil fuel supply of 450 PJ/y. (5MWy / ton nat. U at 35% efficiency).

NONFOSSIL SOURCES AND CONVERSION PROCESSES WORKSHEET

URANIUM AND NUCLEAR

Country: Denmark

WAES SCENARIO CASE (1985): All

WAES SCENARIO CASE (2000): All

Total Resource in Place: 200000 tons

Domestic Uranium Price: ______

Technology Assumptions: ______

	(Units)	1972	1975	1980	1985	1990	1995	2000	2010
1. Ultimate Recoverable Uranium Resource:	10^3ton	200	200	200	200	200	200	200	200
2. Reserve of Uranium Proved to Date:	10^3ton	16	16	16					
3. Potential Production of Uranium:	10^3ton/y	-	-	-	0.5	1	1	1.5	
4. Expected Production of Uranium:	10^3ton/y	-	-	-	0.5	1	1	1	
5. Reserve/Production Ratio:									
6. Maximum Installed Capacity (GWe)*:		-	-	-	1.4	3.0	4.3	5.7	
7. Capital Cost ($/KWe):					600				
8. Operating Cost (¢/Kwhr)**:					0.5				

*Indicate assumed load factor and LWR or HWR.

**Excluding capital charges.

6) Load factor 70% assumed for LWR. Electric efficiency reduced because of district heat production.

8) Operating cost include uranium and fuel element production as follows:

	¢/kWh
Uranium (25$/lb	0.21
Enrichment	0.13
Fabrication	0.05
Waste treatment	0.08
Pu	- 0.07
Fuel cost	0.40
Personel etc.	0.12
Operating cost	0.52

INDIGENOUS SUPPLY ASSESSMENT WORKSHEET

CRUDE OIL

Country: Denmark

WAES SCENARIO CASE (1985): All

WAES SCENARIO CASE (2000): All

Total Resource in Place: $0.6 \cdot 10^9$ TOE

Domestic Oil Price: $12.5/B

Technology Assumptions:

Crude Oil:
Offshore x
Onshore

	(Units)	1972	1975	1980	1985	1990	2000	2010	2025
(1) Ultimate Recoverable Resource:	10^6TOE		80	80	80	80	80	80	80
(2) Discoveries:			20	10	10	10	10	0	0
(3) Reserves Proved to Date:	10^6TOE	20	40	50	60	70	80	80	80
(4) Potential Production:	10^6TOE	0.10	0.20	1.0	1.5	1.7	1.4	0.9	0.4
(5) Expected Annual Production 10^6TOE	heavy								
	light	0.10	0.20	1.0	1.5	1.7	1.4	0.9	0.4
(6) Cumulative Production:	10^6TOE	0	0.4	3	10	18	33	45	55
(7) Remaining Proved Reserves:	10^6TOE	20	40	47	50	52	47	35	25
(8) Reserve/Production Ratio:	y	100	200	50	40	30	30	40	60
(9) Host Government Take ($/unit of production):			12.5% of production value plus 40% income tax						
(10) Capital Cost ($/unit of daily capacity):			$300/ton/y = $\$1.5 \cdot 10^4$/B						
(11) Operating Cost ($/unit of production):			$40/ton = $6.7/B						

(1) Recoverable Ressource: Estimated to 2 x recoverable reserve 1975. This might be optimistic but is in line with the present maximum estimate of recoverable reserves[1].

---

(2) All reserves assumed proven by 2000.

(10) Investment per unit capacity. Total exploration cost 1962-70 $\$30x10^6$ of which 50% is ascribed to oil. Investment in production equipment 1970-72 $\$30x10^6$. Resulting production capacity approximately 150000 t/y: $\$45x10^6/150000$ t/y = \$300/t/y.

(11) Estimated operation cost 1973 $\$4x10^6$ for production of 150000 t. $\$6x10^6/150000$ t = \$40/t.

INDIGENOUS SUPPLY ASSESSMENT WORKSHEET

NATURAL GAS

Country: Denmark

WAES SCENARIO CASE (1985): All

WAES SCENARIO CASE (2000): All

Total Resource in Place: $0.2 \cdot 10^9$ TOE

Domestic Gas Price:

Technology Assumptions:

Natural Gas:
Offshore x
Onshore

	(Units)	1972	1975	1980	1985	1990	2000	2010	2025
(1) Ultimate Recoverable Resource:	10^6TOE	80	80	80	80	80	80	80	80
(2) Discoveries:	10^6TOE		10	10	20	10	10	0	0
(3) Reserves Proved to Date:	10^6TOE	20	30	40	60	70	80	80	80
(4) Potential Production:	10^6TOE	0	0	0	2.0	2.4	1.6	0.8	0
(5) Expected Annual Production:	10^6TOE	0	0	0	2.0	2.4	1.6	0.8	0
(6) Cumulative Production:	10^6TOE	0	0	0	0	11	32	43	49
(7) Remaining Proved Reserves:	10^6TOE	20	30	40	60	59	48	37	31
(8) Reserve/Production Ratio:	y	-	-	-	25	25	30	45	-
(9) Host Government Take ($/unit of production):		12.5% of production value plus 40% income tax							
(10) Capital Cost ($/unit of daily capacity):		∿ $ 300/TOE/y = $1.5 \cdot 10^4$ $/B/d							
(11) Operating Cost ($/unit of production):		∿ $ 40/TOE = $ 6.7/B							

(1) Resource estimate uncertain. Based on upper estimates of recoverable reserves.

---

(2) All reserves assumed proven, by 2000. Strong exploration effort assumed in the early 1980s.

(5) Production assumed initiated 1985.

(10) Capital cost as for oil (wellhead gas).

(11) Operating cost as for oil. Very uncertain estimate.

10.1 OVERVIEW AND SUMMARY

The projected energy supply potential in Finland is described by fuel and by scenario in the summary table. The Finnish energy supply is in principle simple, because Finland has few major indigenous energy resources. Still, indigenous fuels account for a notable share of all commercial fuels, because of small or middle scale and decentralized production and utilization of waste materials, waste heat, and peat. If the price of imported energy continues to rise, utilization of indigenous resources will become more intensive, although the potential above the present level is not very great. The GNP growth rate will be positively correlated with indigenous energy production, requiring investments for new processes that can use indigenous fuels. Although the government policy response will have an effect on all indigenous energy production, it will be most visible in the case of nuclear energy. However, except for the choice between nuclear and coal as the principal replacement fuel, different policy options have not been analyzed in the scenarios.

10.2 CONCLUSIONS

Since Finland has no high-grade fossil fuel resources, it has to have other kinds of responses to the possible shortage of oil. It is often difficult and a matter of accounting to distinguish between energy conservation and indigenous energy production. In many industrial processes waste materials and by-products can be used as energy sources. For example, cascading processes permit waste heat and other nonstorable energy sources to be used instead of exogenous fuels. In space heating the demand for exogenous fuels can be decreased by insulation and better utilization of solar radiation, which in itself is not sufficient to be the primary energy source.

The major new indigenous energy source is nuclear, although the entire fuel cycle is not indigenous.

Another nearly new indigenous energy source is peat. Its potential supply is estimated to be close to one-half of the assumed nuclear energy supply by the year 2000, depending on the scenario.

10.3 COMMENTS ON EACH RESOURCE

10.3.1 Fossil Fuels

Fossil fuels provided about 70 percent of the national energy supply in the base year. Although their share will be decreasing, they remain an important component of the imported materials and goods. The Finnish economy is very open and dependent on foreign trade; the ratio of imports to GNP is close to 30 percent. The negative balance of payments is a serious and permanent problem of the economy. Thus a critical factor in the well-being of the nation is the quantity and value of goods and services that can be exported to balance the necessary imports. This is why bilateral trade in general is advantageous to an economy like that of Finland.

An essential part of the imported energy in Finland is included in bilateral trade agreements with the Soviet Union and Poland. These agreements are usually made on a long-term basis, thus securing energy supply and permitting long-term planning in the indigenous energy sector. For a country with no fossil fuel resources, bilateral trade can be considered a partial substitute for indigenous energy supply. However, since the prices in these agreements are tied to the world market prices, this kind of energy supply is not shielded from price rises in world markets.

Oil is expected to remain the major fossil fuel. The existing Finnish oil refineries can supply the demand for oil products up to the year 1985. Even after that the major part of oil is assumed to be refined in Finland.

After the base year natural gas was introduced into the Finnish energy market. Finland and the Soviet Union have a trade agreement for the importation of natural gas.

10.3.2 Peat

Peat is one of the low-calorific fossil fuels. The thermal value is 1.0 Gcal/m^3, or approximately 3 Gcal/tonne. The peat resources in Finland are very large, the largest in the world except those in the Soviet Union and Canada. The estimated resources are close to 20×10^9 m^3, which is equivalent to 2×10^9 tonnes of oil. This is nearly 100 times larger than the present yearly total primary energy demand in Finland. There are, however, many technical, geographic, environmental, and financial obstacles to exploiting these resources. The economically recoverable reserves are much smaller than the ultimately recoverable resources.

The production target for 1985 is 10–15 million m^3, which will not cover more than 5 percent of the primary energy demand. The production phase is not, however, the critical factor in peat utilization. The problems rise more in the consumption processes. While peat is a very cheap fuel (the price per thermal unit is about half that of heavy fuel oil), the investments required for peat-fueled processes are comparatively high. Thus the amount of peat that can be used does not depend so much on the production capacity as on the degree to which it can replace other fuels in heat- and electricity-generating processes.

At the moment peat can be used economically only in rather large plants sufficiently close to the production fields. But even a moderate change in oil price may change the situation, and peat utilization technology is also developing rapidly.

Peat is used mainly to provide district and industrial heat and to cogenerate electricity. At present peat does not seem suitable for separate electricity generation.

Projections of peat utilization in different scenarios are shown in Table 10.1.

10.3.3 Nuclear

At the moment four nuclear reactors with a total capacity of 2.2 GWe are either under construction or in operation in Finland. This capacity is rather large compared to the total population of 4.7 million. Because the growth rate of demand for electricity has decreased, the nuclear program will be slowing down. The supply of nuclear energy will be mainly determined by the estimated demand for electricity. In principle, nuclear energy is economically attractive to Finland since there are no major indigenous energy sources and the transportation costs for imported fossil fuels are high. However, the Finnish economy is characterized by a permanent shortage of capital, and thus the nuclear energy supply can be affected by financial constraints. It is difficult to estimate what effect public reaction might have on the nuclear program, but it might be expected that the public would respond more favorably to extending the present plants than to establishing new reactor sites.

At this moment it seems likely that neither the front end nor the back end of the fuel cycle will be taken care of indigenously.

The scenarios for year 2000 include nuclear reactors for the combined generation of district heat and electricity.

Projections of nuclear installations in operation for different scenarios are shown in Table 10.2.

10.3.4 Hydroelectric

The technically available unused hydro energy potential is about 6 TWh/year, which could increase the present level of 12 TWh by 50 percent. To utilize this potential, however, would require a great number of very small-scale hydro power plants, which would not be economical under present and projected circumstances. The estimated average unit investment cost is about twice that of nuclear power. In addition, conservation considerations hinder use of this supply source. Thus, even in the case of high energy importation prices, the remaining hydro potential has not been included in electricity generation. The hydro energy supply is constant in all scenarios.

Table 10.1
Projections of peat utilization in the WAES scenarios (10^6 m^3)

1985		2000		
low growth	high growth	C-1	C-2	D-7/D-8
11	13.5	48	45	25

Table 10.2
Projections of nuclear installations in the WAES scenarios (GWe)

	1985	2000		
	all scenarios	C-2	C-1	D-7/D-8
Nuclear Electric	2.2	5.2	4.2	2.65
Combined Nuclear Electric and District Heating	—	2.0	1.0	1.00
Total	2.2	7.2	5.2	3.65

10.3.5 Wastes

Waste materials have traditionally been used for energy production, especially in wood processing industries. The trend has been to increase the utilization of wastes, although there may be less waste material as raw materials are used more efficiently.

Urban wastes have not yet been used very much, but they seem to be suitable for district heating.

10.4 ENERGY SUPPLY WORKSHEETS

The following worksheets have been included (see Section 1.2.1 of Chapter 1 for the definitions of line items):

Energy supply summary

Nuclear electric (3)

Peat (3)

Hydroelectric (1)

Other indigenous fuels (2)

WAES NATIONAL ENERGY SUPPLY STUDIES, 1972-2000
ENERGY SUPPLY SUMMARY

Country: Finland

Scenario Assumptions	1972	1985					2000			
WAES CASE	Base Year	A	B	C	D	E	C-1	C-2	D-7	D-8
GNP 1976-85; 1985-2000		High	Low	High	Low	High	High	High	Low	Low
Oil/Energy Price*		11.50-17.25	11.50-17.25	11.50	11.50	11.50-7.66	11.50-17.25	11.50-17.25	11.50	11.50
National Policy Response							Vig	Vig	Vig	Vig
Principal Replacement Fuel							Coal	Nuclear	Coal	Nuclear
Energy Supply Units 10^{15} J										
Crude Oil										
Natural Gas										
Coal										
Nuclear		149	149	149	149	149	355	490	248	248
Hydro-electric	41	47	47	47	47	47	47	47	47	47
Peat	1	55	45	55	45	55	195	180	100	100
Other Indigenous	150	137	126	137	126	137	166	161	142	142
Solar Electric										
Total Energy Supply	192	388	367	388	367	388	763	878	537	537

*1975 US dollars per barrel of Arabian light crude oil, fob Persian Gulf.
All Figures are primary production prior to any processing or conversion.

NONFOSSIL SOURCES AND CONVERSION PROCESSES WORKSHEET

NUCLEAR ELECTRIC

Country: Finland

WAES SCENARIO CASE (1985): ACE

WAES SCENARIO CASE (2000): 1

Total Resource in Place: ____________

Domestic Uranium Price: ____________

Technology Assumptions: ____________

	(Units)	1972	1975	1980	1985	1990	2000	2010	2025
(1) Potential Production of Uranium.									
(2) Expected Production of Uranium*:									
(3) Yield of Power Plant:	Twh	-	-		14.7		34.5		
(4) Maximum Potential Output:									
(5) Capital Cost ($/unit of daily capacity):	$/Kw		700						
(6) Cumulative Capital Investment.									
(7) Operating Cost ($/unit of production):	c/Kwh		.3		.4				

*Political potential production.

NONFOSSIL SOURCES AND CONVERSION PROCESSES WORKSHEET

NUCLEAR ELECTRIC

Country: Finland

WAES SCENARIO CASE (1985): ACE

WAES SCENARIO CASE (2000): 2

Total Resource in Place:

Domestic Uranium Price:

Technology Assumptions:

	(Units)	1972	1975	1980	1985	1990	2000	2010	2025
(1) Potential Production of Uranium.									
(2) Expected Production of Uranium*:									
(3) Yield of Power Plant:	Twh	–	–		14.7		47.6		
(4) Maximum Potential Output:									
(5) Capital Cost ($/unit of daily capacity):	$/Kw								
(6) Cumulative Capital Investment.									
(7) Operating Cost ($/unit of production):	c/Kwh		.3		.4				

*Political potential production.

NONFOSSIL SOURCES AND CONVERSION PROCESSES WORKSHEET

NUCLEAR ELECTRIC

Country: Finland

WAES SCENARIO CASE (1985): BD

WAES SCENARIO CASE (2000): 78

Total Resource in Place: ______

Domestic Uranium Price: ______

Technology Assumptions: ______

	(Units)	1972	1975	1980	1985	1990	2000	2010	2025
(1) Potential Production of Uranium.									
(2) Expected Production of Uranium*:									
(3) Yield of Power Plant:	Twh	–	–		14.7		24.1		
(4) Maximum Potential Output:									
(5) Capital Cost ($/unit of daily capacity):	$/Kw		700						
(6) Cumulative Capital Investment.									
(7) Operating Cost ($/unit of production):	c/Kwh		.3		.4				

*Political potential production.

NONFOSSIL SOURCES AND CONVERSION PROCESS WORKSHEET

Hydroelectric ______
Coal Gasification ______
Coal to Syncrude ______
Gas Liquifaction ______
Regasification ______
Other* (specify): Peat

Country: Finland

WAES SCENARIO CASE (1985): ACE

WAES SCENARIO CASE (2000): 1

Total Resource in Place: 2000 Mtoe

Technology Assumptions: ______

	(Units)	1972	1975	1985	1990	2000	2010	2025
(1) Yield:	Mtoe	.02	.04	1.35		4.8		
(2) Maximum Potential Output				2.0				
(3) Capital Cost ($/unit of daily capacity):								
(4) Cumulative Capital Investment:								
(5) Operating Cost ($/unit of production):	$/toe		50.					

*Specify other relevant non-fossil energy source or conversion process, i.e., geothermal, solar, wind, etc.

NONFOSSIL SOURCES AND CONVERSION PROCESS WORKSHEET

Hydroelectric ________
Coal Gasification ________
Coal to Syncrude ________
Gas Liquifaction ________
Regasification ________
Other* (specify): Peat

Country: Finland

WAES SCENARIO CASE (1985): ACE

WAES SCENARIO CASE (2000): 2

Total Resource in Place: 2000 Mtoe

Technology Assumptions: ________

	(Units)	1972	1975	1985	1990	2000	2010	2025
(1) Yield:	Mtoe	.02	.04	1.35		4.5		
(2) Maximum Potential Output				2.0				
(3) Capital Cost ($/unit of daily capacity):								
(4) Cumulative Capital Investment:								
(5) Operating Cost ($/unit of production):			50.					

*Specify other relevant non-fossil energy source or conversion process, i.e., geothermal, solar, wind, etc.

NONFOSSIL SOURCES AND CONVERSION PROCESS WORKSHEET

Hydroelectric ______
Coal Gasification ______
Coal to Syncrude ______
Gas Liquifaction ______
Regasification ______
Other* (specify): Peat

Country: Finland

WAES SCENARIO CASE (1985): BD

WAES SCENARIO CASE (2000): 78

Total Resource in Place: 2000 Mtoe

Technology Assumptions: ______

	(Units)	1972	1975	1985	1990	2000	2010	2025
(1) Yield:	Mtoe	.02	.04	1.1		2.5		
(2) Maximum Potential Output				2.0				
(3) Capital Cost ($/unit of daily capacity):								
(4) Cumulative Capital Investment:								
(5) Operating Cost ($/unit of production):			50.					

*Specify other relevant non-fossil energy source or conversion process, i.e., geothermal, solar, wind, etc.

NONFOSSIL SOURCES AND CONVERSION PROCESS WORKSHEET

Hydroelectric x
Coal Gasification ______
Coal to Syncrude ______
Gas Liquifaction ______
Regasification ______
Other* (specify): ______

Country: Finland

WAES SCENARIO CASE (1985): ABCDE

WAES SCENARIO CASE (2000): 1278

Total Resource in Place: 18 Twh

Technology Assumptions: ______

	(Units)	1972	1975	1985	1990	2000	2010	2025
(1) Yield:	Twh	10.2	12.0	12.0		12.0		
(2) Maximum Potential Output						18.0		
(3) Capital Cost ($/unit of daily capacity):								
(4) Cumulative Capital Investment:								
(5) Operating Cost ($/unit of production):								

*Specify other relevant non-fossil energy source or conversion process, i.e., geothermal, solar, wind, etc.

NONFOSSIL SOURCES AND CONVERSION PROCESS WORKSHEET

Hydroelectric ____________
Coal Gasification ____________
Coal to Syncrude ____________
Gas Liquifaction ____________
Regasification ____________
Other* (specify): Other indigenous fuels

Country: Finland

WAES SCENARIO CASE (1985): BD

WAES SCENARIO CASE (2000): 78

Total Resource in Place: ____________

Technology Assumptions: ____________

	(Units)	1972	1975	1985	1990	2000	2010	2025
(1) Yield:	Mtoe	3.8	3.4	3.1		3.5		
(2) Maximum Potential Output								
(3) Capital Cost ($/unit of daily capacity):								
(4) Cumulative Capital Investment:								
(5) Operating Cost ($/unit of production):								

*Specify other relevant non-fossil energy source or conversion process, i.e., geothermal, solar, wind, etc.

NONFOSSIL SOURCES AND CONVERSION PROCESS WORKSHEET

Hydroelectric ______
Coal Gasification ______
Coal to Syncrude ______
Gas Liquifaction ______
Regasification ______
Other* (specify): Other indigenous fuels

Country: Finland

WAES SCENARIO CASE (1985): ACE

WAES SCENARIO CASE (2000): 12

Total Resource in Place: ______

Technology Assumptions: ______

	(Units)	1972	1975	1985	1990	2000	2010	2025
(1) Yield:	Mtoe	3.8	3.4	3.4		4.1		
(2) Maximum Potential Output								
(3) Capital Cost ($/unit of daily capacity):								
(4) Cumulative Capital Investment:								
(5) Operating Cost ($/unit of production):								

*Specify other relevant non-fossil energy source or conversion process, i.e., geothermal, solar, wind, etc.

11 France

11.1 INTRODUCTION

The share of oil in the French energy supply increased between 1960 and 1972 from 31.4 percent to 67 percent as the competitive position of this fossil fuel vis-à-vis coal accelerated the decline of solid fuel extraction (coal decreased from 54.7 percent in 1960 to 17 percent in 1972).

The low price of oil, compared with other fuels, also held back the development of replacement fuels such as nuclear.

As a consequence, the dependence of France on foreign supply increased from 41 percent in 1960 to about 75 percent in 1972.

Due to the new situation created by the 1973 oil crisis and the consequent steep price increase of oil and natural gas, the present French energy policy can be characterized by a number of constraints and implications:

- The need to assure the security of supply by diversification of the sources of imported fuels.
- The need to develop replacement sources that allow increased independence and that have now become competitive (such as nuclear energy for electricity generation, which is now being rapidly developed).
- The need to promote foreign trade to reduce the deficit in the balance of payments.
- The need to encourage a lower degree of importation by promoting an energy conservation policy and exploitation of the indigenous energy supply (nuclear, hydraulic, new forms of energy).

While energy demand continues to increase due to the growth rate of the economy, despite an active energy conservation policy, exploitation of the indigenous fossil fuel energy resources will remain rather limited, at least from now to 1985.

11.2 COAL

Although some sources suggest a much higher level of coal reserves, the recognized resources amount to 1.4 billion tons, from which 0.8 billion are economically recoverable. As in other European countries, extraction of coal from new mines will in the future depend on the price evolution of competitive forms of energy but will in any case, in France, be very limited before 1985.

Roughly 20 million tons will be extracted yearly from the national mines over the next decade. This extraction rate is expected to be roughly maintained into the future unless there is some drastic change of economic conditions. The major portion of the coal used in the steel industry must be imported, the indigenous coal being mainly used as steam coal.

According to the forecast, at least 50 percent of the coal demand will have to be satisfied by imports during the next decade.

11.3 OIL

Because there are practically no resources of oil on the national territory (less than 1 percent of the demand in 1975), the sources of imports will be, according to the national energy policy, diversified. Important oil research activity goes on in the Channel and in the Mediterranean Sea, but oil in 2000 is still expected to be limited.

11.4 NATURAL GAS

The indigenous secured reserves of natural gas amount to 300 billion m^3, and the annual production in 1975 reached about 10 billion m^3, a level that will be maintained up to 1983 but will probably decrease progressively thereafter.

In 1972, about 50 percent of the natural gas consumed was imported; in 1985, it is expected that the degree of dependence on foreign supply will reach at least 75 percent.

As with oil, government policy is to diversify the foreign sources of imports by encouraging long-term contracts.

From the point of view of the security of the gas supply, the geographical distribution of gas reserves in the world is apparently more favorable for a diversification of supply than in the case of oil, and the rigidity of the gas "chain" from the producer to the consumer is much stronger due to transport constraints.

On the other side, the cost of gas transport and storage is higher than that for oil.

Natural gas will in the future be used mainly in industry and for heating in the residential and commercial sectors and practically not at all in power plants. By 2000, natural gas supplies will be largely depleted.

11.5 HYDRAULIC

Some 80 to 85 percent of the economically recoverable hydraulic resources have already been exploited.

In 1972, the yearly production reached 58 TWh, and it is expected that the remaining capacity will be developed up to 63 TWh by 1985 and about 70 TWh by the year 2000.

France has in operation a tidal plant with a yearly production of 0.5 TWh.

None of these figures takes account of the possible construction of an important new tidal plant at Mont Saint Michel in western France, with a yearly production estimated at 25–27 TWh. For the time being, the financial impact of the project has postponed its realization, although it has not been abandoned.

11.6 NUCLEAR

The situation with regard to indigenous resources of uranium is, compared to that with fossil fuel reserves, more favorable for the time being; at least 100,000 tons of uranium (up to $30 per pound of U_3O_8) can be considered as assured and probable. Due to long-term contracts with foreign suppliers (Gabon, Niger), the expectation for the security of demand satisfaction is high.

The above-mentioned figure of 100,000 tons must be considered conservative, the real efforts of exploration having been undertaken only recently.

The French nuclear program in the next decade will be based on light-water reactors (pressured water), the needed enrichment sources for the fuel being provided by long-term contracts with the United States and by the EURODIF plant now under construction in France, with the participation of Belgium, Italy, and Spain. The yearly output of this gas diffusion plant will reach about 11 million separative work units (SWUs) after 1980. A new plant (COREDIF) is being planned, with possible participation of foreign partners.

The LWR fuel reprocessing will take place in a plant that will be in operation in the near future (final capacity 800 tons/year).

The French nuclear program started with some natural uranium, gas-graphite-moderated plants, but this line has now been abandoned in favor of the light-water reactors. Also, as part of the attempt to assure the future fuel supply, the breeder reactor technique was developed, and the 250 MWe Phénix plant has been operated successfully for two years. A new commercial breeder plant (1200 MWe) will be ordered in the near future for commissioning in 1981–1982.

The safety and other problems connected with nuclear energy are being investigated very intensively.

11.7 OIL SHALE

Oil shale deposits in France are far from negligible. For the Parisian region alone, at a depth of about 50 meters recoverable reserves are estimated to be about 1 billion tons of oil.

Nevertheless, the total financial and environmental impact—confirmed by the hesitation in some more favored countries—will contribute to postponing indefinitely the possible exploitation of this resource.

11.8 GEOTHERMAL ENERGY

Low- and medium-temperature geothermal energy is rather abundant in France, mainly in the "Basin Parisien," the "Basin Aquitarius," Alsace, and the "coulor Rhodinien."

The maximum yearly output of these geothermal fields is estimated to be between 15 and 20 MTOE. This energy source will be used mainly for heating water, for district heating, and for some industrial uses.

Because most of the geothermal fields lie under dense urban areas, the development of geothermal energy could be significant before the year 2000, and this energy might then provide about 10 MTOE.

11.9 SOLAR ENERGY

Solar energy in France could be used mainly for space and water heating. Two techniques have been tested for some years already: Trombe's technique and the flat collector technique. These techniques

are briefly described in the French contribution to the third WAES technical report, *Supply-Demand Integration to the Year 2000* (The MIT Press, 1977).

Although the costs of these techniques are fairly high at the present (at least for the flat collector), one can reasonably assume that these costs will decline if large-scale production is developed.

In that case, solar heating could be developed rather fast, and this energy source could provide about 8 MTOE.

11.10 HEAT RECOVERY

The recovery of waste heat (for district heating and industrial low-temperature heat) is an important possibility for the future.

The potential of such an energy source is directly linked to the level of industrial energy consumption and the amount of electricity generated by thermal plants.

Technical, financial, and institutional constraints could greatly impede development of this source. But if these constraints could be overcome in a satisfying way, waste heat could provide energy equivalent to 10–13 MTOE.

11.11 ENERGY SUPPLY WORKSHEETS

The following worksheets appear in this section (see Section 1.2.1 of Chapter 1 for the definitions of line items):

Energy supply summary

Coal

Natural gas

Uranium and nuclear

Hydroelectric

Geothermal and district heating

Solar

WAES NATIONAL ENERGY SUPPLY STUDIES, 1972-2000
ENERGY SUPPLY SUMMARY

Country: France

Scenario Assumptions	1972	1985					2000			
							C-1'		D-7'	
WAES CASE	Base Year	A	B	C	D	E	C-1	C-2	D-7	D-8
GNP 1976-85; 1985-2000		High	Low	High	Low	High	High	High	Low	Low
Oil/Energy Price*		11.50-17.25	11.50-17.25	11.50	11.50	11.50-7.66	11.50-17.25	11.50-17.25	11.50	11.50
National Policy Response		Vig	Vig	Vig	Res	Res	Vig	Vig	Vig	Vig
Principal Replacement Fuel							Coal	Nuclear	Coal	Nuclear
Energy Supply **Units** MTOE										
Crude Oil	2	–	–	–	–	–	–	–	–	–
Natural Gas	8	7	7	7	7	7	5	5	5	5
Coal	22	13	13	13	13	13	13	13	13	13
Nuclear **	1	53	46	60	55	40	85–105	145	80–88	121
Hydro-electric	11	14	14	14	13	14	15	15	15	15
Geothermal	–	3	4	3	3	–	10–30	10	10–30	10
Solar Thermal	–	3	4	3	3	–	10–30	10	10–30	10
Solar Electric	–	3	4	3	3	–	10–30	10	10–30	10
Total Energy Supply	44	90	84	97	91	74	148	188	131–143	164

*1975 US dollars per barrel of Arabian light crude oil, fob Persian Gulf.
All Figures are primary production prior to any processing or conversion.

Footnotes:

** Equivalences: 1 TWh = 0.22 MTOE.

***Including heat recovery.

INDIGENOUS SUPPLY ASSESSMENT WORKSHEET

COAL

Country: France

Coal:
Surface Mined ____
Deep Mined x

WAES SCENARIO CASE (1985): all

WAES SCENARIO CASE (2000): all

Total Resource in Place: ____

Domestic Coal Price: ____

Technology Assumptions: ____

Million Tons	1972	1975	1980	1985	1990	2000	2010	2025
(1) Ultimate Recoverable Resource:		31,000						
(2) Reserves Proved to Date:		1,400	1,300	1,200	1,100	900		
(3) Potential Production:		800	700	600	500	300		
(4) Expected Annual Production:		20	20	20	20	20		
(5) Cumulative Production:		20	120	220	320	520		
(6) Remaining Proved Reserves:		780	680	580	480	280		
(7) Reserve/Production Ratio:		40/1	35/1	30/1	25/1	15/1		
(8) Primary Transportation* —Capital Cost ($/unit of daily capacity): —Operating Cost ($/unit of daily capacity):								
(9) Capital Cost ($/unit of daily capacity):								
(10) Operating Cost ($/unit of production):								

*From mine mouth to first processing stage.

INDIGENOUS SUPPLY ASSESSMENT WORKSHEET

NATURAL GAS

Country: France

Natural Gas:
Offshore ____
Onshore x

WAES SCENARIO CASE (1985): all

WAES SCENARIO CASE (2000): all

Total Resource in Place: 300 billion m^3

Domestic Gas Price: ____

Technology Assumptions: ____

Billions m^3 (Units)	1972	1975	1980	1985	1990	2000	2010	2025
(1) Ultimate Recoverable Resource:		300						
(2) Discoveries:		–	–	–	–	–		
(3) Reserves Proved to Date:		300	250	201	151	83		
(4) Potential Production:								
(5) Expected Annual Production:		10	10	8	7.5	5.5		
(6) Cumulative Production:		10	50	107	156.5	222.5		
(7) Remaining Proved Reserves:		290	240	193	143.5	77.5		
(8) Reserve/Production Ratio:		30/1	25/1	25/1	20/1	15/1		
(9) Host Government Take ($/unit of production):								
(10) Capital Cost ($/unit of daily capacity):								
(11) Operating Cost ($/unit of production):								

NONFOSSIL SOURCES AND CONVERSION PROCESSES WORKSHEET

URANIUM AND NUCLEAR

Country: France

WAES SCENARIO CASE (1985): C

WAES SCENARIO CASE (2000): C-2

Total Resource in Place: ______

Domestic Uranium Price: ______

Technology Assumptions: ______

	(Units)	1972	1975	1980	1985	1990	1995	2000	2010
1. Ultimate Recoverable Uranium Resource:	tons		2.10^{5}	2.10^{5}					
2. Reserve of Uranium Proved to Date:	"		1.10^{5}	1.10^{5}					
3. Potential Production of Uranium:	"		3.10^{3}	3.10^{3}					
4. Expected Production of Uranium:									
5. Reserve/Production Ratio:			1/30	1/30		(1)			
6. Maximum Installed Capacity (GWe)*:			2.5	20	45	75	100	125	
7. Capital Cost ($/KWe):	(at 1.1.76)			440 (2)	480 (3)				
8. Operating Cost (¢/Kwhr)**:	(fuel....... (operation..			0.40 0.35	0.48 0.34				

*Indicate assumed load factor and LWR or HWR.

**Excluding capital charges.

1 dollar = 5 francs

(1) After 1985 the development of the breeder line is expected. As base load the load factor is 0.7 in later operation, around 2000 (with the older plants) will be about 0.6.

(2) 900 MWe units.

(3) 1 300 MWe units (included new development costs).

NONFOSSIL SOURCES AND CONVERSION PROCESS WORKSHEET

Hydroelectric x
Coal Gasification ____
Coal to Syncrude ____
Gas Liquifaction ____
Regasification ____
Other* (specify): ____

Country: France

WAES SCENARIO CASE (1985): all

WAES SCENARIO CASE (2000): all

Total Resource in Place: ____

Technology Assumptions: ____

	(Units)	1972	1975	1980	1990	2000	2010	2025
(1) Yield:								
(2) Maximum Potential Output		58	58	61	64	70		
(3) Capital Cost ($/unit of daily capacity):								
(4) Cumulative Capital Investment:								
(5) Operating Cost ($/unit of production):								

*Specify other relevant non-fossil energy source or conversion process, i.e., geothermal, solar, wind, etc.

NONFOSSIL SOURCES AND CONVERSION PROCESS WORKSHEET

Hydroelectric ______
Coal Gasification ______
Coal to Syncrude ______
Gas Liquifaction ______
Regasification ______
Other* (specify): Geothermal and district heating

Country: France

WAES SCENARIO CASE (1985): C, D

WAES SCENARIO CASE (2000): C-1, D-7

Total Resource in Place: 15-20 MTOE/year

Technology Assumptions: ______

	(Units)	1972	1975	1980	1990	2000	2010	2025
(1) Yield:								
(2) Maximum Potential Output	MTOE			0.5	2	10		
(3) Capital Cost (1975 $) ($/unit of daily capacity):				$600/TOE/year capacity				
(4) Cumulative Capital Investment: (1975 $, billions)				0.3	1.2	6		
(5) Operating Cost (1975 $) ($/unit of production):				$120/TOE/year				

*Specify other relevant non-fossil energy source or conversion process, i.e., geothermal, solar, wind, etc.

Note: Geothermal energy is counted on a delivered energy basis. To allow for consistent comparisons with other energy sources, one should consider that geothermal energy will replace oil burned for heating with an average efficiency of 70 percent.

NONFOSSIL SOURCES AND CONVERSION PROCESS WORKSHEET

Hydroelectric ______
Coal Gasification ______
Coal to Syncrude ______
Gas Liquifaction ______
Regasification ______
Other* (specify): Solar

Country: France

WAES SCENARIO CASE (1985): C,D

WAES SCENARIO CASE (2000): C-1, D-7

Total Resource in Place: ______

Technology Assumptions: ______

	(Units)	1972	1975	1980	1990	2000	2010	2025
(1) Yield:								
(2) Maximum Potential Output	MTOE			0.5	1	1.5		
(3) Capital Cost (1975 $) ($/unit of daily capacity):				$1500/TOE/year capacity				
(4) Cumulative Capital Investment:	(1975 $, billions)			.75	1.5	11.5		
(5) Operating Cost ($/unit of production):								

*Specify other relevant non-fossil energy source or conversion process, i.e., geothermal, solar, wind, etc.

Note: See footnote for geothermal energy worksheet.

German Federal Republic

12.1 SUMMARY

In 1975 about 66 percent of the primary energy consumed was imported. This indicates the deficiency of economically recoverable energy resources in the Federal Republic of Germany. Although there is a considerable resource base consisting of lignite and in particular hard coal, coal's share of total energy has declined. Hard coal production has declined by some 40 percent since the late 1950s. Even the fourfold rise in crude oil prices in 1973 did not significantly increase coal production again.

Except for coal—hard coal and lignite—fossil energy resources are very small. Indigenous oil and natural gas, which provided only about 8 percent of the primary energy consumed in 1975, have reached or already passed peak production.

The economic and technical potential for nonfossil, nonnuclear natural energy sources such as hydro, geothermal, solar, and wind is not very promising. The economically recoverable hydro power potential is almost fully tapped; the geothermal activity usable with expected technology is quite low. The climate permits only small-scale prospects for solar energy, and the integration of large-scale wind energy utilization requires uneconomically high investments, given the existing system and prospective conditions.

Non-resource-based sources such as urban waste have only a relatively small potential.

This supply study was carried out at the Energy Institute at the University of Cologne by Dieter Schmitt and Paul H. Suding.

Because uranium deposits in the Federal Republic of Germany add up to only about 5000 metric tons, uranium requirements have to be met entirely by imports. Nuclear energy, therefore, cannot be regarded as a genuine indigenous energy source before the fast breeder comes into large-scale use.

The scenarios to 1985 (Table 12.1 and Figure 12.1) show considerable growth in energy supply. But this growth is almost entirely caused by expanding nuclear generation of electricity. The other sources are estimated to provide a total output in 1985 similar to that of 1975. A small decrease in oil production is balanced by marginal increases in gas, coal, hydro power, and production from other sources. Among the scenarios only the policy assumption makes a difference: the restrained policy for hard coal (expiration of the Verstromungsgesetze, the coal combustion acts, in the early 1980s) is assumed to cause a drop in hard coal production from 94 to 80 $\times$ 10^6 MTCE. The expansion of nuclear electricity generation depends largely on the demand for electricity, assuming that nuclear energy has cost advantages in the production of base-load electricity. In 1985, 40 percent of electricity is assumed to be generated in nuclear power stations in the high economic growth case and 34 percent in the low growth case.

The growth of the energy supply in the period between 1985 and 2000 (Table 12.1 and Figure 12.1) is almost entirely caused by nuclear development, as in the earlier period. While hydrocarbon production is decreasing, small additional contributions from lignite, solar, and other sources compensate for this

Table 12.1
Actual and expected indigenous primary energy supply of the Federal Republic of Germany in 1975, 1985, and 2000 (10^6 metric tons of coal equivalent)

		1985					2000			
Source	1975	A	B	C	D	E	C-1	C-2	D-7	D-8
Crude oil	8.3	5.7	5.7	5.7	5.7	5.7	3.8	3.8	2.3	2.3
Natural gas	20.7	22	22	22	22	22	10	10	10	10
Lignite	33.3	35	35	35	35	35	36.5	36.5	36.5	36.5
Hard coal	92.8	94	94	94	80	80	112	94	100	80
Hydro	5.2	6.0	6.0	6.0	6.0	6.0	6.8	6.8	6.8	6.8
Solar and others	1.7	6.3	5.4	6.3	5.4	6.3	13.6	13.6	10.2	10.2
Nuclear	7.0	71.8	51.4	70.5	50.3	69.1	120.0	271.0	120.0	204.0
Total	169.0	240.8	219.5	239.5	204.4	124.1	302.7	435.7	285.8	349.8

decrease. Hard coal production varies among the cases as a function of coal policy to 1985 and the nuclear development assumptions. Low nuclear means high coal and vice versa. The nuclear contribution in the low nuclear cases C-1 and D-7 is arbitrarily fixed to be 50 GW. This level is meant to represent a "technological and social drawbacks scenario"; in the high nuclear cases where there is no drawback, nuclear power production is assumed to be determined widely by electricity consumption. Given favorable cost developments in these cases, up to 70 percent of electricity can be generated from nuclear energy by year 2000.

12.2 ASSUMPTIONS

12.2.1 Overview

The WAES methodology requires estimating supply and demand independently from each other under the same scenario assumptions. The assumptions for this study are summarized in Table 12.2. For supply development the most important assumptions are the government's policy response and the price assumption, but economic growth also has some impact. The following sections describe in more depth the specific impact of individual scenario variables on energy supply.

Table 12.2
Scenario assumptions

1985	A	B	C	D	E
Annual economic growth, 1976–1985 (%)	4.5	2.5	4.5	2.5	4.5
Oil price (real 1975 $ per bbl)	17.25	17.25	11.50	11.50	7.66
National energy price level (1975 = 100)	130	130	100	100	90
National policy response	vig	vig	vig	restr	restr
2000	**C-1**	**C-2**	**D-7**	**D-8**	**D-3**
Annual economic growth, 1976–2000 (%)	3.4	3.4	2.2	2.2	2.2
Oil price (real 1975 $ per bbl)	17.25	17.25	11.50	11.50	17.25
National energy price level (1975 = 100)	150	150	100	100	150
National policy response	vig	vig	vig	vig	vig
Nuclear capacity	60 GW	high	50 GW	high	50 GW

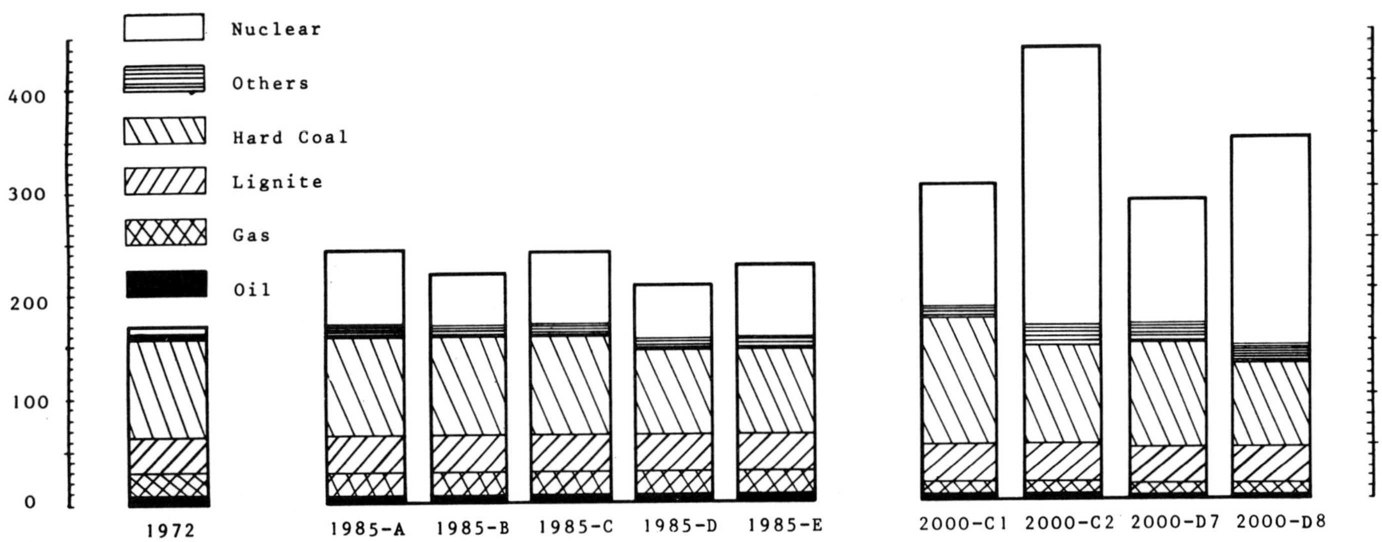

Figure 12.1
Indigenous energy supply in the Federal Republic of Germany, 1972–2000 (units: 10^6 metric tons hard coal equivalent).

12.2.2 Economic Growth

Supply and demand can generally be estimated independently. However, since nuclear generating capacity depends heavily on the individual country's demand for electricity, the demand estimate must be taken into account. The main determinant of electricity consumption is economic growth. Moreover, since capital is a major determinant of the development of a supply source, economic growth has a secondary influence on energy supply because it determines capital formation.

12.2.3 Oil Price

The overarching market share and the substitutivity of oil and oil products have made crude oil the price leader. Since oil may lose its price leadership function before 2000, we have used the scenario variable "oil price" as an analogue for the world market price of primary energy for the period between 1985 and 2000.

We have assumed that the distinction between oil price (1975–1985) and primary energy price (1985–2000) makes a difference in the internal price of energy. For example, a 50 percent higher crude oil price FOB Persian Gulf to 1985 does not necessarily imply a 50 percent higher national energy price level, whereas a 50 percent higher world market price of primary energy can quite plausibly lead to a 50 percent higher internal energy price. Thus the internal price level in Cases A and B (1985) is assumed to be lower than the price level in Case C (2000).

12.2.4 National Policy Response

Policy is assumed to have a decisive impact on indigenous energy supply for both coal and nuclear.

We have interpreted a restrained coal policy to mean that no follow-up act will replace the Verstromungsgesetze (coal combustion acts) when they expire in the 1980s. The vigorous policy assumes a follow-up act, keeping hard coal production at 94×10^6 MTCE.

The nuclear/coal variable for the year 2000 scenarios is only partly affected by policy. Technological and social drawbacks must also be weighed. High coal is plausible when these drawbacks constrain nuclear energy because low nuclear leads to a need for a high capacity to generate electricity from coal. In the case of high coal, mining is assumed to include three additional new deep coal mines, each with a production capacity of 6×10^6 MTCE per year.

Existing import restrictions for hard coal are expected to be removed, if a reasonable indigenous production is assured.

12.3 ENERGY SUPPLY FROM INDIGENOUS AND QUASI-INDIGENOUS SOURCES

To survey energy supply prospects completely, it is appropriate to review not only the indigenous energy resource base and production but also existing long-term supply contracts, in particular those for natural gas and the nuclear fuel cycle. Worksheets with supporting data are gathered at the end of this chapter.

12.3.1 Fossil Nonnuclear Sources

Oil

The Bundesanstalt für Geowissenschaften und Rohstoffe (BGR) estimates the ultimate recoverable oil resources in Germany as of 1 January 1975 to be 175×10^6 TOE, of which 75×10^6 TOE were proved under actual economic conditions (reference 1). At the end of 1975 the proved reserves had decreased by only 3.3×10^6 TOE to 71.7×10^6 TOE in spite of a production of 5.7×10^6 TOE due to secondary recovery. The reserves proved as of 1 January 1976 including the production of 1975 rose therefore to 77.4×10^6 TOE, according to Niedersächsisches Landesamt für Bodenforschung (NLBF), the official agency of the state of Lower Saxony (see reference 2).

Additional oil fields are not likely to be found in the Federal Republic of Germany. Since even the considerable onshore and offshore exploration efforts since 1973 have not been successful, future production will probably rely entirely on known fields and will be determined by the success of recovery technology.

A higher oil price is assumed to slow the exhaustion of the oil fields for some years. This accounts for the difference between Cases C and D. Oil production is estimated to fall from 5.7×10^6 TOE in 1975 to 3.8×10^6 TOE in 1985 and to 2.5 or 1.5

Table 12.3
Long-term contracts for natural gas ($10^9 m^3$)

	1980	1985	1990	1995	2000
The Netherlands	29.4	29.4	27[a]	3[a]	—
Norway	12.3	12.3	12.3	6[a]	—
U.S.S.R.	10.4	10.4	10.4	3[a]	3[a]
Iran	—	5.5	5.5	5[a]	5[a]
Total	52.1	57.6	54	17	8

Source: Ruhrgas AG.
[a]Approximate.

$\times 10^6$ TOE in 2000, depending on the oil price assumption.

Natural Gas

BGR estimates the ultimately recoverable natural gas reserves (associated and unassociated) as of 1975 to be 615×10^9 m^3, of which 315×10^9 m^3 were proved. The average calorific value of this gas is about 7700 kcal/m^3. Thus 615×10^9 m^3 equals about 3.25×10^9 BOE (references 1 and 2).

Additional significant gas fields are not likely to be found. Even the offshore activities have not proved promising. Thus production is estimated to remain at its present level (20×10^9 m^3) through the 1980s and to decline slowly beyond 1990 (references 1 and 3).

Long-term contracts exist for natural gas deliveries to West Germany from The Netherlands, Norway, the Soviet Union, and from Iran via the U.S.S.R. (Table 12.3), but most existing contracts expire in the early 1990s. The Netherlands, the main contractors, has already clearly indicated that new contracts cannot be hoped for (reference 3).

Hard Coal

According to BGR estimates the geologically measured resources amount to about 230×10^9 metric tons (1 metric ton = 7 Gcal); of this amount about 24 have been proved recoverable under technical conditions and 6 have been proved under economic conditions as of 1975 (reference 1).

A large fraction of the German coal is usable as coking coal. The coking coal market shows substitutivity characteristics rather different from the other coal markets. Therefore, the scenario assumption "oil price" does not necessarily determine the future world market price for coking coal. In addition, current contracts and political constraints tie the German iron and steel industry closely to German coking coal production regardless of differences in the world market price. Moreover, the European Community regulations for the use of community coal in the European iron and steel industry balance contingent price differences in the world market price. As a result, coking coal production depends strongly on policies and demand development, that is, on the growth and structure of the iron and steel industry. For coking coal the political conditions in our scenarios are constant through 2000. This leads to the assumption that the production level for coking coal will be about 45×10^6 MTCE per year. The difference in coking coal demand due to different growth assumptions in the scenarios is not sufficient to change German coal production considerably.

Given the scenario assumptions about the energy price level and the WAES preliminary working assumption about the availability of other sources at given prices, the price of coal—other than coking coal—would not climb sufficiently even to keep production at the present level. But for steam coal used to generate electricity, the government's energy policy has a significant impact on production (reference 4). Labor policy, regional policy, and long-term energy supply policy will keep the production of hard coal at today's level by influencing the electric industry to burn coal. A steam coal production of 30 to 33×10^6 metric tons per year is ensured by political measures (embodied in the Verstromungsgesetze) including financial aid for coal and licensing reservations for oil- and gas-fired power stations. However, the act stipulates that financial aid is to expire in the early 1980s, and the extension of this steam coal regulation seems to be less certain than the extension of the coking coal regulations.

The hard coal production estimate must therefore take into account a coking coal production of about 45 million metric tons, a steam coal production sensitive to a now uncertain policy, and a production for other uses of about 15 million tons. For this study the steam coal production through 1985 depends on whether the energy policy is restrained or vigorous.

Under the restrained policy further steam coal subsidies would be eliminated, whereas under a vigorous policy the current system would be extended. The restrained policy with a constant or decreasing oil price is assumed to cause coal production to decline to about 80×10^6 metric tons. A vigorous policy with either higher or constant oil prices is assumed to keep coal production at its present level through 1985.

For the period 1985 to 2000 high nuclear is assumed to result in coal production corresponding to the 1985 level (94 and 80 × 10^6 metric tons in Cases C-2 and D-3). Low nuclear/high coal, on the other hand, is assumed to result in governmental activities to extend coal production. The addition of three mines, each producing about 6 × 10^6 metric tons per year, seems plausible and possible from an environmental, labor force, and political point of view. In the high coal cases C-1 and D-7 for year 2000, therefore, the production levels are 112 and 100 × 10^6 metric tons, respectively. Since the lead time for the opening of a deep coal mine is ten years, the mines would have to be started in 1990 to be open by 2000.

Lignite

According to a BGR estimate, the geologically measured resources account for 52 × 10^9 metric tons of lignite (1 metric ton = 1.9 Gcal), of which 35 × 10^9 are proved recoverable under current 1975 technical conditions and 10 × 10^9 under 1975 economic conditions (reference 1). The regional concentration of the lignite production (Table 12.4) coincides almost exactly with an organizational centralization. The Rhenanian surface mines are operated by the Rheinische Braunkohle AG, which is owned by the Rheinisch-Westfählisches Elektrizitätswerk AG, the largest electricity producer in Germany. About 90 percent of the lignite is burned in this company's power plants. The cost of the lignite makes it reasonably favorable for use in power plants or for gasification. Demand, therefore, would allow higher production. Limits are set by environmental conditions, however, since lignite is extracted in densely populated areas. New space-intensive surface mining requires programs to restore the environment as well as to demolish and resettle villages. This can only be done in steps and requires long-range planning and approval by regional authorities. Because of the organizational unity and the long-range planning, the production plans up to 1990 and 2000 are currently determined and are insensitive to changes in the economic conditions. Only very tight supply in the years to come could open another deep open mine, which could hardly go into operation before 2000.

The expected annual production in 2000 is not assumed to vary with the scenario (Table 12.4). About 90 percent of the actual production is extracted in the Rhineland, and this percentage will rise in the future since the Bavarian and Hessian production will be exhausted before 2000. Helmstedt in Lower Saxony is likely to extend its production, depending on what decision is made about the production of lignite with high salt content.

Since the production share of lignite with higher calorific value is rising, total production is increasing slightly from 33.3 × 10^6 MTCE 1975 to 35 × 10^6 in 1985 and to 36.5 × 10^6 MTCE in 2000.

If nuclear energy becomes the main fuel for electricity generation and if advanced technology such as the high temperature reactor (HTR) becomes available, gasification of lignite with HTR will replace natural gas imports by 2000. In the high nuclear cases, therefore, some 19 × 10^6 MTCE are assumed as net input in gasification (see the supply-demand integration study in the third WAES technical report).

12.3.2 Nonnuclear, Nonfossil Sources

Nonnuclear, nonfossil energy sources include the natural nonfossil sources as well as renewable sources. Explicit estimates of these sources are given only for solar and hydroelectric. All others are added together under the category "other" which consists mainly of urban waste conversion. Wind, biomass, geothermal, and orbital solar are presumed to provide marginal or no contributions.

Hydroelectric

By the end of 1975 the installed hydro power plants in the Federal Republic of Germany were providing up to 5109 MW, of which about 2.3

Table 12.4
Actual and expected lignite production in German provinces (metric tons)

	1975	1985	1990	1995	2000
Rhineland	107.4	120	120	120	120
Helmstedt	4.9	5.0	5.5	8.0	8.0
Hessia	3.1	2.9	1.2	1.0	0
Bavaria	8	3.5	0	0	0
Total	123.4	131.4	126.7	129.0	128.0

Source: Rheinbraun AG, DEBRIV.

GW were from river water, 0.3 GW from storage, and 2.5 GW from pump storage systems. The pump storage systems do not add net energy, but they help to economize peak load problems and provide reserve capacity in case the normal capacity fails. For energy supply, the river water and pure storage systems are more relevant.

The theoretical potential capacity for hydro power given by the amount of water drain and the difference in geodetic altitudes amounts to about 11 GW. The technical potential production is determined by the transformation technology and by the load factor.

About 75 percent of the technical hydro power potential was used up by 1976. Since the economic potential is somewhat lower, only very small extensions of the hydroelectric capacity and production are believed possible (reference 6). The estimate is constant for all scenario cases. Hydroelectric capacity excluding pump storage systems is assumed to be as high as 2.7 GW. With an average load duration of 5500 hr/year this capacity could produce about 14,850 GWh annually.

Capacity for pump storage systems (existing, under construction, and planned capacity) is about 3.3 MW in 1985. If this capacity is maintained through 2000, the output not diminished by the electricity consumption for pumping purposes could amount to about 4000 GWh per year. Pump storage capacity can certainly be much higher by 2000 if this technology is used extensively to provide reserve capacity.

Solar

The geographical position of West Germany in the global climatic system allows only limited use of solar energy. Of the numerous technically possible utilization systems only a few are regarded as options for Germany.

In the WAES time frame, only collector systems are estimated to play even a minor (though growing) role. In the residential sector collector systems can be used, when combined with an additional central heating system, to produce warm water for direct use in heating systems. A marginal economic potential of solar is estimated to be 278×10^{15} joules (reference 6). In the scenarios a more conservative estimate of solar use is assumed: 100×10^{15} joules in the high and 50×10^{15} joules in the medium oil price case.

Of all other solar energy systems only decentralized solar cells are potentially of importance, but current production projections are too low and too uncertain to be included in the WAES cases (reference 6). The cost of the cells must decline dramatically for this application to be economically feasible.

Wind

Large-scale use of wind energy in the Federal Republic of Germany would require major reconstruction of electricity generation, storage, and grids—an effort that seems inappropriate in all the scenarios.

Small-scale units may be put into operation at remote places, but the overall production is too small to be included explicitly. Therefore, the contribution from wind energy is included in the category "other."

12.3.3 Nuclear

To determine nuclear estimates for 1985 and 2000 (high and low) we depart somewhat from the WAES principle of estimating supply and demand independently. Nuclear energy depends almost entirely on electricity consumption, both in volume and in load distribution over time. For 2000 better load management is assumed so that almost 80 percent of electricity is produced in base load (more than 6000 hours per year). Since hydro and lignite also contribute to base-load electricity production, the maximum nuclear contribution is about 70 percent. For 1985 the maximum nuclear contribution is assumed to be 40 percent. Nuclear cannot fuel the same share of the base load (about 65 percent of total load in 1985), since nuclear must be introduced into an existing system and has long lead times. The exact share of nuclear in base load is determined by the availability of competitive fuels such as hydro power and lignite (reference 7).

The demand for electricity and the cost of generating it are the main determinants of the nuclear contribution, unless public policy restraints imposed for safety reduce its contribution. These restraints depend on the technological solution of safety problems and the public acceptance of such solutions and the risks. The low and high estimates for nuclear in 2000 are meant to reflect such unrestrained (high) and restrained (low) cases. The numbers chosen may have more meaning after the following overview.

Installed and Planned Nuclear Power Capacity
In March 1975 the installed nuclear capacity in the Federal Republic of Germany was about 3.5 GW. The 12.8 GW now under construction and the additional 13.7 GW projected would bring 1985 capacity to 30 GW. Consequently, the government's initial estimate of 45–50 GW has been revised downward. Since the lead time for the construction of a new unit is six years or more the maximum capacity possible by 1985 is about 35 GW. By 2000, engineering knowhow and construction capacity will not be limiting factors in the case of light water reactors.

For advanced reactor technology engineering knowhow and construction capacity will still be constraints in 2000. One 300 MW prototype high-temperature reactor that has been under construction since 1974 will not go into operation before 1980. By the end of 1976 the projected construction time had increased from 61 to 87 months, and the total cost of the project had risen about 60 percent. Another high-temperature prototype for the combined production of heat and electricity is planned for construction beginning around 1980, but its construction depends on the experience gained in developing the first prototype. The construction of the fast breeder prototype SNR-300 has also been delayed by about 20 months and will not be completed before 1981. Construction costs for the fast breeder reactor have also risen, because of additional security requirements and inflation. The next step in the development of the fast breeder will be the 1300 MW breeder SNR-2, which can take advantage of the experience gained from the SNR-300 and the French Phénix. Fusion has still not been successfully demonstrated. The next R&D step, the implementation of the Joint European Torus (JET), has not been decided yet. Fusion is therefore not taken into consideration in the scenarios.

Nuclear Fuel Cycle
Uranium deposits in the Federal Republic of Germany are insignificant. They will probably not allow a continuous production but may serve as a short-term stockpile (Table 12.5).

Common companies and interests in foreign companies plus long-term contracts are assumed to cover about 60 percent of the demand for natural and enriched uranium. Some 25,000 metric tons of natural uranium will be imported from Canada, South Africa, and France within the next ten years, but by 1979 the supply contracted for at the end of 1976 will no longer meet projected demand (reference 8). By 1979 long-term contracts for separative work units of uranium enrichment will also be insufficient to meet demand. These gaps, of course, occur only if the current nuclear plans materialize. In the 1980s between 1100 and 1500 metric tons per year of separative work units have been contracted from the United States and between 600 and 800 metric tons per year have been contracted from the Soviet Union. Contracted deliveries from URENCO start in 1978 and grow to 1230 metric tons per year by the mid 1980s; those from EURODIF start in 1981 and plateau at 130 metric tons per year by the mid 1980s (reference 8). Both contracts supply peak quantities in the early 1980s. There are plans now to build an enrichment facility in West Germany in addition to an existing small (250 metric ton) facility. Siting activities in Germany are being conducted by the Anglo-Dutch-German URENCO group.

For the back end of the nuclear fuel cycle, the federal government has developed a waste management program that integrates reprocessing, recycling of fissile material, waste treatment, and waste storage. Although the first stage of this system may take six to eight years and must be available by

Table 12.5
Known uranium deposits in the Federal Republic of Germany

Location	Production Cost ($/lb U_3O_8)	Quantity (1000 metric tons U_3O_8)
Menzenschwand	10	1
Geroldsau	12	0.3
Tischenreuth	12–16	0.2–0.5
Stockheim	19	2.0
Mähring	19	0.2–0.5
Rudolfstein	20	0.05
Total	under 21	3.75–4.35

Source: BMFT (Ministry of Research and Technology).

about 1985, siting disputes had not been settled even by the end of 1976. Three sites in Lower Saxony, each located upon virgin underground salt formations, have been determined to be appropriate for the currently planned storage technology. Politicians have announced their willingness to decide among these, but growing public concern about radioactive material, plutonium disuse, and final storage of high-level waste has delayed their decision.

High and Low Estimates in 2000

The technological, economic, and social prospects are sufficiently uncertain to consider nuclear a scenario variable. If the safety problems can be solved technically, if the reprocessing capacity can be established in time, if the public accepts nuclear as a necessary option, and if residents near plant sites accept the risks, then the private cost and the consumption profile allow nuclear to satisfy up to 70 percent of the electricity demand. Given the electricity consumption estimated in Cases C-2 and D-8 (see the third WAES technical report), nuclear power capacities of about 120 or 80 GW, respectively, would be required to produce 70 percent of the electricity.

In the low nuclear case an arbitrary production capacity of about 50 GW in 2000 is assumed. Development is assumed hampered by a delay in the construction of reprocessing facilities, together with major opposition from the concerned public as well as a temporary stop in licensing, all leading to a delay in the realization of power plants planned through 1985. To reach the assumed 50 GW in 2000 these problems must finally be overcome. The low nuclear case also assumes that technical difficulties will delay the use of waste heat from nuclear stations for residential and industrial purposes until after 2000.

12.4 SUPPLY BEYOND THE SCENARIO ASSUMPTIONS

Supply is not very sensitive to changes in the scenario assumptions for price and growth. Even higher prices and higher economic growth rates than those assumed in the scenarios would not bring much additional indigenous supply into markets.

Hydrocarbons and hydroelectric have reached the technical potential limits. Coal production is either environmentally constrained or politically determined.

If environmental obstacles are overcome, early recovery from an additional lignite surface mine (Hembach II) could increase lignite production by some 50×10^6 metric tons (some 10 million TOE). Tentative estimates of potential additional hard coal production are half that order of magnitude. Major politically caused increases beyond 120×10^6 MTCE per year are hardly imaginable.

The solar contribution estimate for the supply scenarios is conservative, but even a threefold expansion of the high solar case would result in as little as 7 MTOE. Neither solar nor geothermal energy has the physical potential to contribute significantly to West Germany's energy supply.

The prospects for a considerable expansion of wind use depend on economic conditions beyond the scope of this study.

Because indigenous supply is so inelastic, quasi-indigenous supplies (nuclear; long-term contracts for gas) and the conversion technologies to increase substitutivity among fuels are becoming more important.

Additional options for energy supply technologies such as chemical transportation of heat by ADAM and EVA and hydrogen economy are frequently discussed, but those technologies have little chance of being developed before 2000.

12.5 REFERENCES

1. Bundesanstalt für Geowissenschaften und Rohstoffe. *Die künftige Entwicklung der Energienachfrage und deren Deckung: Perspektiven bis zum Jahre 2000. Abschnitt III: Das Angebot von Energierohstoffen* (Hannover, 1976).
2. Hans-Ulrich Hark. "Die Erdöl- und Erdgasexploration in der Bundesrepublic Deutschland," Erdöl und Kohle, 8 August 1976, pp. 336–345.
3. Klaus Liesen. "Der Beitrag des Erdgases in der Energieversorgung der Bundesrepublic Deutschland," Glückauf No. 3, 1976, pp. 136–142.
4. *Gesamtverband des deutschen Steinkohlenbergbaus, Jahresbericht Steinkohle 1975/1976* (Essen, 1976).
5. Joachim Leuschner. "Tagebau- und Landschaftsplanung im Rheinischen Braunkohlenrevier." In: *Festschrift zur Verleihung des Krupp-Energie-Preises* (Essen, 1975).
6. Bundesministerium für Forschung und Technologie, *Energiequellen für morgen? Nichtfossile-nichtnukleare Primärenergiequellen* (Frankfurt am Main, 1976).
7. Bundesministerium für Forschung und Technologie. *Künftiger Bedarf an elektrischer Energie in Abhängigkeit von wirtschafts- und gesellschaftspolitischen Entwicklungen und dessen Deckung, insbesondere mit Hilfe der Kernenergie.* Forschungsbericht K 76-03, Kernforschung (Eggenstein-Leopoldshafen, 1976).

8. Antwort der Bundesregierung auf die kleine Anfrage betr. Brennstoffkreislauf, friedliche Nutzung der Kernenergie (Drucksache 7/5763).

12.6 ENERGY SUPPLY WORKSHEETS

This section includes worksheets in the following order (see Section 1.2.1 of Chapter 1 for the definitions of line items):

Crude oil

Natural gas

Coal (3)

Nuclear electric

INDIGENOUS SUPPLY ASSESSMENT WORKSHEET

CRUDE OIL

Country: Fed. Rep. of Germany

Crude Oil:
Offshore ______
Onshore x

WAES SCENARIO CASE (1985): A, B, C, D, E

WAES SCENARIO CASE (2000): C/D

Total Resource in Place: 440×10^6 t

Domestic Oil Price: ______

Technology Assumptions: ______

	(Units)	1972	1975	1980	1985	1990	2000	2010	2025
(1) Ultimate Recoverable Resource:	10^6t		175	175	175	175	175		
(2) Discoveries:									
(3) Reserves Proved to Date:	10^6t		75	79	90	107/90	120/103		
(4) Potential Production:	10^3 BD		115	95	75	75/50	50/20		
(5) Expected Annual Production 10^3BDOE	heavy								
	light	142	115	95	75	75/50	50/30		
(6) Cumulative Production: from 1975	10^6t		6	31	53	70/68	95/88		
(7) Remaining Proved Reserves:	10^6t		69	48	37	37/27	25/15		
(8) Reserve/Production Ratio:			11.5	10	10	10	10		
(9) Host Government Take ($/unit of production):									
(10) Capital Cost ($/unit of daily capacity):									
(11) Operating Cost ($/unit of production):									

INDIGENOUS SUPPLY ASSESSMENT WORKSHEET

NATURAL GAS

Country: Fed. Rep. of Germany

WAES SCENARIO CASE (1985): A, B, C, D, E

WAES SCENARIO CASE (2000): C, D

Total Resource in Place: 615 x 10^9 m^3

Domestic Gas Price: ______

Technology Assumptions: ______

Natural Gas:
Offshore ______
Onshore x

	(Units)	1972	1975	1980	1985	1990	2000	2010	2025
(1) Ultimate Recoverable Resource:	$10^9 m^3$		615						
(2) Discoveries:									
(3) Reserves Proved to Date:	$10^9 m^3$		315	315	365	415	450	450	
	10^9 BOE		1.7	1.7	1.9	2.2	2.4	2.9	
(4) Potential Production:	$10^9 m^3/a$		20.0		20.0	20.0	9	5	
	10^6 BDOE		.29	.29	.29	.29	.13	.08	
(5) Expected Annual Production:	$10^9 m^3$	17.8	18.4	20.0	20.0	20.0	9	5	
	10^6 BDOE		.26	.29	.29	.29	.13	.08	
(6) Cumulative Production: from 1975	$10^9 m^3$		18.4	115	165	215	365	400	
	10^9 BOE		.097	.607	.871	1.14	1.93	2.11	
(7) Remaining Proved Reserves:	$10^9 m^3$		300	200	200	200	85	50	
	10^9 BOE		1.58	1.06	1.06	1.06	.45	.26	
(8) Reserve/Production Ratio:			16	10	10	10	9.4	10	
(9) Host Government Take ($/unit of production):									
(10) Capital Cost ($/unit of daily capacity):									
(11) Operating Cost ($/unit of production):									

INDIGENOUS SUPPLY ASSESSMENT WORKSHEET

COAL

Country: Fed. Rep. of Germany

WAES SCENARIO CASE (1985): A,B,C

WAES SCENARIO CASE (2000): C1/C2

Total Resource in Place: 230×10^9 t

Domestic Coal Price: ______

Technology Assumptions: ______

Coal:
Surface Mined ______
Deep Mined Hard Coal

	(Units)	1975	1980	1985	1990	2000	2010	2025
(1) Ultimate Recoverable Resource:	(10^9 t)	24.2	24.2	24.2	24.2	24.2	24.2	24.2
(2) Reserves Proved to Date:	(10^9 t)	6	6	6	6	7	8	8
	(10^9 BOE)	28.8	28.8	28.8	28.8	33.6	38.4	38.4
(3) Potential Production:	(10^6 t)	94	94	94	94	112	118	124
(4) Expected Annual Production:	(10^6 t)	92.8	94	94	94	112/94		
	(10^6 BOE)	445	451	451	451	538/451		
(5) Cumulative Production:	(10^6 BOE)	445	2685	4940	7195	12200/11705		
(6) Remaining Proved Reserves:	(10^9 t)	5.9	5.4	5.o	4.5	4.5/4.6		
	(10^9 BOE)	28.4	26.1	23.9	21.6	21.4/21.9		
(7) Reserve/Production Ratio:		63	58	53	48	40 / 49		
(8) Primary Transportation* —Capital Cost ($/unit of daily capacity): —Operating Cost ($/unit of daily capacity):								
(9) Capital Cost ($/unit of daily capacity):	($/BDOE)	7.4	8.0	8.6	9.0	9.5		
(10) Operating Cost ($/unit of production):	($/BDOE)	9.2	9.2	9.2	9.5	9.8		

*From mine mouth to first processing stage.

INDIGENOUS SUPPLY ASSESSMENT WORKSHEET

COAL

Country: Fed. Rep. of Germany

WAES SCENARIO CASE (1985): D,E

WAES SCENARIO CASE (2000): D7/D8

Total Resource in Place: 230×10^9 t

Domestic Coal Price: ______

Technology Assumptions: ______

Coal:
Surface Mined ______
Deep Mined Hard Coal

	(Units)	1975	1980	1985	1990	2000	2010	2025
(1) Ultimate Recoverable Resource:	(10^9t)	24.2	24.2	24.2	24.2	24.2	24.2	24.2
(2) Reserves Proved to Date:	(10^9t)	6	6	6	6	7	8	8
	(10^9BOE)	28.8	28.8	28.8	28.8	33.6	38.4	38.4
(3) Potential Production:	(10^6t)	94	94	94	94	112	118	124
(4) Expected Annual Production:	(10^6t)	92.8	94	80	86/80	100/80		
	(10^6BOE)	445	451	384	412/384	480/384		
(5) Cumulative Production:	(10^6BOE)	445	2685	4680	6680/6600	11100/10440		
(6) Remaining Proved Reserves:	(10^9t)	5.9	5.4	5.0	4.6/4.6	4.7/4.8		
	(10^9BOE)	28.4	26.1	24.1	22.1/22.2	22.5/23.2		
(7) Reserve/Production Ratio:		63	58	63	54/58	47/60		
(8) Primary Transportation* —Capital Cost ($/unit of daily capacity): —Operating Cost ($/unit of daily capacity):								
(9) Capital Cost ($/unit of daily capacity):	($/BDOE)	7.4	8.0	8.6	9.0	9.5		
(10) Operating Cost ($/unit of production):	($/BOE)	9.2	9.2	9.2	9.5	9.8		

*From mine mouth to first processing stage.

INDIGENOUS SUPPLY ASSESSMENT WORKSHEET

COAL

Country: Fed. Rep. of Germany

Coal:
Surface Mined Lignite
Deep Mined ______

WAES SCENARIO CASE (1985): A, B, C, D, E

WAES SCENARIO CASE (2000): C,D

Total Resource in Place: 52×10^9 t

Domestic Coal Price: ______

Technology Assumptions: ______

	(Units)	1975	1980	1985	1990	2000	2010	2025
(1) Ultimate Recoverable Resource:	10^9 t	35	35	35	35	35	35	35
(2) Reserves Proved to Date:	10^9 t	10.1	10.1	10.1	10.1	11	12	13
	10^9 BOE	12.8	12.8	12.8	12.8	14.o	15.2	16.5
(3) Potential Production:	10^6 t	130	130	132	127	128	130	130
(4) Expected Annual Production:	10^6 t	123.4	130	132	127	128		
	10^6 BOE	157	164	168	164	172	(rising calor. value)	
(5) Cumulative Production:	10^6 BOE	157	964	18o7	2647	4326		
(6) Remaining Proved Reserves:	10^9 t	10.1	9.3	8.6	8.0	7.6		
	10^9 BOE	12.8	11.8	10.9	10.2	9.7		
(7) Reserve/Production Ratio:		82	72	65	62	56		
(8) Primary Transportation* —Capital Cost ($/unit of daily capacity): —Operating Cost ($/unit of daily capacity):								
(9) Capital Cost ($/unit of daily capacity):	($/BDOE)	5.4	6.0	6.3	6.5	7.0		
(10) Operating Cost ($/unit of production):	($/BOE)	4.0	4.0	4.0	4.0	4.0		

*From mine mouth to first processing stage.

NONFOSSIL SOURCES AND CONVERSION PROCESSES WORKSHEET

NUCLEAR ELECTRIC

Country: Fed. Rep. of Germany

WAES SCENARIO CASE (1985): A, B, C, D, E

WAES SCENARIO CASE (2000): C, D

Total Resource in Place: 5000 tons

Domestic Uranium Price: ______

Technology Assumptions: ______

	(Units)	1972	1975	1980	1985	1990	2000	2010	2025
(1) Potential Production of Uranium.	t	0	0	250	250	250	250		
(2) Expected Production of Uranium*:	t	0	0	0	100	100	100		
(3) Yield of Power Plant:	%	31	31	31	31	31	31	31	
(4) Maximum Potential ~~Output~~: Capacity	GWe		3.5	15	35	63	120	150	
(5) Capital Cost ($/unit of daily capacity):	$/KWe 1)		885	885	885	885	885	885	
(6) Cumulative Capital Investment.	10^9 $		3.1	13.3	31	56	106	133	
(7) Operating Cost ($/unit of production):	¢/KWh 1)		.35	.35					

*Political potential production.

1) Estimates as of 1976 for LWR, on stream in 1981; Operating cost incluede maintenance cost.

13.1 INTRODUCTION

This report considers the future Italian production of major energy resources: fossil fuels—natural gas, oil, coal, bitumens, and asphalt—and nonfossil fuels—hydro power, geothermal power, wood, solid waste (for electricity generation), and uranium. The potential for utilizing solar energy in space heating, cooling, and sanitary water heating is analyzed in *Energy Supply-Demand Integrations to the Year 2000: Global and National Studies* (Third Technical Report of the Workshop on Alternative Energy Strategies, MIT Press, 1977).

The results are summarized in the worksheets at the end of this chapter. Most of the detailed calculations and assumptions concerning the supply estimates are footnoted on the worksheets; only the general framework for each source is indicated in the text.

13.2 INFLUENCE OF SOCIOECONOMIC SCENARIOS UPON ITALIAN DOMESTIC SUPPLY ALTERNATIVES

The Italian economy produced approximately 16 percent of its primary energy requirements in 1975. Because Italy's energy self-sufficiency is very low, vigorous development of the limited domestic energy sources poses no great conflict for the allocation of funds and is the implicit strategy in many resource areas. To this extent the development of Italian domestic supplies is independent of the scenario.

However, nuclear power is an important exception to this general trend. The expansion of nuclear power is a basic energy policy decision in the WAES scenarios for 2000. This energy conversion system is not limited by the wealth of mineral resources and is highly capital intensive. Moreover, in contrast to other resources, which are relatively independent of the demand for energy, the amount of nuclear power likely to be installed depends on long-term economic growth rates and the national energy policy (choice of energy mix and extent of savings) adopted.

To a certain extent all the nonnuclear resources may suffer a reduction in funding under a "vigorous nuclear" scenario; however, since many of these domestic sources are themselves quite limited, the possible differences in development are likely to be seen only in the short to intermediate term, with approximately equal development by the 1990s.

In general these supply scenarios are "surprise-free." That is, the possibility of exceptional resource discoveries or the introduction of technological breakthroughs (such as nuclear fusion) have not been considered. In any case this assumption appears reasonable because for most energy sources the lead time from the demonstration of a scientific discovery to its commercialization on a wide scale must be measured in decades.

13.3 FOSSIL FUELS

13.3.1 Natural Gas and Petroleum

Two methods of estimating future oil and gas discoveries and production have been used. The first method, a financial or investment approach (see notes on gas and oil worksheets), was used to forecast a production minimum and maximum through 1985; the second method, incorporating specific geologic hypotheses of new discoveries, provided estimates for 1985 and year 2000. For natural gas the second technique gives 250.0×10^9 m^3 of new discoveries between 1977 and year 2000, slightly more than the 196.0×10^9 m^3 of known reserves in 1975. For petroleum a similar hypothesis leads to additions of 40.0×10^6 tons from 1976 to year 2000, compared to 45.0×10^6 tons existing reserves in 1975. (For comparison, the Malossa field has estimated reserves of 40×10^6 tons of oil and 50×10^9 m^3 of natural gas.)

The two methods give comparable results for the year 1985 (see Table 13.1). The geologic hypotheses are used in the WAES worksheets since they provide continuity to 2000. The supply estimates are also consistent with current plans for hydrocarbon exploration and development.

Recently (1976) a field of approximately 100 ×

The group working on this report included Azienda Municipale Nettezza Urbana (Milano), Azienda Servizi Municipalizzati (Brescia), CISE, EGAM, ENI, IEFE, Istituto Internazionale per le Ricerche Geotermiche, Montedison, SNAM, Università di Napoli.

10^6 tons of heavy oil was discovered offshore in the Adriatic Sea near the region of the Marches. The technology required for production is under study; however, it is not certain that costs will permit production under WAES price assumptions, and these reserves are not included in the worksheets.

In addition to the domestic supply of natural gas, Italy has secured four long-term contracts for importation from the U.S.S.R., Libya, The Netherlands, and Algeria. These contracts should be fully operational by 1985 (the first three by 1977–1978) providing for the level of imports shown in Table 13.2. The total available supply of natural gas in 1985 is estimated to be approximately 43.0 × 10^9 m^3 or 35.5 × 10^6 TOE. If Algerian gas is liquefied and shipped to Italy, instead of being transported by pipeline, approximately 25 percent of the gas will be consumed in liquefaction.

Estimates given in the supply study are based upon planned exploration expenditures and geologic expectations. If, in the context of oil prices above $17.25/bbl or "extra vigorous" energy policy, the exploration and development expenditures are substantially increased, the likelihood of finding more oil and gas increases.

For example, let us assume that in the next fifteen years an additional amount of money is annually invested equal to the capital investment in 1974 in oil and gas exploration and development ($90 million). In this case an additional productive capacity from 75,000 to 150,000 BDOE is possible. (An average exploration and development capital cost, from well to refinery sites, between $9000 and $10,000 per BDOE is used, consistent with European offshore oil and gas costs estimated by WAES and others.) This corresponds to maximum annual production from 3.7 to 7.4 × 10^6 TOE, as compared to the 1972 oil and gas production of 12 × 10^6 TOE.

The additional hydrocarbons are assumed to be Italian property, but not necessarily discovered within Italy. However, after the discovery of the deep oil- and gas-bearing sediments in connection with the Malossa field, the possibility that all the additional production could be of Italian origin is not to be excluded.

With prices higher than $17.25/bbl it is also likely that the heavy oil field discovered in the Adriatic near the Marches would be produced. The recoverable reserves are estimated at 100 × 10^6 TOE.

13.3.2 COAL

There are relatively small quantities of coal in Italy, and they are of the low-heat-content lignite type. Most of the lignite is located in the Sulcis area in Sardinia and has a high sulfur content (6 percent); a small amount of low-sulfur (∿ 1 percent) reserves are located in the mainland.

Recently the estimates of lignite reserves at Sulcis have been revised upward from 40.0 to 150.0 × 10^6 tons of proven reserves (from 17.6 to 66 × 10^6 TOE) with 132.0 to 176.0 × 10^6 TOE of potential resources (reference 1, p. 56). However, because of the high sulfur content (equivalent to 16 percent sulfur in fuel oil) and given the low heat content (4400 kcal/kg) and the uncertain future extraction costs, there is considerable controversy over future production. The Sulcis production, 200,000 tons (88,000 TOE) per year, was suspended in 1972 due to high costs, and Sulcis lignite is currently not considered competitive with imported fuel oil (reference 2, p. 34).

Despite these uncertainties there are plans to reopen the Sulcis mine, bringing capacity up to 1 to 2 × 10^6 tons per year (reference 1, p. 57). It is estimated that up to 2 × 10^6 tons, or 880,000

Table 13.1
Estimated natural gas and petroleum production

	Geologic Hypothesis (10^6 TOE)	Financial-Investment Hypothesis (10^6 TOE)	
		Minimum	Maximum
Natural Gas			
1985 production	14.0	5.8	15.4
2000 production	7.2	—	—
Petroleum			
1985 production	3.0	1.4	3.8
2000 production	1.7	—	—

Table 13.2
Importation of natural gas

Import Country	Natural Gas ($10^9 m^3$)	(10^6 TOE)
Libya	3.0	2.5
The Netherlands	6.0	4.9
U.S.S.R.	7.0	5.8
Algeria (4) and others	10.0	8.3
Total	26.0	21.5

TOE, per year can be consumed locally at the power plant at Port Vesme and at the nearby aluminum plant, thereby minimizing transportation costs. With mechanization of the Sulcis mine, production costs have been estimated at about $21 (1975 U.S.$) per ton, sufficiently low to be produced under all WAES price scenarios. In addition, initial government funding has been approved for reopening the mine. Two million tons annual capacity, the amount that can be utilized locally, is assumed as the production for all WAES cases in 1985. Only under the "high coal" scenario for year 2000 (Cases C-1 and D-7) is capacity expanded beyond this level. Under these conditions with proven reserves of 150.0×10^6 tons, production is likely to reach 4.0×10^6 tons per year.

13.3.3 Bitumens and Heavy Asphalts

In the central zone of Sicily near Caltanissetta are outcroppings of the Tripoli formation, a diatomite of deep Miocene origin, that is occasionally accompanied by layers of bitumen. The bitumen layers are quite thin, from 20 centimeters to several meters, and highly irregular, following the tectonic structures. They are located in a mountainous zone and covered by considerable overburden in addition to being in frequent contact with chalky sulfurous faces. Because of the thinness and irregularity of the bitumens and the presence of sulfurous material, an in situ process does not appear feasible. Conventional mining would be extremely costly because of the extensive overburden relative to the small quantities of bitumens. Production is not anticipated even at petroleum prices in the range $20 to $25 (1975 U.S.$) per barrel oil equivalent.

In 1953 an estimate of recoverable reserves of the Tripoli bitumens was made without considering these geological and technical production limits, resulting in a gross misrepresentation of the situation (reference 4). Unfortunately this estimate of recoverable reserves has been uncritically reported in several international publications, giving the impression of substantial reserves on the order of 10^9 TOE (reference 2, p. 139).

Near the town of Ragusa, in another part of Sicily, is a deposit of limestone with bitumens. Total reserves are estimated at 10^7 TOE. The concentration, however, is quite low (about 4 percent), and the limestone is mined as the primary product with the bitumens and heavy asphalt being the by-products. All of this asphalt is used for road construction. Under WAES energy price assumptions, production for energy use is unlikely.

Finally, there are deposits of asphalt in the Abruzzi region, but their quantity is negligible.

13.4 NONFOSSIL FUEL SOURCES

13.4.1 Hydro power

It has been estimated that hydro power is economically and technically feasible for generating an additional 4 to 5×10^9 kWhe, thus bringing the 1975 total of 44.7×10^9 kWhe up to a maximum of 50.0×10^9 kWhe. Using the 1975 ratio of kWh to hydroelectric capacity (44.7×10^9 kWhe to 15,000 MWe), this results in an additional capacity of 1700 MWe at the winter peak. This figure has been used as additional capacity by 1985 in all scenario cases. By year 2000, in the strong diversification or high coal scenarios, an additional 5×10^9 kWhe is added, bringing the total production to 55.0×10^9 kWhe.

In the case of higher energy prices or more vigorous policy, it may be possible to slightly increase hydroelectric production. The total existing and maximum technically feasible is estimated to be 64.6×10^9 kWhe (reference 4, p. 4). However, production over 64.6×10^9 kWhe is difficult to imagine because of the large economic and social costs involved. For example, many potential dam sites are unacceptable because (1) local topography requires a very large area reservoir, increasing evaporation and capital costs; (2) communities are located within the site area; (3) the land slopes too gently with no great difference in height (if water is piped to a point where there is a significant difference in height, large losses in friction may be incurred); or (4) the capacity of the site is limited by the capacity of the downstream system which can accept oscillations in the water level only within certain limits, except at the huge expense of modifying the entire downstream river system. Even if a dam is constructed, there is increasing competition for use of the water for irrigation and urban purposes.

On the other hand, substantial improvements in

the hydrogeologic resources, primarily in terms of better erosion control, would result in more gradual and useful drainage; they would also reduce the amount of solid matter present in the water, which alone could increase hydroelectric power by 2 to 4 percent. A comprehensive study is required to assess the possible increases in hydroelectric power from such an improvement program.

13.4.2 Pumped Storage

Although not strictly an energy source, pumped storage is included here since this form of energy storage permits a reduction in conventional capacity, thereby changing the allocation of the supply and utilization of the electricity generation system.

The capacity for adding pumped storage is quite significant. In addition to the 2500 MWe already installed, 17,500 MWe of additional pumped storage is estimated as economically and technically feasible. (The *Programma Energetico Nazionale* includes plans to add 4500 MW between 1976 and 1983. Between 1983 and 1991, ENEL intends to construct another 4000 MW. Beyond 1991, 9000 MW are estimated as feasible, and an additional capacity of 500 MW/year is given as a possible policy, bringing the total additional capacity to 17,500 MW by 2010.) This storage, together with other forms of delivering peak power such as gas turbine generators, is used to cover fluctuations in the demand; normally it will not exceed 10–12 percent of the total installed capacity of the system. Accordingly, in the supply-demand integration it is added in proportion to the electricity demand growth; the worksheets presented in this study include only the maximum estimated capacity (Case E).

13.4.3 Geothermal Energy

A distinction is made between vigorous and restrained resource development. The restrained and high nuclear cases (D, E, C-2, D-8) assume that the capacity of existing geothermal fields will be expanded from 417 MWe in 1976 to 550 MWe in 1985 and that this capacity will be maintained through 2000. New discoveries are not expected to be very large—200 MWe new field capacity by 1985 and 150 MWe more by year 2000, for a total installed capacity of 750 and 900 MWe, respectively.

Nonelectric use of geothermal energy, currently 1800 TOE, is hypothesized to increase to 8600 TOE in 1985 and to 172,000 TOE by 2000. Possible nonelectric use of geothermal energy includes utilization in agriculture, animal husbandry, aquaculture, space heating, recreational and health facilities, and light industry.

In the vigorous energy policy scenarios through 1985 and in the strong diversification of coal for 2000, existing fields are assumed to be further developed and expanded to 800 MWe in 1985 and to 1500 MWe in 2000. Additions from new fields are estimated at 300 MWe in 1985 and another 200 MWe in 2000, a total of 1100 and 2000 MWe, respectively. A demand of 0.034×10^6 TOE is hypothesized for nonelectrical uses of geothermal energy in 1985, and 1.2×10^6 TOE in 2000.

In the case of an "extra vigorous" energy policy, more research and development on alternative energy forms, such as hot dry rocks, is expected. If a method for utilizing the energy in hot dry rocks is developed within the next fifteen to twenty years, exploitation by 2000 is likely in Italy, given the wide distribution of sites with abnormally high geothermal gradients. In the year 2000 an electrical capacity of as much as 1500 MWe from hot dry rocks could then be assumed. This figure indicates only a possible beginning of the development of this source, since even assuming the first demonstration of commercial feasibility by 1990, considerable time and effort will be needed for the exploration, drilling, fracturing, and integration of hot rock power into the existing electrical grid. The development of an economical means to utilize the heat trapped in rocks and sediments (less than 10 km deep) could substantially alter the long-term Italian supply situation, and sustained research efforts should therefore be made in this field. For example, if it is assumed that 0.1 percent of Italian territory is underlayed by hot rock with a producible rock sediment of 1 km, an average temperature of 400°C, a minimum usable temperature of 200°C, a heat capacity of 0.20 cal/°C/g, and an average rock density of 2.65, then the volume of rock is $0.3 \times 10^6\ \text{km}^2 \times 0.001 \times 1\ \text{km} = 3.0 \times 10^{17}\ \text{cm}^3$, and the total usable heat is 0.33×10^{17} kcal. With a 10 percent efficiency for conversion of total usable heat to electricity, this represents 3700×10^9 kWhe, equal to 30 times the 1972 demand.

13.4.4 Wood

The production of wood used for fuel has remained stable since 1969; its average output is 0.5×10^6 TOE per year. Production is expected to decline because of the trend to increase the production of trees for lumber and pulp, uses of wood that are more profitable than energy production.

13.4.5 Uranium

Uranium reserves are modest; total proven reserves are given as 2500 tons of U_3O_8 at prices up to $30/lb. There are plans to mine uranium oxide near Bergamo at the rate of 150 tons per year beginning in 1980. An additional deposit has been identified in the high Val Seriana area and is currently under evaluation. An increase in the proven reserves of 1000 tons has been hypothesized for the period between 1986 and 2000.

In the upper Lazio region, north of Rome, a fairly extensive but low concentration (200 to 500 ppm) deposit of uranium is being evaluated. Preliminary and partial estimates of reserves have been set at 600 tons of uranium (reference 5, p. 49). However, this deposit is not included in the estimate of resources.

13.4.6 Urban Solid Waste

Milan and Genoa generate part of their electricity from municipal solid waste, without treatment or sorting. Current capacity is 25 MWe, and it could be increased to 100 MWe by 1985. This estimate is based on the known historic relation between income and per capita solid urban waste production in Italy. Transportation and handling costs for collection of solid waste limit consideration of the technology to cities with more than 200,000 inhabitants. At lower per capita incomes, solid waste is not self-combusting because of its lower cellulose content and higher proportion of vegetable matter. For this reason the computation to 1985 was extended to cities in northern and central Italy. By the year 2000 it is expected that per capita income will increase sufficiently to permit the technology to spread to large parts of southern Italy as well. The capacity estimates for year 2000 are 360 MWe and 500 MWe for the low and high economic growth scenarios, respectively.

13.5 NUCLEAR POWER GENERATION

Only the nuclear part of the electricity supply is indicated here; a complete discussion of electricity supply will be found in the supply-demand integration report.

13.5.1 Nuclear Power Plants in Operation by 1985

As in other countries there is considerable debate concerning the Italian nuclear power program. A realistic estimate of maximum additional nuclear capacity for 1985 is 10,000 MWe. This corresponds to what remains of total additional capacity required to meet the maximum projected demand after subtracting (1) the new capacity for hydroelectric, geothermal, and pumping plants and (2) the capacity of conventional thermoelectric plants already under construction. (The maximum demand occurs in Case E. In all cases a ratio of reserve capacity to power required at the winter peak of 1 to 4 is used. For complete details see the supply-demand integration report.) Because of the long lead time needed for the construction of nuclear power plants, and to avoid the risk of having insufficient installed capacity, 10,00 MWe has been hypothesized for all cases in 1985.

13.5.2 Nuclear Power Plants in 2000

The maximum nuclear capacity, achieved under the assumptions of Case C-2, is hypothesized as 65,000 MWe. If 25,000 MWe are operational by 1990, this policy implies putting into service an average of four 1000 MWe nuclear plants each year from 1990 to 2000. This means having approximately twenty-four nuclear power plants in various stages of construction each year from 1990 through 1995 and is considered to be what the Italian nuclear power industry can reasonably manage. In the low economic growth/vigorous nuclear and the high economic growth/vigorous coal scenarios, the nuclear program is kept at substantial levels (40,000 and 30,000 MWe, respectively), ensuring more than enough nuclear construction for industrial "take-off" and possible exportation of components.

13.6 REFERENCES

1. *Programma Energetico Nazionale,* 1975.

2. *World Survey of Energy Resources 1974,* U.S. National Committee of the World Energy Conference.

3. "Tripoli Bituminosi di Sicilia." *Rivista Mineraria Siciliana* Nr. 22/23, Luglio-Ottobre 1953.

4. Pier Antonio Rolla and Francesco Veronese. "Indagine sulla Risorse Idrolettriche Italiane." *Quaderni dell'Istituto di Ricerca sulle Acque* Nr. 7, 1973.

5. Locardi Enzo. Notizario del CNEN, anno 20, nr. 7, July 1974.

13.7 ENERGY SUPPLY WORKSHEETS

The following worksheets are included (see Section 1.2.1 for definition of terms):

Energy supply summary

Natural gas

Coal (2)

Nuclear electric

Crude oil

Hydroelectric (2)

Geothermal (2)

Uranium and nuclear

Urban solid waste

Wood

WAES NATIONAL ENERGY SUPPLY STUDIES, 1972-2000
ENERGY SUPPLY SUMMARY

Country: Italy

Scenario Assumptions	1972	1985					2000			
WAES CASE	Base Year	A	B	C	D	E	C-1	C-2	D-7	D-8
GNP 1976-85; 1985-2000		High	Low	High	Low	High	High	High	Low	Low
Oil/Energy Price*		11.50-17.25	11.50-17.25	11.50	11.50	11.50-7.66	11.50-17.25	11.50-17.25	11.50	11.50
National Policy Response		Vig	Vig	Vig	Res	Res	Vig	Vig	Vig	Vig
Principal Replacement Fuel							Coal	Nuclear	Coal	Nuclear
Energy Supply **Units**		10^6 tons equivalent								
Crude Oil	1.1	2.9	2.9	2.9	2.9	2.9	1.7	1.7	1.7	1.7
Natural Gas	11.0	14.0	14.0	14.0	14.0	14.0	7.2	7.2	7.2	7.2
Coal	0.7	0.9	0.9	0.9	0.9	0.9	1.8	0.9	1.8	0.9
Nuclear (1)	0.8	14.4	14.4	14.4	14.4	14.4	43.5	93.9	29.0	57.9
Hydro-electric (1)	9.5	11.2	11.2	11.2	11.2	11.2	12.2	11.2	12.2	11.2
Geothermal (1)	0.6	1.6	1.6	1.6	1.1	1.1	2.9	1.4	2.9	1.4
Solar Thermal	0	1.5	1.2	1.0	0.7	0.2	10.5	5.3	8.4	4.2
Solar Electric	0	0	0	0	0	0	0	0	0	0
Miscellaneous(2)	0.5	0.6	0.5	0.6	0.5	0.6	1.8	1.1	1.6	0.9
Total Energy Supply	24.2	47.1	46.7	46.7	45.7	45.3	81.6	122.7	64.8	85.4

*1975 US dollars per barrel of Arabian light crude oil, fob Persian Gulf.
All Figures are primary production prior to any processing or conversion.

(1) Given in terms oil equivalent necessary to produce the same quantity of electricity by conventional thermoelectric plants, using 2260 Kcal/kWh.

(2) Includes wood, urban solid wastes (for electricity) and geothermal heat (for non-electrical uses).

INDIGENOUS SUPPLY ASSESSMENT WORKSHEET

COAL

Country: Italy

Coal:
Surface Mined ______
Deep Mined X

WAES SCENARIO CASE (1985): All Cases

WAES SCENARIO CASE (2000): C_1 and D_7

Total Resource in Place: -

Domestic Coal Price: Equal to international coal price

Technology Assumptions: Modernization and mechanization of the Sulcis mine

	1972	1975	1980	1985	1990	2000	2010	2025
(1) Ultimate Recoverable Resource: 10^6 ton	110 [1)]	400 [2)]	400	400	400	400	400	
(2) Reserves Proved to Date: 10^6 ton	33 [1)]	150 [2)]	150	150	150	150	150	
(3) Potential Production:	-	-	-	-	-	-	-	
(4) Expected Annual Production: 10^6 ton	1.5 [3)]	0	0	2.0	4.0	4.0	4.0	
(5) Cumulative Production: 10^6 ton	0	0	0	0	10.0	50.0	90.0	
(6) Remaining Proved Reserves: 10^6 ton	31.5	150.0	150.0	150.0	140.0	100.0	60.0	
(7) Reserve/Production Ratio:	21			75	35	25	23	
(8) Primary Transportation* —Capital Cost ($/unit of daily capacity):				N.A.				
—Operating Cost ($/unit of daily capacity):				N.A.				
(9) Capital Cost ($/unit of daily capacity):				N.A.				
(10) Operating Cost $/toe ($/unit of production):				~50 [4)]				

*From mine mouth to first processing stage.

(1) Survey of World Energy Resources 1974 p. 76
(2) Revised estimates given in Programma Energetico Nazionale p. 56
(3) Sulcis mine was closed mid 1972
(4) Estimates given by EGAM ($ 1975)

INDIGENOUS SUPPLY ASSESSMENT WORKSHEET

COAL

Country: Italy

Coal:
Surface Mined ______
Deep Mined x

WAES SCENARIO CASE (1985): All Cases

WAES SCENARIO CASE (2000): C_2 and D_8

Total Resource in Place: -

Domestic Coal Price: Equal to international coal price

Technology Assumptions: Modernization and mechanization of the Sulcis mine

	1972	1975	1980	1985	1990	2000	2010	2025
(1) Ultimate Recoverable Resource: 10^6 ton	110 [1)]	400 [2)]	400	400	400	400	400	
(2) Reserves Proved to Date: 10^6 ton	33 [1)]	150 [2)]	150	150	150	150	150	
(3) Potential Production:	-	-	-	-	-	-	-	
(4) Expected Annual Production: 10^6 ton	1.5 [3)]	0	0	2.0	2.0	2.0	2.0	
(5) Cumulative Production: 10^6 ton	0	0	0	0	10.0	30.0	50.0	
(6) Remaining Proved Reserves: 10^6 ton	31.5	150.0	150.0	150.0	140.0	120.0	100.0	
(7) Reserve/Production Ratio:	21			75	70	60	50	
(8) Primary Transportation* —Capital Cost ($/unit of daily capacity):				N.A.				
—Operating Cost ($/unit of daily capacity):				N.A.				
(9) Capital Cost ($/unit of daily capacity):				N.A.				
(10) Operating Cost \$/toe ($/unit of production):				~ 50 [4)]				

*From mine mouth to first processing stage.

(1) Survey of World Energy Resources 1974 p. 76
(2) Revised estimates given in Programma Energetico Nazionale p. 56
(3) Sulcis mine was closed mid 1972
(4) Estimates given by EGAM (\$ 1975)

NONFOSSIL SOURCES AND CONVERSION PROCESSES WORKSHEET

NUCLEAR ELECTRIC

Country: Italy

WAES SCENARIO CASE (1985): All Cases

WAES SCENARIO CASE (2000): C_2 (Other cases footnoted)

Total Resource in Place: -

Domestic Uranium Price: -

Technology Assumptions: -

	(Units)	1972	1975	1980	1985	1990	2000	2010	2025
(1) Potential Production of Uranium.									
(2) Expected Production of Uranium*:									
(3) Yield of Power Plant:									
(4) Maximum Potential Output: (Capacity)	GW_e	0.6	0.6	1.4	10.6	~25.0	65.0[1)]		
(5) Capital Cost 2) ($/unit of daily capacity):	$/$kW_e$		600		>600		>600		
(6) Cumulative Capital Investment.	$ 10^6		0		>6,000		>38,700		
(7) Operating Cost 2) ($/unit of production):	Mils/kWh_e		6		>6		>6		

*Political potential production.

(1) See supply demand integration report for complete description of nuclear program. The nuclear capacity for cases C_1, D_7 and D_8 is respectively : 30, 20, and 40 GW_e.

(2) Costs as given in Programma Energetico Nazionale ($ 1975)

INDIGENOUS SUPPLY ASSESSMENT WORKSHEET

NATURAL GAS

Country: Italy

Natural Gas:
Offshore x
Onshore x

WAES SCENARIO CASE (1985): All Cases

WAES SCENARIO CASE (2000): All Cases

Total Resource in Place: -

Domestic Gas Price: (1)

Technology Assumptions: Maximum development, in keeping with field conservation practices.

	(Units)	1972	1975	1980	1985	1990	2000	2010	2025
(1) Ultimate Recoverable Resource:		-	-	-	-	-	-	-	-
(2) Discoveries: (2)	10^9 m^3	-	0			250.0		-	-
(3) Reserves Proved to Date: (2)	10^9 m^3	137.8	196.0		356.0		446.0		
(4) Potential Production:	10^9 m^3	-	-	-	-	-	-	-	-
(5) Expected Annual Production:	10^9 m^3	13.3	14.5		17.0		8.7		
(6) Cumulative Production:	10^9 m^3		14.5		151.0		341.0		
(7) Remaining Proved Reserves:	10^9 m^3	137.8	181.5		205.0		105.0		
(8) Reserve/Production Ratio:		10.4	12.5	12.0	12.0	12.0	12.0		
(9) Host Government Take ($/unit of production): (3)	Mils/ m^3		3.0						
(10) Capital Cost ($/unit of daily capacity):	$ 10^3/ bdoe		8.0	>8.0	>8.0	>8.0	>8.0		
(11) Operating Cost (4) ($/unit of production):	Mils/ m^3		7.0	7.0	7.0	>7.0	>7.0		

(1) In 1975 the natural gas prices were : 18.74 Lit/m^3 for domestic use, 39.76 Lit/m^3 for industrial use and 51.60 Lit/m^3 for use by utilities. The average price was approximately 34.0 Lit/m^3 or

---

70% of the value of petroleum. In 1985 with approximately two thirds of the natural gas being supplied from imports, the average price is expected to be approximately that of an equivalent amount of crude oil at international prices.

(2) Given the new areas of hydrocarbon exploration, deep onshore and offshore sediments in connection with the Malossa field, it is difficult to predict the future discovery rate. The 250 x 10^9 cubic meters of new discoveries hypothesized for the 1977-2000 period can be considered an optimistic result of exploration efforts. Most of this, 160 x 10^9 cubic meters is assumed to be discovered in the next decade, in the worksheet.

(3) There is an Italian government royalty of 9% of the value for onshore gas and 5% for offshore gas. Assuming a future ratio of 60% offshore and 40% onshore,results in an average royalty of 6,6%. Only the value for 1975 is shown.

(4) Operating costs are given as 5 Lire/m^3 equal to 6.9 mils per cubic meter at 720 Lire/$.

INDIGENOUS SUPPLY ASSESSMENT WORKSHEET

CRUDE OIL

Country: Italy

Crude Oil:
Offshore x
Onshore x

WAES SCENARIO CASE (1985): All Cases

WAES SCENARIO CASE (2000): All Cases

Total Resource in Place: -

Domestic Oil Price: WAES Petroleum Prices

Technology Assumptions: Maximum development, in keeping with field conservation practices.

	(Units)	1972	1975	1980	1985	1990	2000	2010	2025
(1) Ultimate Recoverable Resource:									
(2) Discoveries: (1)	10^6 ton		40.0 (1975–2000)						
(3) Reserves Proved to Date: (1)	10^6 ton		45.0		65.0	-	85.0		
(4) Potential Production:									
(5) Expected Annual Production	heavy	1.0	1.0		0.6		0.3		
	light	0.1	0.1		2.3		1.4		
(6) Cumulative Production: (2)	10^6 ton		3.5		31.0	-	64.4		
(7) Remaining Proved Reserves:	10^6 ton		41.5		34.0		20.6		
(8) Reserve/Production Ratio:			37.4	12.0	12.0	12.0	12.0		
(9) Host Government Take ($/unit of production): (4)									
(10) Capital Cost ($/unit of daily capacity):	\$ 10^3 bdoe		8.0	>8.0	>8.0	>8.0	>8.0		
	\$ 10^3 ton/year		400.0	>400.0	>400.0	>400.0	>400.0		
(11) Operating Cost (5) ($/unit of production):	\$/bbl.		1.0	>1.0	>1.0	>1.0	>1.0		
	\$/ton		7.0	>7.0	>7.0	>7.0	>7.0		

(1) Given the new areas of hydrocarbon exploration, deep onshore and offshore sediments in connection with the Malossa field, it is quite difficult to predict the future discovery and production

rates. The 40.0 x 10^6 tons of oil hypothesized for the 1975-2000 period, is approximately equivalent to the remaining proved reserves to date and can be considered an optimistic result of exploration efforts.

(2) Purely for indicative purposes the production rates for years 1985 and 2000 are calculated assuming the above discovery is uniformly distributed over the period.

(3) This change is due to bringing on stream of the Malossa field, heavy petroleum remaining only in Sicily.

(4) 8.5% of value of crude.

(5) Operating costs are given as 5000 Lire/ton or $ 7.00/ton.

NONFOSSIL SOURCES AND CONVERSION PROCESS WORKSHEET

Hydroelectric x
Coal Gasification
Coal to Syncrude
Gas Liquifaction
Regasification
Other* (specify):

Country: Italy

WAES SCENARIO CASE (1985): All Cases

WAES SCENARIO CASE (2000): C_1 and D_7

Total Resource in Place: 341×10^9 kWh_e, gross theoretical capacity 1)

Technology Assumptions: -

	(Units)	1972	1975	1985	1990	2000	2010	2025
(1) Yield:	%	78	78	~78		~78		
(2) Maximum Potential Output 2)	10^9 kWh_e	44	44.7	50	52	55		
(3) Capital Cost ($/unit of daily capacity):	$/kW	-	340	>340	>340	>340		
(4) Cumulative Capital Investment:	$ 10^6			840	1180	1680		
(5) Operating Cost ($/unit of production):	Mils/kWh_e	-	2	>2	>2	>2		

*Specify other relevant non-fossil energy source or conversion process, i.e., geothermal, solar, wind, etc.

(1) <u>Survey of World Energy Resources</u> 1974

(2) Maximum production, technically feasible is estimated as 65×10^9 kWh_e

(3) Estimates of average capital cost for 1972-1982 period, provided by ENEL

NONFOSSIL SOURCES AND CONVERSION PROCESS WORKSHEET

Hydroelectric X
Coal Gasification
Coal to Syncrude
Gas Liquifaction
Regasification
Other* (specify):

Country: Italy

WAES SCENARIO CASE (1985): All Cases

WAES SCENARIO CASE (2000): C_2 and D_8

Total Resource in Place: 341×10^9 kWh_e, gross theoretical capacity 1)

Technology Assumptions: -

	(Units)	1972	1975	1985	1990	2000	2010	2025
(1) Yield:	%	78	78	~78		~78		
(2) Maximum Potential Output 2)	10^9 kWh_e	44	44.7	50	50	50		
(3) Capital Cost ($/unit of daily capacity):	$/kW	-	340	>340	>340	>340		
(4) Cumulative Capital Investment:	$ 10^6			>840	>840	>840		
(5) Operating Cost ($/unit of production):	Mils/kWh_e	-	2	2	>2	>2		

*Specify other relevant non-fossil energy source or conversion process, i.e., geothermal, solar, wind, etc.

(1) Survey of World Energy Resources 1974

(2) Maximum production, technically feasible is estimated as 65 x 10^9 kWh_e

(3) Estimates of average capital cost for 1972-1982 period, provided by ENEL

NONFOSSIL SOURCES AND CONVERSION PROCESS WORKSHEET

Hydroelectric ______
Coal Gasification ______
Coal to Syncrude ______
Gas Liquifaction ______
Regasification ______
Other* (specify): Geothermal Power

Country: Italy

WAES SCENARIO CASE (1985): A,B,C

WAES SCENARIO CASE (2000): C_1, D_7

Total Resource in Place: -

Technology Assumptions: -

	(Units)	1972	1975	1980	1985	2000	2010	2025
(1) Yield:	%	11	11		11	11		
(2) Maximum Potential 1) Output (Capacity)	MW_e	395	408		1100	2000		
(3) Capital Cost 2) ($/unit of daily capacity):	$\$/kW_e$	-	180		>180	>180		
(4) Cumulative Capital Investment:	$\$ 10^6$	-			>125	>290		
(5) Operating Cost 3) ($/unit of production):	$Mils/kWh_e$	-	8		>8	>8		

*Specify other relevant non-fossil energy source or conversion process, i.e., geothermal, solar, wind, etc.

(6) Non-electric uses	10^6 toe	0	~0		~0	0.9		

(1) Estimates provided by E. Barbier, M. Fanelli and L. Taffi of International Institute for Geothermal Research, Pisa. Above capacities exclude possible contribution for geothermal hot rocks, which is given in the supply demand integration report.

(2) Includes power plant costs only. Di Mario and Leardini, *Acqua-Aria*, October 1974 p. 546.

(3) Refers to a plant with atmospheric discharge.

NONFOSSIL SOURCES AND CONVERSION PROCESS WORKSHEET

Hydroelectric ______
Coal Gasification ______
Coal to Syncrude ______
Gas Liquifaction ______
Regasification ______
Other* (specify): Geothermal Power

Country: Italy

WAES SCENARIO CASE (1985): D, E

WAES SCENARIO CASE (2000): C_2, D_8

Total Resource in Place: -

Technology Assumptions: -

	(Units)	1972	1975	1980	1985	2000	2010	2025
(1) Yield:	%	11	11		11	11		
(2) Maximum Potential 1) Output (Capacity)	MW_e	395	408		750	900		
(3) Capital Cost 2) ($/unit of daily capacity):	$/$kW_e$	-	180		>180	>180		
(4) Cumulative Capital Investment:	$ 10^6	-			>62	>88		
(5) Operating Cost 3) ($/unit of production):	Mils/kWh_e	-	8		>8	>8		

*Specify other relevant non-fossil energy source or conversion process, i.e., geothermal, solar, wind, etc.

(6) Non-electric uses	10^6 toe	0	~0		~0	0.2		

(1) Estimates provided by E. Barbier, M. Fanelli and L. Taffi of International Institute for Geothermal Research, Pisa. Above capacities exclude possible contribution for geothermal hot rocks, which is given in the supply demand integration report.

(2) Includes power plant costs only. Di Mario and Leardini, _Acqua-Aria_, October 1974 p. 546.

(3) Refers to a plant with atmospheric discharge.

NONFOSSIL SOURCES AND CONVERSION PROCESSES WORKSHEET

URANIUM AND NUCLEAR

Country: Italy

WAES SCENARIO CASE (1985): All Cases

WAES SCENARIO CASE (2000): All Cases

Total Resource in Place: -

Domestic Uranium Price: $ 15-30/lb.

Technology Assumptions: -

	(Units)	1972	1975	1980	1985	1990	1995	2000	2010
1. Ultimate Recoverable Uranium Resource:		-	-	-	-	-	-	-	
2. Reserve of Uranium Proved to Date:			1500	1500 1)	2500 2)	2500	2500	2500	
3. Potential Production of Uranium:		-	-	-	-	-	-	-	
4. Expected Production of Uranium:		0	0	150 1)	250 2)	100	100	0	
5. Reserve/Production Ratio:									
6. Maximum Installed Capacity (GWe)*:	(See Nuclear Worksheets)								
7. Capital Cost ($/KWe):									
8. Operating Cost (¢/Kwhr)**:									

*Indicate assumed load factor and LWR or HWR.

**Excluding capital charges.

(1) The Val Seriana deposit in the Novazza area (near Bergamo) of 1500 tons U_3O_8 is to be brought into production in 1979, producing 150 t/yr through 1989.

(2) It is possible that a second deposit in the Novazza area, the Val Rendena site, will go into production about 1985 at 100 t/yr. Current estimates of reserves, still undergoing evaluation, are given

---

as 1000 tons U_3O_8. In addition to the Novazza deposit, there is an extensive, low concentration deposit of volcanic uranium, typically in the form of carnotite, in the northern Lazio region. The concentration is in the range of 200 to 450 parts per million. A strongly radioactive zone has been discovered in Sardenia and geologic surveys are underway. Because of the preliminary nature of these surveys no hypothesis of reserves or production is made.

NONFOSSIL SOURCES AND CONVERSION PROCESS WORKSHEET

Hydroelectric ______
Coal Gasification ______
Coal to Syncrude ______
Gas Liquifaction ______
Regasification ______
Other* (specify): Urban Solid Waste (for electricity generation)

Country: Italy

WAES SCENARIO CASE (1985): All Cases

WAES SCENARIO CASE (2000): C_1 and C_2 (D_7 and D_8 footnoted)

Total Resource in Place: -

Technology Assumptions: -

	(Units)	1972	1975	1985	1990	2000	2010	2025
(1) Yield:								
(2) Maximum Potential Output (Capacity)	MW_e		25	100		500[1)]		
(3) Capital Cost 2) ($/unit of daily capacity):	$/$kW_e$		110	>110		>110		
(4) Cumulative Capital Investment:	$ 10^6		0	>8		>52		
(5) Operating Cost 3) ($/unit of production):	Mils/kWh_e		10	>10		>10		

*Specify other relevant non-fossil energy source or conversion process, i.e., geothermal, solar, wind, etc.

(1) 360 MW_e for cases D_7 and D_8

(2) Additional capital cost for burning solid waste

(3) Additional operating cost for burning waste

NONFOSSIL SOURCES AND CONVERSION PROCESS WORKSHEET

Hydroelectric ______
Coal Gasification ______
Coal to Syncrude ______
Gas Liquifaction ______
Regasification ______
Other* (specify): Wood

Country: Italy

WAES SCENARIO CASE (1985): All Cases

WAES SCENARIO CASE (2000): All Cases

Total Resource in Place: -

Technology Assumptions: -

	(Units)	1972	1975	1985	1990	2000	2010	2025
(1) Yield:								
(2) Maximum Potential Output	10^6 toe	0.5	0.5	0.4	0.3	0.2		
(3) Capital Cost (1) ($/unit of daily capacity):		N.A.						
(4) Cumulative Capital Investment:		N.A.						
(5) Operating Cost ($/unit of production):		N.A.						

*Specify other relevant non-fossil energy source or conversion process, i.e., geothermal, solar, wind, etc.

(1) Cost data not available in above form. Minimum price to recover total costs at 15% discount rate is approximately $ .10/kg, for firewood without transport costs.

14 Japan

14.1 OVERVIEW AND SUMMARY OF DOMESTIC ENERGY SUPPLY

Because the domestic energy supply in Japan does not meet demand, Japan's policy has been to secure imports by sustaining good relations with foreign suppliers. Although Japan was trying to increase domestic energy supply even before the oil crisis, this effort has produced no substantial gain so far. Since the oil crisis, Japan has emphasized the need to lessen its dependence on oil by diversifying the import sources.

This emphasis is clearly reflected in the governmental policy, which was published in October 1975 in the report of the Energy Advisory Committee to the government. This policy calls for lessening dependence on oil, developing domestic energy resources, and diversifying energy resources through vigorous introduction of nuclear energy and the import of LNG and steam coal.

We interpret the WAES scenario variable "restrained policy" to mean the minimum governmental action, which in our case implies a continued high dependence on imported oil. However, strong governmental action to lessen this dependence is the current Japanese energy policy. The vigorous policy in Japan takes the form of "vigorous nuclear" and "vigorous diversification" through strong governmental action. In practice, we have no choice but to adopt a vigorous supply policy. Whether even the vigorous policy will permit the expected production level to be realized is our great concern.

Including nuclear energy as domestic energy, our domestic supply of energy amounted to 45.4×10^{13} kcal in 1972, which is 13 percent of the total energy supply in that year. The main sources of domestic energy were coal and hydroelectricity.

In 1985 the domestic energy supply will be increased to about 95×10^{13} kcal in all cases and will account for about 13 percent of the total primary energy supply in the high economic growth case and 17 percent in the low economic growth case. The main sources of this increase will be the expansion of nuclear energy and hydroelectricity.

In the year 2000 the domestic supply of energy will depend significantly on the capacities of nuclear power plants. If we assume high nuclear capacity, the domestic supply will be about 240 to 250×10^{13} kcal and will account for 20 percent of the total primary energy supply in Case C-2 and 28 percent in Case D-8. If we assume low nuclear capacity, domestic supply will remain about 190×10^{13} kcal and is expected to provide about 15 percent of the total supply in Case C-1 and 23 percent in Case D-7.

Our forecast for total primary energy supply including imports is summarized in the first supply study worksheet found at the end of this chapter.

14.2 CONCLUSIONS AND DISCUSSION

1. Because domestic fossil fuels are expected to remain limited, we are concerned with the availability of fossil fuel imports and with the expansion of nuclear energy.
2. Efforts to search for oil and gas in the Japanese offshore will be intensified; however, we cannot make any reliable quantitative forecasts of these sources at this time.
3. Expanding nuclear energy supplies might seem to be the only realistic way to lessen our dependence on energy imports; however, because of political and environmental considerations, this possibility is subject to great uncertainty.
4. For fossil fuels, the type of fuel and the import source need to be as diversified as possible to cope with the possible leveling off of oil supply. In this context, we should consider greater use of imported steam coal and the possibility of importing fossil fuel from Communist areas.

14.3 COMMENT ON EACH RESOURCE

14.3.1 Coal

Coal production in Japan has decreased rapidly in the last ten years in spite of heavy financial aid and other governmental actions to support this industry. In 1975 about 20 million tonnes were produced. The "New Coal Policy" of 1975 calls for maintaining at least this production level for the next ten years.

To realize this target we need a vigorous governmental policy, which implies the mandatory use of steam coal to generate electricity, the setting of coal prices by the government, and a substantial amount of financial aid. Without such a vigorous policy, coal production will decline further.

It may be possible to maintain the annual production level of coal reserves at about 20 million tonnes until 2000. However, because it will be difficult to get the necessary labor, achieving this production level is uncertain.

14.3.2 Oil and Gas

We produced about 3.7 million kl (oil equivalent) of oil and gas in 1973, mostly from onshore sites. This level would be the maximum onshore production level in the future. Additional production will come from offshore development. We expect about an additional 10 million kl of oil and gas; about 75 percent of this will be gas and the rest oil by 1985.

The feasibility of further developing offshore resources depends on political considerations as well as the uncertainty surrounding the existence of such resources. We have done little so far to develop the offshore area, but there is some evidence that this area contains large amounts of oil and gas. To more accurately predict these resources, we need more vigorous exploration. At the moment, however, it is difficult to estimate the potential offshore reserves.

14.3.3 Hydro Power

The installed hydroelectric capacity at the end of 1974 was 21.2 million kW, which produced about 85 billion kWh. With a vigorous policy we expect to add about 7 million kW by 1985. This requires developing about 30 percent of the now undeveloped hydro power at a cost of under 160 yen/kWh at 1973 prices.

Our reserve of hydro power is estimated to be about 50×10^6 kW, and we believe that about 80 percent of these reserves will be developed by 2000. This figure does not include pump-storage hydro power, which produces almost no additional energy but will be increasingly important in covering peak load demand.

14.3.4 Geothermal Energy

The installed geothermal power plant capacity at the end of 1972 was about 0.3 million kW. With a vigorous policy we expect to add about 2 million kW by 1985 and another 3 million kW by 2000.

Because most of this resource is located in a national park, the main problem in developing it is preserving the environment.

14.3.5 Solar Energy

The Japanese government published its development plan for new energy, the so-called Sunshine Project, in March 1974. This project called for completing new technologies for industrial application of solar energy by the year 2000. The major thrust of these plans is directed toward developing solar furnaces, solar energy power generation, and solar air conditioning units. To generate solar power the construction of a 1 or 2 million kW power plant is planned. For air conditioning by solar energy, a community-wide air conditioning system is planned by 2000.

At the moment it is very difficult to estimate the amount of solar energy that can be expected. This estimate depends on the future development of the required technologies and the world energy situation.

Table 14.1
Range of annual average demand for main fuel cycle components

	1975–1980	1981–1985	1986–1990	1991–1995	1996–2000
Annual uranium demand (tonnes, U metal)	2000	3500	7000	10000	14000
Annual separative work requirement (SWU, tonnes)	1500	2500	4500	6500	8000
Annual fabrication demand (tonnes, U metal)					
LWR	500	1000	1500	2000	2500
ATR (Pu)	—	110	200	400	600
Annual reprocessing requirement (tonnes, U metal)					
LWR	200	450	1000	1500	2000
ATR (Pu)	—	—	50	150	450
Fissile plutonium stock (tonnes)	10	30	70	150	180

But before solar energy is used for large-scale industrial purposes such as thermal power plants, it may be used on a relatively large scale in the residential sector.

14.3.6 Nuclear Energy

Nuclear energy is one of the most important sources in our efforts to reduce our dependence on oil imports. The capacity of installed nuclear power plants is about 7.0 million kW, and the capacity under construction amounts to about 15 million kW.

The October 1975 government energy report predicts that a 49 million kW nuclear power plant will be in operation by 1985. We estimate, however, that by 1985 the maximum capacity will be about 30 million kW, even with the vigorous policy. We believe that this lower estimate is more likely because of the current commitment and the lead time, which becomes longer as the complexity of getting approval to construct nuclear power plants increases. We estimate that in 1985 the installed capacity will range from a low of 25 million kW to a high of 30 million kW under the vigorous policy. For 2000 we estimate the installed capacity of nuclear power plants to be 75 million kW under the vigorous nuclear policy or 120 million kW under the vigorous diversification policy.

Table 14.1 is a summary of the average demand for the main fuel cycle components in Japan for the WAES high nuclear case (120 million kW in 2000).

14.4 DIVERSIFICATION OF ENERGY

Our energy supply policy is to diversify energy type as much as possible. This policy has three main phases: (1) vigorous introduction of nuclear energy, (2) vigorous development of domestic fossil fuel and other resources, and (3) vigorous importing of LNG and steam coal.

The first two phases have already been discussed. To implement the third phase we need vigorous efforts by government and industries to realize the infrastructure necessary to receive and use imported LNG and steam coal; this infrastructure includes a coal center (coal harbor), LNG loading harbor facilities, and both coal and LNG power plants.

With vigorous governmental and industrial action we expect to be able to import about 30 million tons of LNG and 7 to 8 million tons of steam coal by 1985. Although these figures should be close to the maximum volume, we do not exclude the possibility of exceeding these levels. The levels that will be achieved depend on how quickly and in what depth the real world energy situation is understood as well as on the efforts of government and industry.

14.5 Energy Supply Worksheets

This section includes summary tables and worksheets in the following order (see Section 1.2.1 of Chapter 1 for the definitions of line items):

Energy supply summary

Crude oil

Natural gas

Coal

Nuclear (2)

Hydroelectric

Geothermal

WAES NATIONAL ENERGY SUPPLY STUDIES, 1972-2000
ENERGY SUPPLY SUMMARY

Country: JAPAN

Scenario Assumptions	1972	1985					2000			
WAES CASE	Base Year	A	B	C	D	E	C-1	C-2	D-7	D-8
GNP 1976-85; 1985-2000		High	Low	High	Low	High	High	High	Low	Low
Oil/Energy Price*		11.50-17.25	11.50-17.25	11.50	11.50	11.50-7.66	11.50-17.25	11.50-17.25	11.50	11.50
National Pólicy Response		Vig	Vig	Vig	Res	Res	Vig	Vig	Vig	Vig
Principal Replacement Fuel							Coal	Nuclear	Coal	Nuclear
Energy Supply Units 10^{13}Kcal	"	"	"	"	"	"	"	"	"	"
Crude Oil (Incl. imports of Prod.	261.6	467.2	351.4	492.3	372.8	527.4	924.9	890.9	502.6	490.7
Natural Gas	4.0	48.4	45.6	48.4	45.6	48.4	84.0	84.0	84.0	84.0
Coal	57.2	91.2	73.6	91.2	73.6	91.2	136.4	113.2	106.9	83.6
Nuclear	2.3	45.1	45.1	45.1	45.1	45.1	105.0	167.0	105.0	150.2
Hydro-electric	21.3	29.1	28.4	29.1	28.4	29.1	44.4	44.4	42.1	42.1
Geothermal	-	1.0	1.0	1.0	1.0	1.0	4.8	4.8	4.8	4.8
Solar Thermal	-	-	-	-	-	-	-	-	-	-
Solar Electric	-	-	-	-	-	-	-	-	-	-
(In which imported)	(301.0)	(585.9)	(449.7)	(611.0)	(471.1)	(646.0)	(1105.5)	(1048.3)	(653.7)	(618.5)
Total Energy Supply	346.4	682.0	545.1	707.1	566.5	742.2	1,299.5	1,304.3	845.4	855.4

*1975 US dollars per barrel of Arabian light crude oil, fob Persian Gulf.
All Figures are primary production prior to any processing or conversion.

INDIGENOUS SUPPLY ASSESSMENT WORKSHEET

CRUDE OIL

Country: JAPAN

WAES SCENARIO CASE (1985): } All cases
WAES SCENARIO CASE (2000): }

Total Resource in Place: ________

Domestic Oil Price: ________

Technology Assumptions: ________

Crude Oil:
Offshore ✓
Onshore ✓

	(Units)	1972	1975	1980	1985	1990	2000	2010	2025
(1) Ultimate Recoverable Resource:	10^6 tonnes		260						
(2) Discoveries:									
(3) Reserves Proved to Date:		3.6	70						
(4) Potential Production:		0.7	0.7	1.0	4.0	4.0	4.0		
(5) Expected Annual Production	heavy / light	0.7	0.7	1.0	4.0	4.0	4.0		
(6) Cumulative Production:									
(7) Remaining Proved Reserves:									
(8) Reserve/Production Ratio:									
(9) Host Government Take ($/unit of production):									
(10) Capital Cost ($/unit of daily capacity):									
(11) Operating Cost ($/unit of production):									

INDIGENOUS SUPPLY ASSESSMENT WORKSHEET

NATURAL GAS

Country: JAPAN

WAES SCENARIO CASE (1985): } All cases

WAES SCENARIO CASE (2000): }

Total Resource in Place: ______

Domestic Gas Price: ______

Technology Assumptions: ______

Natural Gas:
Offshore ✓
Onshore ✓

	(Units)	1972	1975	1980	1985	1990	2000	2010	2025
(1) Ultimate Recoverable Resource:	$10^9 m^3$		540						
(2) Discoveries:									
(3) Reserves Proved to Date:		11.0	130						
(4) Potential Production:	$10^9 m^3$	2.5	2.5	2.5	7.0	8.0	8.0		
(5) Expected Annual Production:	$10^9 m^3$		2.5	2.5	7.0	8.0	8.0		
(6) Cumulative Production:									
(7) Remaining Proved Reserves:									
(8) Reserve/Production Ratio:									
(9) Host Government Take ($/unit of production):									
(10) Capital Cost ($/unit of daily capacity):									
(11) Operating Cost ($/unit of production):									

INDIGENOUS SUPPLY ASSESSMENT WORKSHEET

COAL

Country: JAPAN

Coal:
Surface Mined ✓
Deep Mined ✓

WAES SCENARIO CASE (1985): } All cases
WAES SCENARIO CASE (2000): }

Total Resource in Place: $20{,}250 \times 10^6$t.

Domestic Coal Price: ______

Technology Assumptions: ______

	1972	1975	1980	1985	1990	2000	2010	2025
(1) Ultimate Recoverable Resource: 10^6t		3,180						
(2) Reserves Proved to Date:		1,026						
(3) Potential Production:	28	20	20	20	20	20		
(4) Expected Annual Production:	28	20	20	20	20	20		
(5) Cumulative Production:								
(6) Remaining Proved Reserves:								
(7) Reserve/Production Ratio:								
(8) Primary Transportation* —Capital Cost ($/unit of daily capacity):								
—Operating Cost ($/unit of daily capacity):								
(9) Capital Cost ($/unit of daily capacity):								
(10) Operating Cost ($/unit of production):								

*From mine mouth to first processing stage.

NONFOSSIL SOURCES AND CONVERSION PROCESSES WORKSHEET

URANIUM AND NUCLEAR

Country: JAPAN

WAES SCENARIO CASE (1985): ______

WAES SCENARIO CASE (2000): Low Nuclear

Total Resource in Place: ______

Domestic Uranium Price: ______

Technology Assumptions: ______

	(Units)	1972	1975	1980	1985	1990	1995	2000	2010
1. Ultimate Recoverable Uranium Resource:									
2. Reserve of Uranium Proved to Date:	10^3t 15\$/lbU_3O_8		1.1						
	10^3t 15-30\$/lbU_3O_8		6.6						
3. Potential Production of Uranium:		16	4	30	30				
4. Expected Production of Uranium:									
5. Reserve/Production Ratio:									
6. Maximum Installed Capacity (GWe)*:			6.6		25			75	
7. Capital Cost (\$/KWe):									
8. Operating Cost (¢/Kwhr)**:									

*Indicate assumed load factor and LWR or HWR.

**Excluding capital charges.

NONFOSSIL SOURCES AND CONVERSION PROCESSES WORKSHEET

URANIUM AND NUCLEAR

Country: JAPAN

WAES SCENARIO CASE (1985): ______

WAES SCENARIO CASE (2000): High Nuclear

Total Resource in Place: ______

Domestic Uranium Price: ______

Technology Assumptions: ______

	(Units)	1972	1975	1980	1985	1990	1995	2000	2010
1. Ultimate Recoverable Uranium Resource:									
2. Reserve of Uranium Proved to Date:	$10^3 t\ 15\$/lb\ U_3O_8$		1.1						
	$10^3 t\ 15\text{-}30\$/lb\ U_3O_8$		6.6						
3. Potential Production of Uranium:		16	4	30	30				
4. Expected Production of Uranium:									
5. Reserve/Production Ratio:									
6. Maximum Installed Capacity (GWe)*:			6.6		30			120	
7. Capital Cost ($/KWe):									
8. Operating Cost (¢/Kwhr)**:									

*Indicate assumed load factor and LWR or HWR.

**Excluding capital charges.

NONFOSSIL SOURCES AND CONVERSION PROCESS WORKSHEET

Hydroelectric ✓
Coal Gasification ____
Coal to Syncrude ____
Gas Liquifaction ____
Regasification ____
Other* (specify): ____

Country: JAPAN

WAES SCENARIO CASE (1985): } All cases
WAES SCENARIO CASE (2000): }

Total Resource in Place: 49.6×10^6 KW

Technology Assumptions: ____

	(Units)	1972	1975	1980	1985	1990	2000	2010	2025
(1) Yield:									
(2) Maximum Potential Output (Expected Annual Prod.)	10^6KW (10^9Kwh)	20.7 (88.0)	23.5 (92.0)		28.3 (110.0)		40.0 (150.0)		
(3) Capital Cost ($/unit of daily capacity):									
(4) Cumulative Capital Investment:									
(5) Operating Cost ($/unit of production):									

*Specify other relevant non-fossil energy source or conversion process, i.e., geothermal, solar, wind, etc.

NONFOSSIL SOURCES AND CONVERSION PROCESS WORKSHEET

Hydroelectric ________
Coal Gasification ________
Coal to Syncrude ________
Gas Liquifaction ________
Regasification ________
Other* (specify): Geothermal

Country: JAPAN

WAES SCENARIO CASE (1985): All cases

WAES SCENARIO CASE (2000):

Total Resource in Place: 140×10^6 KW

Technology Assumptions:

	(Units)	1972	1975	1980	1985	1990	2000	2010	2025
(1) Yield:									
(2) Maximum Potential Output (Expected Annual Prod.)	10^6KW (10^9Kwh)	0.3 (1.8)	0.45 (2.8)		2.0 (12.0)		5.0 (30.0)		
(3) Capital Cost ($/unit of daily capacity):									
(4) Cumulative Capital Investment:									
(5) Operating Cost ($/unit of production):									

*Specify other relevant non-fossil energy source or conversion process, i.e., geothermal, solar, wind, etc.

15 Mexico

15.1 OVERVIEW

This paper is based on the studies carried out by the Mexican WAES team and is based on certain subjective considerations regarding the energy situation. By no means does it represent the official position of the Mexican Government or of the individual energy agencies. The Mexican participants in WAES are solely responsible for the contents herein.

The following is a brief summary of the WAES cases presented in this report.

Case A, which presumes a high GWP and an increase in oil prices, indicates a high growth rate of energy consumption and represents the upper boundary to the acceleration of development.

Due to the high price of petroleum, exports of surplus oil would aid the economic development of the country.

Until now our energy needs have been satisfied by domestic resources, and they are expected to continue to do so. The availability of oil, gas, and coal for internal consumption has allowed the national energy market to maintain itself practically independent of international oil prices; but in the case of hydrocarbon exports, a high price is an advantage to our country.

Case B assumes slow world economic growth and high oil prices. This would result in low demand for hydrocarbons in a highly competitive market and would limit our level of export. This would result in a low economic growth rate for Mexico, representing the lower boundary regarding the exploitation of our resources.

Case C represents the possibility of exporting a small quantity of petroleum, which would compensate for the coal imports estimated to occur under the assumptions made in these cases.

For Cases D and E, an equilibrium between supply and demand was maintained, in which case Mexico would neither export nor import.

15.2 CONCLUSIONS

The apparent abundance of hydrocarbons, their undisputed advantages as a source of energy, and the simplicity of their use and wide field application have conditioned modern societies to a high degree of dependence on this nonrenewable resource. This dependence is even more significant in countries that found their socioeconomic development on petroleum and is heightened in countries counting on this resource within their territory. Almost 90 percent of Mexico's energy supply depends on hydrocarbons. This situation, until now justified and manageable, is beginning to show signs of possible short-run changes, with considerable consequences.

The first signs were given during the oil crisis of 1973–1974, when the country was forced to import large volumes of petroleum to satisfy national demand. Fortunately, the discoveries in the southeast eliminated the need to import and even made it possible to export about 10 percent of the production. Nevertheless, this experience must be used to prevent similar crises.

It is estimated that world oil production will begin to decline around 1990. In Mexico, however, if we use our energy resources "competently," the decline in production could be deferred until the beginning of the next century. Mexico could then find itself in a very favorable position for the transition from conventional sources (hydrocarbons, coal, uranium) to asymptotic sources (solar, fusion).

Using our energy resources competently means securing a less vulnerable position, increasing the use of those resources not utilized until now to diversify and allow us to cope more flexibly with abnormal situations. In our case, this means substituting coal and uranium for hydrocarbons to produce electricity.

The projected volume of energy required before the turn of the century is of such magnitude that the reserves of hydrocarbons, large as they may be, will not be sufficient to satisfy even the needs of a few decades of the next century.

The hydroelectric resources should be totally developed by the year 2000, yet their contribution will only represent about 20 percent of the total production of electricity.

Mexico's reserves of coal are abundant when compared to the present level of production. Their size, according to some sources, is comparable to that of the hydrocarbons, but in proportion coal utilization is almost 20 times less.

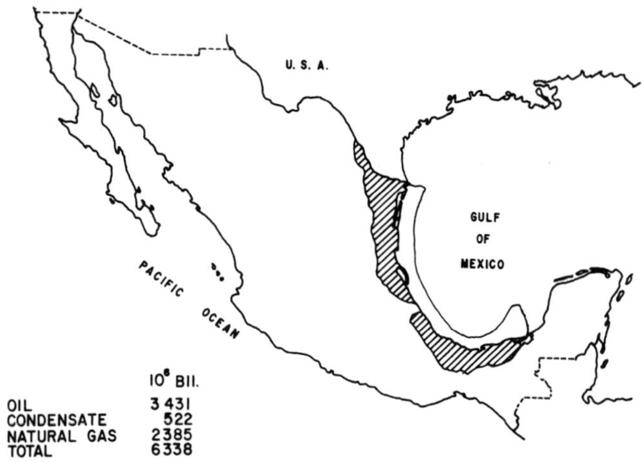

Figure 15.1
Hydrocarbons reserves

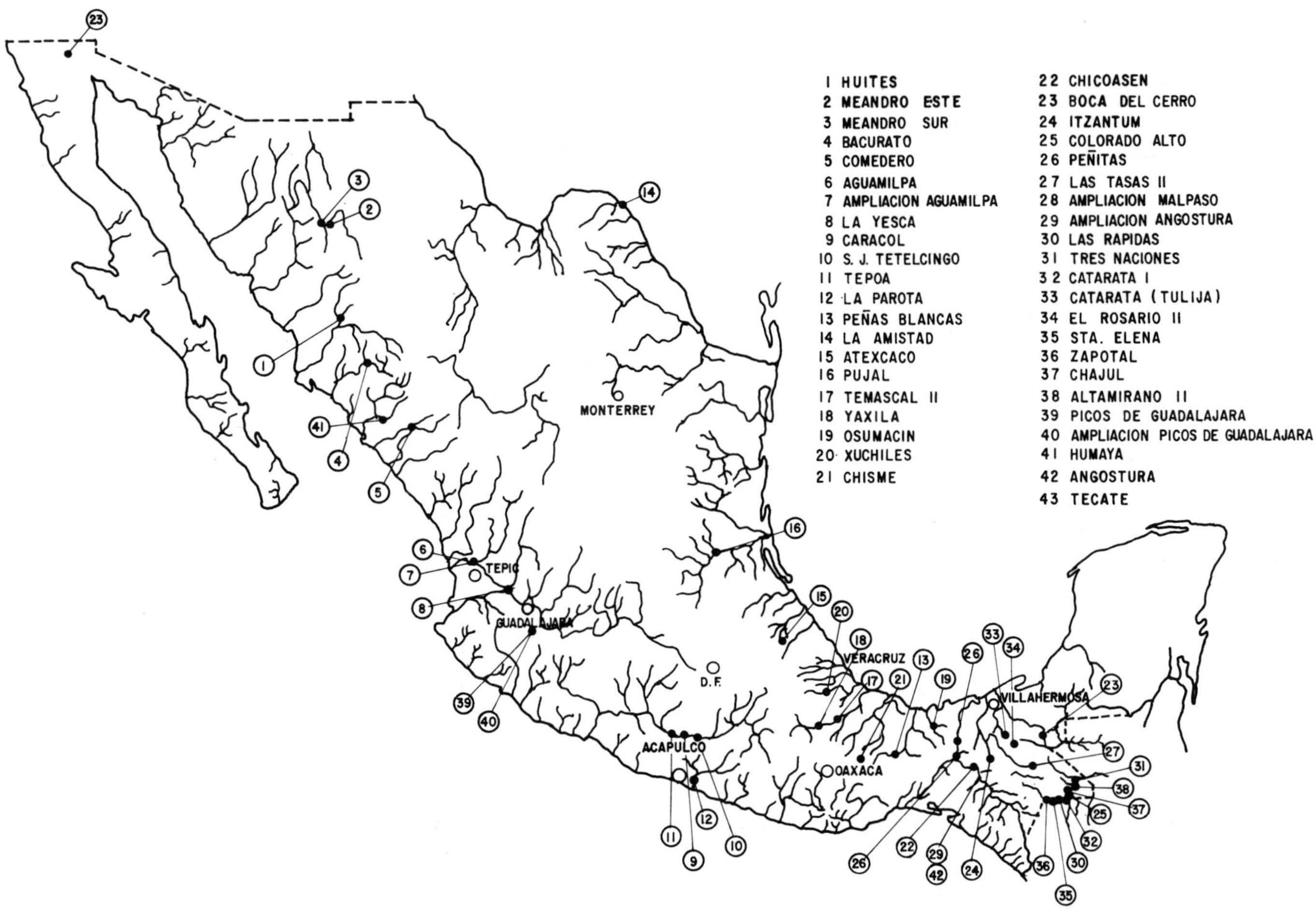

Figure 15.2
Location of viable hydroelectric sources to enter into operation between 1975 and 2000.

Although geothermal energy presents interesting possibilities, it is unlikely to be significant before the turn of the century.

Mexico must rely on more energy to continue its development. Reducing consumption is important, but it will only reduce demand marginally in the current consuming sectors. Moreover, the incorporation of marginal populations will demand more energy. Access of the younger generation, currently composed of marginal consumers, to the productive sector of the country will require still more energy.

15.3 ENERGY SUPPLY

For more than half a century, Mexico's growth has depended heavily on its oil industry. Inevitably, the course of future development will be closely related to the progress of this industry. Mexico has turned to hydrocarbons as the principal energy source—over 90 percent of demand is supplied by hydrocarbons—and for the year 2000 this dependence is still expected to be at least 70 percent.

The main objective of Mexico's strategy has been to produce enough oil and gas to satisfy the country's demand (whose annual growth of 7 percent is above world average) and to maintain a reserves-to-production ratio of 15 to 1.

Reported reserves were 6400 million barrels; however, recent studies based on more modern methods of measurement have changed this assessment to 11,000 million barrels. This new assessment is presently being confirmed by a specialized foreign consultant.

Almost 50 percent of the proven oil reserves are located in the southeastern part of the country. Production in these fields started in 1973, when as much as 390,000 barrels a day were produced. Today this area accounts for 70 percent of national production, and the cumulative production to 1975 of 191.6 million barrels clearly reveals its importance. In the cretacic area of this part of the country, the needed infrastructure is being established with exceptional speed (Figure 15.1).

As a preventive measure, the Mexican government has been making an effort to substitute other primary energy sources for hydrocarbons, to the extent possible. The development of hydroelectricity, restricted by the inconvenient geographical location of available resources, is being actively promoted.

Potential sources of hydraulic energy are located in the southeast, far from consumption centers. Furthermore, the practical potential is calculated to be not more than 60,000 million kWh per year; by the year 2000 this potential is planned to be in operation (Figures 15.2 and 15.3). Other alternatives are coal and uranium.

Solar and geothermal energy are considered interesting resources for the next century, but during this century technical and economic problems limit their contribution.

Coal has been used in Mexico since the end of the last century, but attempts to evaluate the overall coal reserves have been initiated only recently. Estimates of coal reserves cover a wide range. Only in the states of Coahuila, Sonora, and Oaxaca has commercially exploitable coal been found. Of these, the first is the most important; total measured reserves there are 1685 million tons (Figure 15.4).

Uranium and coal could well be the basic resources to reduce dependence on hydrocarbons.

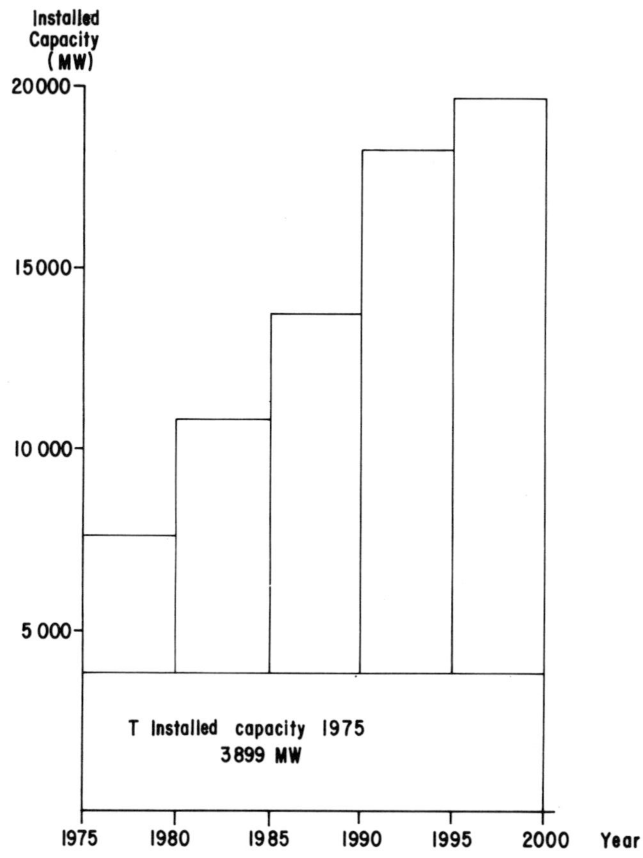

Figure 15.3
Hydroelectric projects, 1975–2000.

At present, two 650 MW nuclear power units are being installed. It is estimated that our measured reserves of 8000 tons of U_3O_8 could feed the two plants during the next thirty years. The main deposits of uranium are found in the states of Tamaulipas and Chihuahua (Figure 15.5).

The tectonic characteristics of Mexico suggest considerable potential geothermal energy. However, with the present technology, only marginal production can be expected during the remainder of this century. A 75 MW electrical geothermal plant now in operation in the Baja California peninsula (Cerro Prieto) has an estimated potential of 400 MW (Figure 15.6). Mexico is rich in solar energy—over 60 percent of its territory is semidesert or desert—and the area of insolation is vast (Figure 15.7). Thus in the next century when solar becomes a reality, Mexico should be in a very favorable position to take advantage of this source of energy.

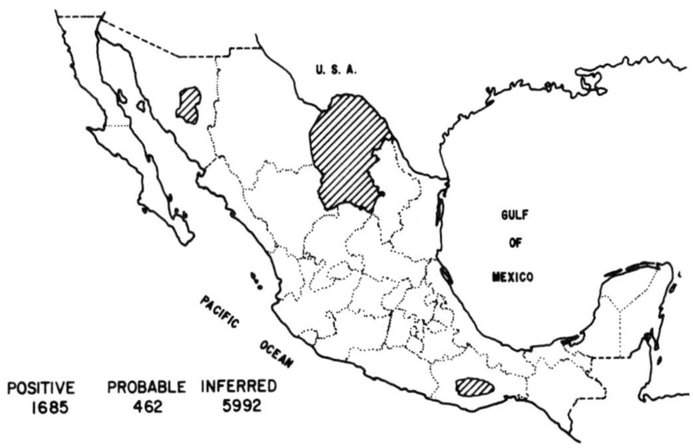

Figure 15.4
Coal reserves (10^6 tons).

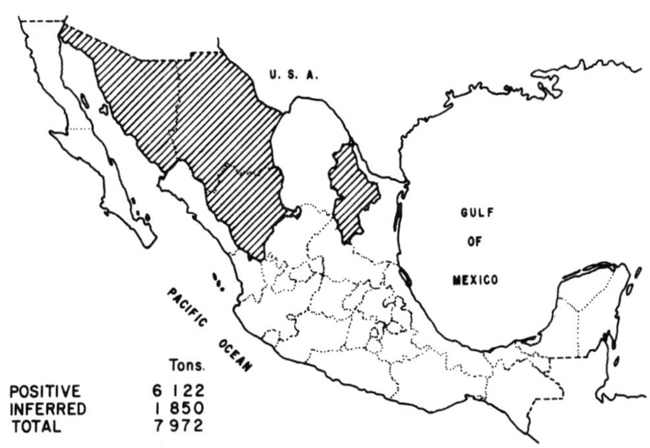

Figure 15.5
Uranium reserves.

The general conclusion is in agreement with the energy history of Mexico. The oil industry has been very vigorous and has dominated the Mexican energy scene. The goal of energy diversification will require considerable efforts. A vital necessity is to incorporate other energy sources to cover increasing national demand and lighten the heavy burden on hydrocarbons.

15.4 ENERGY RESERVES

An energy policy is a direct function of the information regarding the energy sources of the country. This information helps to determine which sources should be restrained, which substitutes would be convenient, which type of energy should be promoted, what export possibilities exist without sacrificing the future, and so on. The greater the certainty of available information, the more successful the policies will be. Unfortunately, very little reliable and complete information on reserves is available.

Without a doubt, we must enhance our knowledge of reserves; this is an easy statement to make, but many practical problems must be overcome before we can do so. First, large investments are needed, and only a small part of the resulting benefits will be seen in the short term. Second, exploration necessarily takes time, so results cannot be expected in a short period.

Despite these problems, efforts must be made to obtain the necessary information, for the lack of it might create serious short-run economic problems, like the one Mexico dramatically faced when it was forced to import crude oil, or the shortcomings in mining facilities needed to satisfy the coal demands of the steel industry.

Figure 15.6
Geothermal resources.

15.5 RESERVES

15.5.1 Oil and Gas

According to the official information of PEMEX, the accumulated hydrocarbon reserves discovered in Mexico from 1938 to 1976 amount to 12,000 million barrels of oil. The reported reserves as of 1976 were 6338 million barrels. (These reserves are currently being reassessed by PEMEX and are now estimated to be 11,000 million barrels.) The reserves are distributed in four areas: the south, the most important, had 41 percent; Poza Rica, 39 percent; the north, 20 percent; and the Angostura, less than 1 percent. These four areas are distributed only among the five land and two offshore provinces that belong to the category of preferred provinces. Of Mexico's remaining forty-three provinces, twenty are considered to have oil possibilities, and twenty-three belong to the category of unexplored provinces.

From the analysis of the provinces' geological characteristics and the results of recent exploration in production areas, we can make an optimistic estimate of the country's reserves of hydrocarbons. In the WAES studies hydrocarbon reserves were estimated at 54,000 million barrels. Consumption in the year 2000 will range between 1675 million barrels and 2063 million barrels. Total production in 1975 was 454 million barrels. For the purposes of comparison, Figure 15.8 contrasts the structure of energy consumption expected in the year 2000 to that of 1975; C-1 is the scenario that assumes low oil consumption, and D-8 is the high oil scenario.

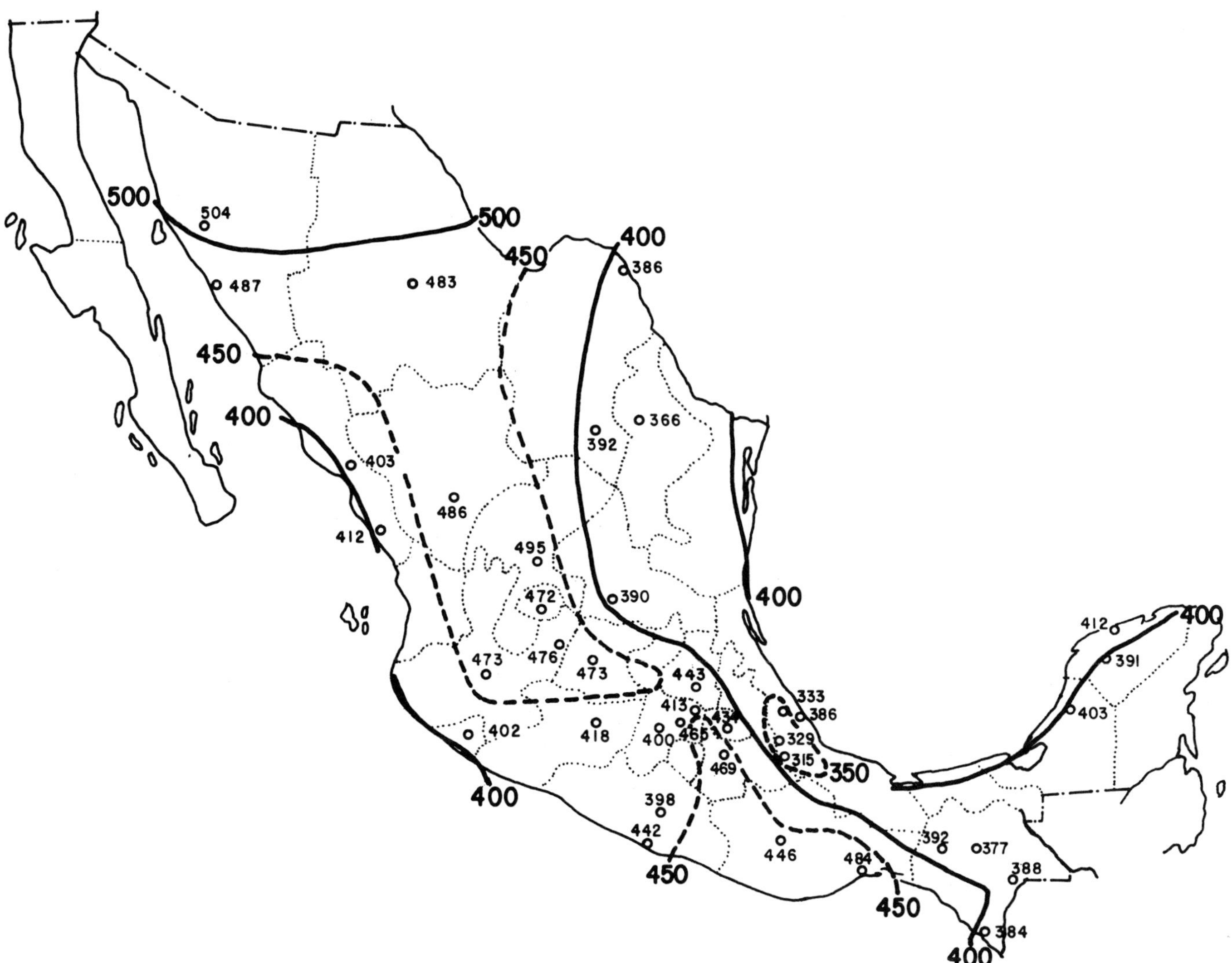

Figure 15.7
Annual average daily radiation (Langleys/day).

15.5.2 Hydroelectric Energy

Hydroelectric energy is a renewable source and, therefore, should be considered not in terms of reserves, but in terms of potential annual output. It also differs from other sources in that studies on it are more complete, to the point of determining its potential very precisely, and so we do not expect significant changes in estimates of its potential.

The hydroelectric potential of the country is calculated at 80,000 GWh average annual output, of which approximately 20 percent is now in use. Because of geological problems, it is most probable that not all potential sources will be fully utilized. However, the goal is total development of the practical sites by the end of the century.

15.5.3 Coal

Attempts to evaluate coal reserves started only twenty-five years ago. Lack of information, as well as explorations limited to only a very small portion of the country, have resulted in evaluations ranging from 1000 million to 15,000 million tons; sometimes these are described as general reserves, sometimes as positive reserves, and other times as geological reserves.

Coal is found in many parts of the country, but the only reserves considered commercially exploitable are found in Coahuila, Oaxaca, and Sonora. The Coahuila area is undoubtedly the most important in the country; its fields hold over 90 percent of the total known reserves, and it is now the only commercially exploited area. Exploration has taken place mostly in the richest areas; it has been fragmentary and only a relatively small percentage of the total territorial extension has been explored.

Following the Coahuila fields in importance is the Mixteca in Oaxaca, though its potential is inferior to what the estimate shows, for only 13 million of the total of 127 million tons are considered positive. This field has been almost unexplored and unexploited; it contains 108 million tons of uncokable coal, which could only be used in generating electricity, and 19 million tons of cokable coal.

Sonora's coal was exploited on a small scale many years ago, but production was discontinued. It is the uncokable anthracitic variety of coal, and its main use would be in the generation of electricity.

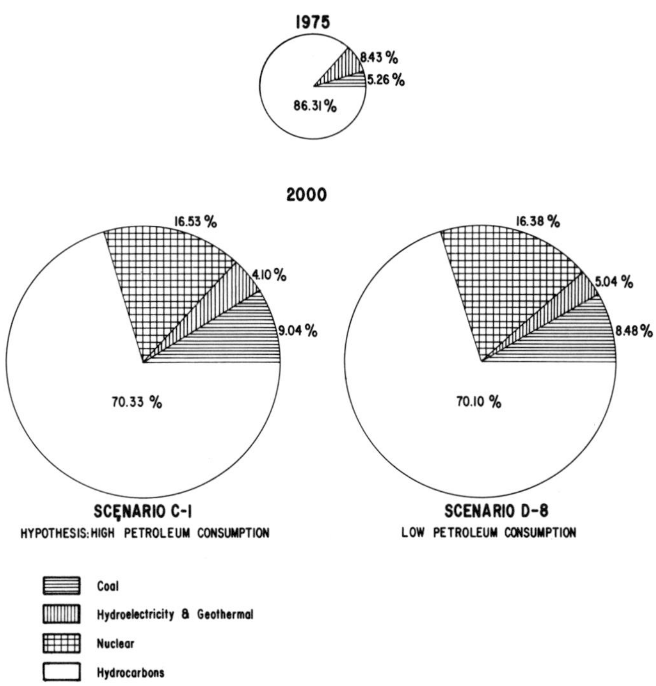

Figure 15.8
Composition of primary energy consumption under hydrocarbon frontier conditions.

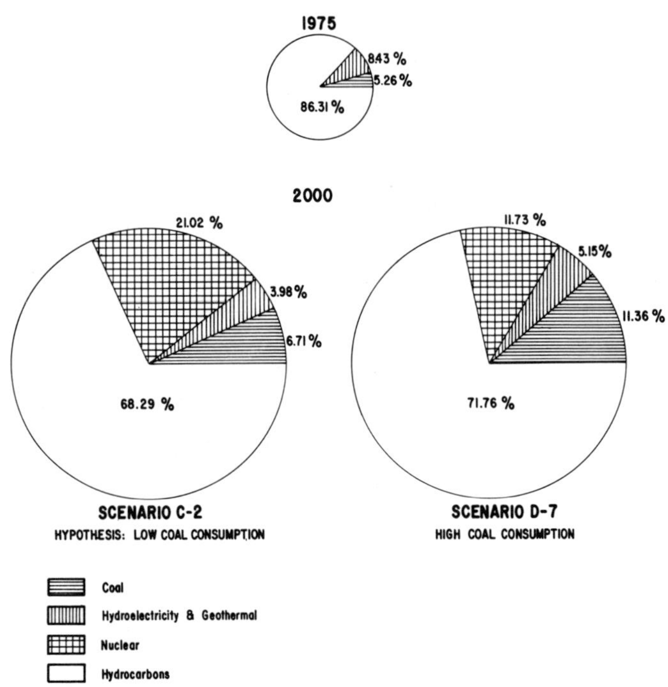

Figure 15.9
Composition of primary energy consumption under coal frontier conditions.

Here again, the total reserves figure of 22 million tons is an overestimate since only 2 million tons fall into the positive category.

In our WAES analysis total coal reserves were considered to be 2394 million tons. Our study concluded that production in the year 2000 would be in the neighborhood of 51 to 70 million tons, compared to the 5.1 million tons produced in 1975. Figure 15.9 shows coal's share of energy in the year 2000 and compares it with that in 1975. Both cases of coal production are presented; C-2 is the low scenario for coal consumption, and D-7 is the high scenario.

15.5.4 Uranium

Less than 1 percent of the country's area has been explored for uranium. In December 1975 positive reserves of about 6000 tons of U_3O_8, plus roughly 2000 tons of inferred reserves, were estimated. Since then, however, the estimates of some experts, based on exploration carried out by the Instituto Nacional de Energía Nuclear, show that Mexico's uranium potential is significant. Using geological information resulting from our study, we calculated uranium resources of 144,000 tons. Installed nuclear power plant capacity was calculated within the limits of 20 GW for Case D-7 and 50 GW for Case C-2.

15.5.5 Geothermal Energy

The potential of the only geothermal producing area in Mexico now in operation is expected to reach at least 3000 GWh per year.

Even though the many manifestations of endogenous heat would indicate considerable potential for geothermal energy, it is generally accepted that this source will contribute slowly and only marginally. The development of a geothermal field takes ten to fifteen years, and even assuming the development of five fields with positive results, production by the end of the century would be only 12,000 to 15,000 GWh per year (about 20 percent of all hydroelectric energy, which by that time is expected to be totally developed).

15.6 TECHNOLOGY

The country is practically self-sufficient in technology for the exploration, exploitation, and processing of petroleum, gas, and coal. The same is true of process technologies. Petróleos Mexicanos (PEMEX) has the qualified personnel and the systems needed to develop the human resources demanded by its growth. But a less favorable situation prevails in the design and manufacture of equipment associated with the energy sector, and the need to import components is significant. Furthermore, there is a chronic lack of specialists in the exploration for and exploitation of energy resources other than hydrocarbons.

In Mexico the institutions responsible for exploring, exploiting, processing, and distributing hydrocarbons, electric utilities, nuclear fuels, and a large part of the coal industry are state-owned. Thus Mexico can coordinate the energy sector.

PEMEX is the country's largest company and relies on the most up-to-date methods for exploration. It is a pioneer in the field of petrochemical plants, especially those for the production of ammonia. It handles all of the petroleum and basic petrochemical industries with local technology and imports engineering only for certain special processes.

The Instituto Mexicano del Petróleo, founded in 1965, is the largest R&D institution in the country. Since its inception, it has established itself as the generator of petroleum technology and engineering to assist PEMEX in the design of its installations.

The Comisión Federal de Electricidad (CFE) is entrusted with providing the country with electrical energy and is self-sufficient in the technologies for operation, planning, and construction. It is self-sufficient in the design of hydroelectric power plants, including dams, but needs to import technology for the design of thermal and nuclear power plants.

CFE operates hydroelectric and thermoelectric (fuel oil, coal, gas, turbine, and diesel) power plants and is currently constructing its first nuclear and its first large coal plant. Also, it is one of the pioneers in geothermal energy; its Cerro Prieto plant, near Mexicali, now has two 37 MW units in operation and a projected total capacity above 400 MW. CFE also has in operation a dual purpose plant in Rosarito, B.C., where in addition to generating electricity, it also desalinates seawater to complement the supply of fresh water to the city of Tijuana.

The geological services provided by the CFE include coal exploration, geothermal prospecting, and the geology of dams. It relies on an experimental engineering unit to provide technology on soil mechanics, rock formation, concrete, and so on, for all its projects.

Recently, the Instituto de Investigaciones Eléctricas was created to carry out R&D in the energy field, including the development of design techniques for electromechanical equipment.

The Instituto Nacional de Energía Nuclear is responsible for the nuclear fuel cycle (prospecting, exploration, mining, fuel manufacture, reprocessing of irradiated fuel), for nuclear safety, and for the development of nuclear energy for peaceful purposes. Its resources have been relatively limited, and until now it could be considered self-sufficient only in uranium exploration. However, it has established small groups of highly qualified personnel in the fields of nuclear fuel manufacture and some specialized fields of nuclear physics and applications of radiation.

Coal is produced by the iron and steel industry, both public and private, on a small scale; total annual output is about 6 million metric tons. The country has sufficient technicians in this field for its low level of production. However, if coal exploitation increases as expected, it could be limited by the lack of specialized personnel.

15.7 COORDINATING ACTIVITIES

The Energy Commission has been a natural forum for examining the opportunities provided by new technologies and for analyzing problems associated with this sector.

With the support of the energy sector, energy-economic studies have been promoted; the potential of the industry for domestic manufacturing of nuclear plant components has been examined, and solar energy research has been supported.

15.8 EQUIPMENT MANUFACTURE

The problem of capital goods in the Mexican energy sector is of great importance. The projected level of equipment purchases will impose an additional severe burden on the already negative balance of payments. The country must take the necessary steps to manufacture its own equipment for exploring, exploiting, processing, and distributing energy.

15.9 ENERGY SUPPLY WORKSHEETS

The following supply worksheets appear in this section (see Section 1.2.1 of Chapter 1 for the definitions of line items):

Energy supply summary

Crude oil (5)

Natural gas (5)

Hydroelectric

Uranium and nuclear (6)

Nuclear electric (4)

Coal (3)

WAES NATIONAL ENERGY SUPPLY STUDIES, 1972-2000
ENERGY SUPPLY SUMMARY

Country: MEXICO

Scenario Assumptions	1972	1985					2000			
WAES CASE	Base Year	A	B	C	D	E	C-1	C-2	D-7	D-8
GNP 1976-85; 1985-2000		High	Low	High	Low	High	High	High	Low	Low
Oil/Energy Price*		11.50-17.25	11.50-17.25	11.50	11.50	11.50-7.66	11.50-17.25	11.50-17.25	11.50	11.50
National Policy Response		Vig	Vig	Vig	Res	Res	Vig	Vig	Vig	Vig
Principal Replacement Fuel							Coal	Nuclear	Coal	Nuclear
Energy Supply Units	Kcal x 10^{12}									
Crude Oil	268	1006	720	827	744	730	1901	1901	1511	1511
Natural Gas	124	297	232	279	255	255	745	745	637	637
Coal	18	70	70	70	70	70	340	260	340	260
Nuclear	-	30	20	20	20	20	622	815	351	502
Hydro-electric	37	51	51	51	51	51	150	150	150	150
Geothermal	-	2	2	2	2	2	4	4	4	4
Solar Thermal	-	-	-	-	-	-	-	-	-	-
Solar Electric	-	-	-	-	-	-	-	-	-	-
Total Energy Supply	447	1456	1095	1249	1142	1128	3762	3875	2993	3064

*1975 US dollars per barrel of Arabian light crude oil, fob Persian Gulf.
All Figures are primary production prior to any processing or conversion.

INDIGENOUS SUPPLY ASSESSMENT WORKSHEET

CRUDE OIL

Country: MEXICO

WAES SCENARIO CASE (1985): A

WAES SCENARIO CASE (2000): ____

Total Resource in Place: ____

Domestic Oil Price: ____

Crude Oil:
Offshore ____
Onshore X

Technology Assumptions: 1975 Technology prevailing

	(Units)	1972	1975	1980	1985	1990	2000	2010	2025
(1) Ultimate Recoverable Resource:									
(2) Discoveries:									
(3) Reserves Proved to Date:	10^9bbl	6.3	7.8	12.9	20.9				
(4) Potential Production:									
(5) Expected Annual Production 10^9bbl	heavy								
	light	0.2	0.3	0.5	0.8				
(6) Cumulative Production: 1938-1972	10^9bbl	3.3	4.0	5.9	9.2				
(7) Remaining Proved Reserves:	10^9bbl	3.0	3.8	7.0	11.7				
(8) Reserve/Production Ratio:		15.0	13.0	15.0	15.0				
(9) Host Government Take ($/unit of production):									
(10) Capital Cost ($/unit of daily capacity):									
(11) Operating Cost ($/unit of production):									

(1) 10^9bbl
(2) 1938-1972

INDIGENOUS SUPPLY ASSESSMENT WORKSHEET

CRUDE OIL

Country: MEXICO

WAES SCENARIO CASE (1985): B

WAES SCENARIO CASE (2000): ______

Total Resource in Place: ______

Domestic Oil Price: ______

Technology Assumptions: 1975 Technology prevailing

Crude Oil:
Offshore ______
Onshore X

	(Units)	1972	1975	1980	1985	1990	2000	2010	2025
(1) Ultimate Recoverable Resource:									
(2) Discoveries:									
(3) Reserves Proved to Date:	10^9bbl	6.3	7.8	11.7	16.6				
(4) Potential Production:									
(5) Expected Annual Production 10^9bbl	heavy								
	light	0.2	0.3	0.4	0.6				
(6) Cumulative Production: 1938–1972	10^9bbl	3.3	4.0	5.7	8.2				
(7) Remaining Proved Reserves:	10^9bbl	3.0	3.8	6.0	8.4				
(8) Reserve/Production Ratio:		15.0	13.0	15.0	15.0				
(9) Host Government Take ($/unit of production):									
(10) Capital Cost ($/unit of daily capacity):									
(11) Operating Cost ($/unit of production):									

(1) 10^9bbl
(2) 1938–1972

INDIGENOUS SUPPLY ASSESSMENT WORKSHEET

CRUDE OIL

Country: MEXICO

WAES SCENARIO CASE (1985): C

WAES SCENARIO CASE (2000): 1 and 2

Total Resource in Place:

Domestic Oil Price: High energy price

Technology Assumptions: 1975 Technology prevailing

Crude Oil:
Offshore
Onshore X

	(Units)	1972	1975	1980	1985	1990	2000	2010	2025
(1) Ultimate Recoverable Resource:									
(2) Discoveries:									
(3) Reserves Pròved to Date:	10^9bbl	6.3	7.8	12.3	18.2	24.3	38.4		
(4) Potential Production:									
(5) Expected Annual Production 10^9bbl	heavy								
	light	0.2	0.3	0.4	0.6	0.8	1.5		
(6) Cumulative Production: 1938-1972	10^9bbl	3.3	4.0	5.8	8.6	12.2	23.6		
(7) Remaining Proved Reserves:	10^9bbl	3.0	3.8	6.5	9.6	12.1	14.8		
(8) Reserve/Production Ratio:		15.0	13.0	15.0	15.0	15.0	10.0		
(9) Host Government Take ($/unit of production):									
(10) Capital Cost ($/unit of daily capacity):									
(11) Operating Cost ($/unit of production):									

(1) 10^9bbl
(2) 1938-1972

INDIGENOUS SUPPLY ASSESSMENT WORKSHEET

CRUDE OIL

Country: MEXICO

WAES SCENARIO CASE (1985): D

WAES SCENARIO CASE (2000): 7 and 8

Total Resource in Place:

Domestic Oil Price: Medium

Technology Assumptions: 1975 Technology prevailing

Crude Oil:
Offshore
Onshore X

	(Units)	1972	1975	1980	1985	1990	2000	2010	2025
(1) Ultimate Recoverable Resource:									
(2) Discoveries:									
(3) Reserves Proved to Date:	10^9bbl	6.3	7.8	11.9	17	22.6	38.4		
(4) Potential Production:									
(5) Expected Annual Production 10^9bbl	heavy								
	light	0.2	0.3	0.4	0.6	0.7	1.2		
(6) Cumulative Production: 1938–1972	10^9bbl	3.3	4.0	5.8	8.3	11.6	21.2		
(7) Remaining Proved Reserves:	10^9bbl	3.0	3.8	6.1	8.7	11.0	17.2		
(8) Reservé/Production Ratio:		15.0	13.0	15.0	15.0	15.0	14.5		
(9) Host Government Take ($/unit of production):									
(10) Capital Cost ($/unit of daily capacity):									
(11) Operating Cost ($/unit of production):									

(1) 10^9bbl
(2) 1938–1972

INDIGENOUS SUPPLY ASSESSMENT WORKSHEET

CRUDE OIL

Country: MEXICO

WAES SCENARIO CASE (1985): E

WAES SCENARIO CASE (2000): ________

Total Resource in-Place: ________

Domestic Oil Price: ________

Technology Assumptions: 1975 Technology prevailing

Crude Oil:
Offshore ________
Onshore X

	(Units)	1972	1975	1980	1985	1990	2000	2010	2025
(1) Ultimate Recoverable Resource:									
(2) Discoveries:									
(3) Reserves Proved to Date:	10^9bbl	6.3	7.8	11.7	16.7				
(4) Potential Production:									
(5) Expected Annual Production 10^9bbl	heavy								
	light	0.2	0.3	0.4	0.6				
(6) Cumulative Production: 1938-1972	10^9bbl	3.3	4.0	5.7	8.2				
(7) Remaining Proved Reserves:	10^9bbl	3.0	3.8	6.0	8.5				
(8) Reserve/Production Ratio:		15.0	13.0	15.0	15.0				
(9) Host Government Take ($/unit of production):									
(10) Capital Cost ($/unit of daily capacity):									
(11) Operating Cost ($/unit of production):									

(1) 10^9bbl
(2) 1938-1972

INDIGENOUS SUPPLY ASSESSMENT WORKSHEET

NATURAL GAS

Country: MEXICO

WAES SCENARIO CASE (1985): A

WAES SCENARIO CASE (2000):

Total Resource in Place:

Domestic Gas Price:

Technology Assumptions: 1975 Technology prevailing

Natural Gas:
Offshore
Onshore X

	(Units)	1972	1975	1980	1985	1990	2000	2010	2025
(1) Ultimate Recoverable Resource:									
(2) Discoveries:									
(3) Reserves Proved to Date:	10^9BOE	3.8	4.5	5.9	7.5				
(4) Potential Production:									
(5) Expected Annual Production:	10^9BOE	0.13	0.16	0.19	0.23				
(6) Cumulative Production:	10^9BOE	1.7	2.2	3.0	4.1				
(7) Remaining Proved Reserves:	10^9BOE	2.1	2.3	2.9	3.4				
(8) Reserve/Production Ratio:		16.0	14.0	15.0	15.0				
(9) Host Government Take ($/unit of production):									
(10) Capital Cost ($/unit of daily capacity):									
(11) Operating Cost ($/unit of production):									

INDIGENOUS SUPPLY ASSESSMENT WORKSHEET

NATURAL GAS

Country: MEXICO

WAES SCENARIO CASE (1985): B

WAES SCENARIO CASE (2000): ________

Total Resource in Place: ________

Domestic Gas Price: ________

Technology Assumptions: 1975 Technology prevailing

Natural Gas:
Offshore ________
Onshore X

	(Units)	1972	1975	1980	1985	1990	2000	2010	2025
(1) Ultimate Recoverable Resource:									
(2) Discoveries:									
(3) Reserves Proved to Date:	10^9 BOE	3.8	4.5	5.6	6.6				
(4) Potential Production:									
(5) Expected Annual Production:	10^9 BOE	0.13	0.16	0.17	0.18				
(6) Cumulative Production:	10^9 BOE	1.7	2.2	3.0	3.9				
(7) Remaining Proved Reserves:	10^9 BOE	2.1	2.3	2.6	2.7				
(8) Reserve/Production Ratio:		16.0	14.0	15.0	15.0				
(9) Host Government Take ($/unit of production):									
(10) Capital Cost ($/unit of daily capacity):									
(11) Operating Cost ($/unit of production):									

INDIGENOUS SUPPLY ASSESSMENT WORKSHEET

NATURAL GAS

Country: MEXICO

WAES SCENARIO CASE (1985): C

WAES SCENARIO CASE (2000): 1 and 2

Total Resource in Place: ________

Domestic Gas Price: High energy price

Technology Assumptions: 1975 Technology prevailing

Natural Gas:
Offshore ______
Onshore X

	(Units)	1972	1975	1980	1985	1990	2000	2010	2025
(1) Ultimate Recoverable Resource:									
(2) Discoveries:									
(3) Reserves Proved to Date:	10^9BOE	3.8	4.5	5.9	7.3	9.8	15.5		
(4) Potential Production:									
(5) Expected Annual Production:	10^9BOE	0.13	0.16	0.19	0.22	0.3	0.58		
(6) Cumulative Production:	10^9BOE	1.7	2.2	3.0	4.0	5.3	9.7		
(7) Remaining Proved Reserves:	10^9BOE	2.1	2.3	2.9	3.3	4.5	5.8		
(8) Reserve/Production Ratio:		16.0	14.0	15.0	15.0	15	10		
(9) Host Government Take ($/unit of production):									
(10) Capital Cost ($/unit of daily capacity):									
(11) Operating Cost ($/unit of production):									

INDIGENOUS SUPPLY ASSESSMENT WORKSHEET

NATURAL GAS

Country: MEXICO

WAES SCENARIO CASE (1985): D

WAES SCENARIO CASE (2000): 7 and 8

Total Resource in Place:

Domestic Gas Price: Medium

Technology Assumptions: 1975 Technology prevailing

Natural Gas:
Offshore
Onshore X

	(Units)	1972	1975	1980	1985	1990	2000	2010	2025
(1) Ultimate Recoverable Resource:									
(2) Discoveries:									
(3) Reserves Proved to Date:	10^9BOE	3.8	4.5	5.7	7.0	9.2	15.5		
(4) Potential Production:									
(5) Expected Annual Production:	10^9BOE	0.13	0.16	0.18	0.2	0.27	0.5		
(6) Cumulative Production:	10^9BOE	1.7	2.2	3.0	4.0	5.2	9.0		
(7) Remaining Proved Reserves:	10^9BOE	2.1	2.3	2.7	3.0	4.0	6.5		
(8) Reserve/Production Ratio:		16.0	14.0	15.0	15.0	15.0	13.0		
(9) Host Government Take ($/unit of production):									
(10) Capital Cost ($/unit of daily capacity):									
(11) Operating Cost ($/unit of production):									

INDIGENOUS SUPPLY ASSESSMENT WORKSHEET

NATURAL GAS

Country: MEXICO

WAES SCENARIO CASE (1985): E

WAES SCENARIO CASE (2000): ________

Total Resource in Place: ________

Domestic Gas Price: ________

Natural Gas:
Offshore ________
Onshore X

Technology Assumptions: 1975 Technology prevailing

	(Units)	1972	1975	1980	1985	1990	2000	2010	2025
(1) Ultimate Recoverable Resource:									
(2) Discoveries:									
(3) Reserves Proved to Date:	10^9BOE	3.8	4.5	5.7	7.0				
(4) Potential Production:									
(5) Expected Annual Production:	10^9BOE	0.13	0.16	0.18	0.2				
(6) Cumulative Production:	10^9BOE	1.7	2.2	3.0	4.0				
(7) Remaining Proved Reserves:	10^9BOE	2.1	2.3	2.7	3.0				
(8) Reserve/Production Ratio:		16.0	14.0	15.0	15.0				
(9) Host Government Take ($/unit of production):									
(10) Capital Cost ($/unit of daily capacity):									
(11) Operating Cost ($/unit of production):									

NONFOSSIL SOURCES AND CONVERSION PROCESS WORKSHEET

Hydroelectric X
Coal Gasification
Coal to Syncrude
Gas Liquifaction
Regasification
Other* (specify):

Country: MEXICO

WAES SCENARIO CASE (1985):

WAES SCENARIO CASE (2000): All scenarios

Total Resource in Place:

Technology Assumptions: Standard engineering practices

	(Units)	1972	1975	1980	1990	2000	2010	2025
(1) Yield:								
(2) Maximum Potential Output	TWH	83	83	83	83	83		
(3) Capital Cost ($/unit of daily capacity):								
(4) Cumulative Capital Investment:								
(5) Operating Cost ($/unit of production):								

*Specify other relevant non-fossil energy source or conversion process, i.e., geothermal, solar, wind, etc.

NONFOSSIL SOURCES AND CONVERSION PROCESSES WORKSHEET

URANIUM AND NUCLEAR

Country: MEXICO

WAES SCENARIO CASE (1985): A

WAES SCENARIO CASE (2000): ______

Total Resource in Place: ______

Domestic Uranium Price: ______

Technology Assumptions: ______

	(Units)	1972	1975	1980	1985	1990	1995	2000	2010
1. Ultimate Recoverable Uranium Resource:	10^3 ton	144.0	144.0	144.0	144.0				
2. Reserve of Uranium Proved to Date:	10^3 ton	3.0	6.0	18.5	31.0				
3. Potential Production of Uranium:									
4. Expected Production of Uranium:	ton			400.0					
5. Reserve/Production Ratio:									
6. Maximum Installed Capacity (GWe)*:				1.0	1.7				
7. Capital Cost ($/KWe):									
8. Operating Cost (¢/Kwhr)**:									

*Indicate assumed load factor and LWR or HWR.

**Excluding capital charges.

NONFOSSIL SOURCES AND CONVERSION PROCESSES WORKSHEET

URANIUM AND NUCLEAR

Country: MEXICO

WAES SCENARIO CASE (1985): B, C, D, E

WAES SCENARIO CASE (2000): ________

Total Resource in Place: ________

Domestic Uranium Price: ________

Technology Assumptions: ________

	(Units)	1972	1975	1980	1985	1990	1995	2000	2010
1. Ultimate Recoverable Uranium Resource:	10^3 ton	144.0	144.0	144.0	144.0				
2. Reserve of Uranium Proved to Date:	10^3 ton	3.0	6.0	18.5	31.0				
3. Potential Production of Uranium:									
4. Expected Production of Uranium:									
5. Reserve/Production Ratio:									
6. Maximum Installed Capacity (GWe)*:				0.9	1.2				
7. Capital Cost ($/KWe):									
8. Operating Cost (¢/Kwhr)**:									

*Indicate assumed load factor and LWR or HWR.

**Excluding capital charges.

NONFOSSIL SOURCES AND CONVERSION PROCESSES WORKSHEET

URANIUM AND NUCLEAR

Country: MEXICO

WAES SCENARIO CASE (1985): C

WAES SCENARIO CASE (2000): 1

Total Resource in Place: ________

Domestic Uranium Price: ________

Technology Assumptions: ________

	(Units)	1972	1975	1980	1985	1990	1995	2000	2010
1. Ultimate Recoverable Uranium Resource:	10^3ton	144.0	144.0	144.0	144.0	144.0		144.0	
2. Reserve of Uranium Proved to Date:	10^3ton	3.0	6.0	18.5	31.0	44.0		85.0	
3. Potential Production of Uranium:									
4. Expected Production of Uranium:	ton			400.0					
5. Reserve/Production Ratio:									
6. Maximum Installed Capacity (GWe)*:				0.7	1.2	15.0		40.0	
7. Capital Cost ($/KWe):									
8. Operating Cost (¢/Kwhr)**:									

*Indicate assumed load factor and LWR or HWR.

**Excluding capital charges.

NONFOSSIL SOURCES AND CONVERSION PROCESSES WORKSHEET

URANIUM AND NUCLEAR

Country: MEXICO

WAES SCENARIO CASE (1985): C

WAES SCENARIO CASE (2000): 2

Total Resource in Place: ________

Domestic Uranium Price: ________

Technology Assumptions: ________

	(Units)	1972	1975	1980	1985	1990	1995	2000	2010
1. Ultimate Recoverable Uranium Resource:	10^3 ton	144.0	144.0	144.0	144.0			144.0	
2. Reserve of Uranium Proved to Date:	10^3 ton	3.0	6.0	18.5	31.0			85.0	
3. Potential Production of Uranium:									
4. Expected Production of Uranium:	ton			400.0					
5. Reserve/Production Ratio:									
6. Maximum Installed Capacity (GWe)*:				0.7	1.2			47.2	
7. Capital Cost ($/KWe):									
8. Operating Cost (¢/Kwhr)**:									

*Indicate assumed load factor and LWR or HWR.

**Excluding capital charges.

NONFOSSIL SOURCES AND CONVERSION PROCESSES WORKSHEET

URANIUM AND NUCLEAR

Country: MEXICO

WAES SCENARIO CASE (1985): D

WAES SCENARIO CASE (2000): 7

Total Resource in Place: ________

Domestic Uranium Price: ________

Technology Assumptions: ________

	(Units)	1972	1975	1980	1985	1990	1995	2000	2010
1. Ultimate Recoverable Uranium Resource:	10^3 ton	144.0	144.0	144.0	144.0			144.0	
2. Reserve of Uranium Proved to Date:	10^3 ton	3.0	6.0	18.5	31.0			85.0	
3. Potential Production of Uranium:									
4. Expected Production of Uranium:	ton			400.0					
5. Reserve/Production Ratio:									
6. Maximum Installed Capacity (GWe)*:				0.7	1.2			20.3	
7. Capital Cost ($/KWe):									
8. Operating Cost (¢/Kwhr)**:									

*Indicate assumed load factor and LWR or HWR.

**Excluding capital charges.

NONFOSSIL SOURCES AND
CONVERSION PROCESSES WORKSHEET

URANIUM AND NUCLEAR

Country: MEXICO

WAES SCENARIO CASE (1985): D

WAES SCENARIO CASE (2000): 8

Total Resource in Place: ______

Domestic Uranium Price: ______

Technology Assumptions: ______

	(Units)	1972	1975	1980	1985	1990	1995	2000	2010
1. Ultimate Recoverable Uranium Resource:	10^3 ton	144.0	144.0	144.0	144.0			144.0	
2. Reserve of Uranium Proved to Date:	10^3 ton	3.0	6.0	18.5	31.0			85.0	
3. Potential Production of Uranium:									
4. Expected Production of Uranium:	ton			400.0					
5. Reserve/Production Ratio:									
6. Maximum Installed Capacity (GWe)*:				0.7	1.2			29.1	
7. Capital Cost ($/KWe):									
8. Operating Cost (¢/Kwhr)**:									

*Indicate assumed load factor and LWR or HWR.

**Excluding capital charges.

NONFOSSIL SOURCES AND CONVERSION PROCESSES WORKSHEET

NUCLEAR ELECTRIC

Country: MEXICO

WAES SCENARIO CASE (1985): C

WAES SCENARIO CASE (2000): 1

Total Resource in Place:

Domestic Uranium Price:

Technology Assumptions:

	(Units)	1972	1975	1980	1985	1990	2000	2010	2025
(1) Potential Production of Uranium.									
(2) Expected Production of Uranium*:									
(3) Yield of Power Plant:	TWH			5.0	8.1		252.4		
(4) Maximum Potential Output:	GW			0.7	1.2		40.0		
(5) Capital Cost ($/unit of daily capacity):									
(6) Cumulative Capital Investment.									
(7) Operating Cost ($/unit of production):									

*Political potential production.

NONFOSSIL SOURCES AND CONVERSION PROCESSES WORKSHEET

NUCLEAR ELECTRIC

Country: MEXICO

WAES SCENARIO CASE (1985): C

WAES SCENARIO CASE (2000): 2

Total Resource in Place: ______

Domestic Uranium Price: ______

Technology Assumptions: ______

	(Units)	1972	1975	1980	1985	1990	2000	2010	2025
(1) Potential Production of Uranium.									
(2) Expected Production of Uranium*:									
(3) Yield of Power Plant:	TWH			5.0	8.1	100.0	330.6		
(4) Maximum Potential Output:	GW			0.7	1.2	15.0	47.2		
(5) Capital Cost ($/unit of daily capacity):									
(6) Cumulative Capital Investment.									
(7) Operating Cost ($/unit of production):									

*Political potential production.

NONFOSSIL SOURCES AND CONVERSION PROCESSES WORKSHEET

NUCLEAR ELECTRIC

Country: MEXICO

WAES SCENARIO CASE (1985): D

WAES SCENARIO CASE (2000): 7

Total Resource in Place: ______

Domestic Uranium Price: ______

Technology Assumptions: ______

	(Units)	1972	1975	1980	1985	1990	2000	2010	2025
(1) Potential Production of Uranium.									
(2) Expected Production of Uranium*:									
(3) Yield of Power Plant:	TWH			5.0	8.1		142.4		
(4) Maximum Potential Output:	GW			0.7	1.2		20.3		
(5) Capital Cost ($/unit of daily capacity):									
(6) Cumulative Capital Investment.									
(7) Operating Cost ($/unit of production):									

*Political potential production.

NONFOSSIL SOURCES AND CONVERSION PROCESSES WORKSHEET

NUCLEAR ELECTRIC

Country: MEXICO

WAES SCENARIO CASE (1985): D

WAES SCENARIO CASE (2000): 8

Total Resource in Place: ________

Domestic Uranium Price: ________

Technology Assumptions: ________

	(Units)	1972	1975	1980	1985	1990	2000	2010	2025
(1) Potential Production of Uranium.									
(2) Expected Production of Uranium*:									
(3) Yield of Power Plant:	TWH			5.0	8.1		203.7		
(4) Maximum Potential Output:	GW			0.7	1.2		29.1		
(5) Capital Cost ($/unit of daily capacity):									
(6) Cumulative Capital Investment.									
(7) Operating Cost ($/unit of production):									

*Political potential production.

INDIGENOUS SUPPLY ASSESSMENT WORKSHEET

COAL

Country: MEXICO

WAES SCENARIO CASE (1985): A,B,E

WAES SCENARIO CASE (2000): ________

Total Resource in Place: ________

Domestic Coal Price: ________

Technology Assumptions: 1975 Technology prevailing

Coal:
Surface Mined ____
Deep Mined X

		1972	1975	1980	1985	1990	2000	2010	2025
(1) Ultimate Recoverable Resource:									
(2) Reserves Proved to Date:	10^6 ton	435	987	1538	2254				
(3) Potential Production:									
(4) Expected Annual Production:	10^6 ton	4	6	9	14				
(5) Cumulative Production:(1)		45	60	98	155				
(6) Remaining Proved Reserves:	10^6 ton	390	927	1440	2240				
(7) Reserve/Production Ratio:		108	160	160	100				
(8) Primary Transportation* —Capital Cost ($/unit of daily capacity): —Operating Cost ($/unit of daily capacity):									
(9) Capital Cost ($/unit of daily capacity):									
(10) Operating Cost ($/unit of production):									

*From mine mouth to first processing stage.

(1) Starting in 1950, 10^6 ton

INDIGENOUS SUPPLY ASSESSMENT WORKSHEET

COAL

Country: MEXICO

WAES SCENARIO CASE (1985): C,D

WAES SCENARIO CASE (2000): 1,7

Total Resource in Place: ______

Domestic Coal Price: ______

Technology Assumptions: 1975 Technology prevailing

Coal:
Surface Mined ____
Deep Mined X

	1972	1975	1980	1985	1990	2000	2010	2025
(1) Ultimate Recoverable Resource:								
(2) Reserves Proved to Date: 10^6 ton	435	987	1538	2254	2394	2394		
(3) Potential Production:								
(4) Expected Annual Production: 10^6 ton	4	6	9	14	25	70		
(5) Cumulative Production: 1950–1972 10^6ton	45	60	98	155	254	728		
(6) Remaining Proved Reserves: 10^6 ton	390	927	1440	2240	2140	1666		
(7) Reserve/Production Ratio:	108	160	160	160	85	24		
(8) Primary Transportation* —Capital Cost ($/unit of daily capacity): —Operating Cost ($/unit of daily capacity):								
(9) Capital Cost ($/unit of daily capacity):								
(10) Operating Cost ($/unit of production):								

*From mine mouth to first processing stage.

(1) Starting in 1950, 10^6ton

INDIGENOUS SUPPLY ASSESSMENT WORKSHEET

COAL

Country: MEXICO

WAES SCENARIO CASE (1985): C,D

WAES SCENARIO CASE (2000): 2,8

Total Resource in Place:

Domestic Coal Price:

Technology Assumptions: 1975 Technology prevailing

Coal:
Surface Mined
Deep Mined X

		1972	1975	1980	1985	1990	2000	2010	2025
(1) Ultimate Recoverable Resource:									
(2) Reserves Proved to Date:	10^6 ton	435	987	1538	2254	2394	2394		
(3) Potential Production:									
(4) Expected Annual Production:	10^6 ton	4	6	9	14	22	51		
(5) Cumulative Production: 1950–1972	10^6 ton	45	60	98	155	244	590		
(6) Remaining Proved Reserves:	10^6 ton	390	927	1440	2240	2150	1804		
(7) Reserve/Production Ratio:		108	160	160	160	100	35		
(8) Primary Transportation* —Capital Cost ($/unit of daily capacity): —Operating Cost ($/unit of daily capacity):									
(9) Capital Cost ($/unit of daily capacity):									
(10) Operating Cost ($/unit of production):									

*From mine mouth to first processing stage.

(1) Starting in 1950, 10^6 ton

The Netherlands

16.1 OVERVIEW AND SUMMARY

Since the Second World War indigenous energy supply in The Netherlands has changed considerably.

Until 1950 indigenous coal accounted for approximately 50 percent of total inland energy consumption. By 1973—the final year in which any Dutch coal was mined—indigenous coal provided less than 2 percent of total energy demand in The Netherlands.

Unlike coal, indigenous oil has never played a substantial role in Dutch energy supply. The resource base is small, and so is the production rate. In 1972 oil provided more than 55 percent of total Dutch energy consumption, but only 4 percent of this oil was supplied from indigenous sources. Although the growth of indigenous oil supply is expected to remain the same, its share of total oil consumption in 2000 will be on the order of 2 percent. Dutch indigenous natural gas is relatively new: production started in 1968. Until then, gas was mainly being produced from coal in city gas factories. Since 1963 penetration of gas increased considerably. By 1972 indigenous natural gas provided for more than 40 percent of total inland energy demand. When production started, policy was directed toward depletion of indigenous gas reserves by 1980. An extensive infrastructure enhanced the growth rate of consumption considerably. Moreover, many large contracts were negotiated for export of Dutch natural gas. Since the oil crisis in 1973 gas depletion policy has been redirected toward long-term continuation of natural gas supply for inland purposes, especially residential use. In addition, Dutch energy policy now also calls for a strategic gas reserve to reduce the vulnerability to oil import disruptions and to preserve indigenous gas as long as possible. This policy leads to an 11 to 15 percent share of indigenous natural gas in total inland energy demand by 2000, corresponding almost completely to residential fuel demand. The share of electricity in final energy demand is expected to rise from approximately 6 percent in 1972 to about 11 percent in 2000 under the WAES assumptions. This electricity is going to be produced by varying percentages of coal, oil, and uranium governed by the WAES assumptions.

This report was prepared by A.C. Sjoerdsma of the Future Shape of Technology Foundation. The data and assumptions are the sole responsibility of the author and do not necessarily reflect official views. The data have been derived from a number of documents and have been supplemented by personal information from a number of specialists.

16.2 COMMENTS ON EACH RESOURCE

16.2.1 Coal

Until 1950 coal was the backbone of Dutch energy supply. Coal provided 70 percent of the total inland energy consumed, and indigenous coal accounted for up to 72 percent of coal consumption. Thus, about 50 percent of total inland energy consumption was then supplied by Dutch coal.

By 1973, the final year in which any Dutch coal was mined, indigenous coal provided less than 2 percent of total inland energy consumption. In 1974 the coal mines in The Netherlands were closed because production was no longer economical.

The total resource of coal in place is estimated at approximately 3.7×10^9 tonnes. Half of this, 1.84 $\times 10^9$ tonnes, could technically be produced, but to be economical, future coal production would require completely new recovery techniques. Since such techniques have not been developed yet, the Dutch assessments of supply through 2000 exclude production of indigenous coal.

16.2.2 Oil

Onshore Oil

Since 1857 small asphalt pits have been exploited in the eastern part of The Netherlands. These were the first indications of possible oil formations. In 1923 the first oil was found in that area. In 1938 oil was discovered accidentally—and to everyone's surprise—when an oil rig was set up at the Educational Exhibition in The Hague to demonstrate drilling in the Dutch East Indies. Under occupation during the war further exploration was enforced in both areas. Actual production started in 1945 in the Schoonebeek concession, and in 1954 in the Rijswijk concession.

Although the total resource in place cannot be

precisely estimated, the quantities are known to be relatively small. In 1975 ultimately recoverable reserves were estimated to be approximately 46 $\times 10^6$ tonnes, about the same as the total produced until 1975 (44 $\times 10^6$ tonnes). The present production rate is only 1.5 $\times 10^6$ tonnes per year.

The Schoonebeek oil contains 6 percent paraffins, 0.9 percent sulphur, and 0.29 percent nitrogen. Its pour point is 10–20° C, and its specific gravity is 0.90–0.93. The high viscosity and high pour point of the Schoonebeek oil make it ideal for experimenting with new extraction and recovery techniques.

It is expected that by 1985 the production rate can be increased slightly to 1.8 $\times 10^6$ tonnes per year, but between 1985 and 2000 onshore oil production in The Netherlands is not expected to increase.

Offshore Oil

Offshore oil in the Dutch part of the North Sea has a very short history. The first offshore legal regulations were formulated in 1965, and the first exploration permit was issued in 1968. The first offshore oil was found in 1970. At present, exploration permits cover 85 percent of the total area of the Dutch continental shelf.

In 1975 the recoverable reserves of offshore oil were estimated at 23 $\times 10^6$ tonnes, including roughly 25 percent natural gas liquids. These small amounts are not being produced now, and there are no plans yet for future production.

16.2.3 Natural Gas

Onshore Natural Gas

The first natural gas in The Netherlands was discovered in 1948 at Coevorden. In 1959 the large Groningen field was discovered at Slochteren, and in 1963 a production concession was granted. Since it became known that the Groningen field is one of the world's largest single gas fields (it covers a surface area of 400 km^2), drilling activity in The Netherlands has become intensive, and as a result a fair number of other smaller gas formations have been discovered in the northern and eastern parts of the country. By the end of 1975, eighty gas fields had been found; fourteen of these are associated with oil in the Schoonebeek and Rijswijk concessions. All these reservoirs except the Groningen field are small because the deep geological formations are intensively disturbed.

During the last few years, exploration activity has declined; growing concern for the environment and for town and country planning has retarded the granting of drilling permits and concessions. In 1974 and 1975 only one new exploratory drilling was performed on the basis of a newly granted drilling permit.

Offshore Natural Gas

The first drilling permit for the Dutch part of the North Sea was granted in 1968. Exploration was first concentrated on the K and L blocks, and later on the A, B, E, F, and Q blocks. Almost all the offshore gas discovered so far has been found in the K and L blocks. A total of forty-six gas reservoirs have been discovered in the North Sea. However, they are all isolated from each other, either horizontally or vertically, by impermeable rock formations; moreover, the deep formations are intensively disturbed and their structure and composition vary rapidly. This is why these gas reservoirs are small in size and content.

By 1976 a total of 130 oil and gas drillings had been performed, and permits covering 85 percent of the total area of the Dutch continental shelf had been granted. The first gas production permit was granted in 1971 (blocks L-10 and L-11). The first offshore gas was landed in 1975 through a 170 km pipeline. A second gas pipeline has been constructed from the K blocks to the shore. Production platforms have been installed in blocks L-10 and K-13.

Natural Gas Reserves

To calculate natural gas reserves the following basic data are needed:

- gross rock volume
- ratio of net to gross reservoir thickness
- porosity
- gas saturation
- gas expansion factor.

These data must be gathered from a sufficient number of exploratory drillings and then processed through interpretation, calculation, and data handling methods.

A distinction must be made between proven and unproven reserves. Proven reserves are those that with 90 percent probability can be produced under existing economic conditions, with existing techniques, and from which the cumulative production is discounted. Unproven reserves are those assumed present and producible but not yet proven with actual drilling; and it is less certain that they can be produced.

These two components add up to the reserves to date with a total production probability of 50 percent. These reserves are presented in Table 16.1.

Cumulative production of natural gas up to 1976 amounted to 448.5 × 10^9 m^3 at 0°C. Ultimately recoverable reserves are estimated at about 3000 × 10^9 m^3 at 0°C, but no probability can be attached to this estimate.

16.3 NATURAL GAS POLICY

When natural gas production in the Groningen field first began in 1963, nuclear energy was expected to become a large-scale substitute for fossil fuels. With the coal crisis of 1958—when oil prices produced a decline in the more expensive mining operations—still fresh in mind, the Dutch government adopted an energy policy calling for depletion of the reserves by 1980. This policy was intended to recover the large investments in the extensive infrastructure necessary to produce, transport, and distribute gas before nuclear energy made gas uneconomical.

In the early 1970s, several factors led to a change in this policy. The obstacles to large-scale application of nuclear energy appeared more formidable than ever. Moreover, further expected natural gas discoveries failed to turn up, and the agreements of Teheran in 1971 as well as the oil crisis of 1973 threw a different light on the world oil situation. A revised Dutch energy policy now called for stretching the natural gas supply as far as possible. In addition, the policy included the formation of a strategic gas reserve to reduce The Netherlands' vulnerability to disruptions in oil imports and to preserve the natural gas reserves as much as possible.

Table 16.1
The Netherlands reserves of natural gas (10^9 m^3 at 0°C)

Source	Proven reserves	Unproven reserves 0°C	Reserves as of 1/1/1976
Groningen	1563.1	190.3	1753.4
Other onshore	95.4	177.3	272.5
Total onshore	1658.5	367.4	2025.9
Offshore	93.7	228.7	322.4
Total	1752.2	596.1	2348.3

Current Dutch energy policy objectives have been formulated as "a guaranteed supply to meet the demand for energy, resulting from the most efficient consumption possible, at the least possible social costs; supply and demand must fit into a more balanced system of international distribution of labour and income, and must be weighted against requirements of ecology, safety and employment." One of the consequences of these energy policy objectives is a slower long-term depletion of the Groningen gas field, thus preserving part of the Dutch gas reserves into the next century.

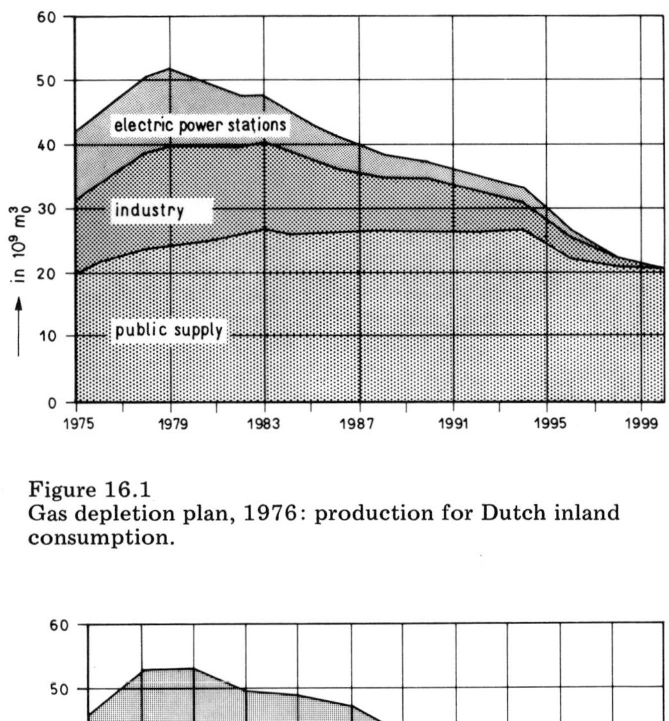

Figure 16.1
Gas depletion plan, 1976: production for Dutch inland consumption.

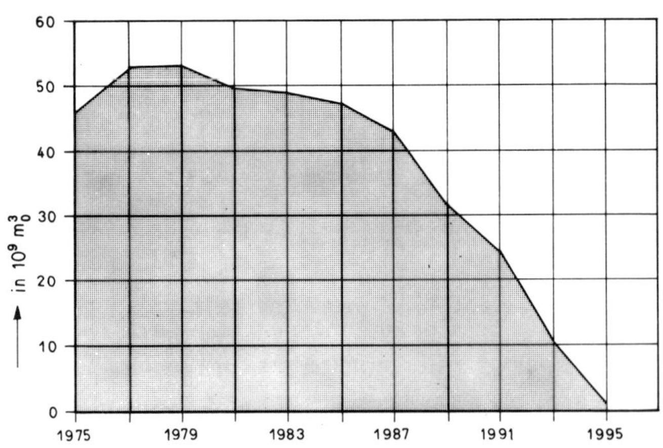

Figure 16.2
Gas depletion plan, 1976: production of Dutch natural gas for export.

The 1976 gas depletion policy is based on the following priorities:

- High priority for residential use (93 percent of all homes use gas).
- Priority for high-grade industrial applications (e.g., chemical feedstock).
- No new contracts with electric power stations.
- Selective new contracts for underfiring of other large boiler installations.
- No extension of export contracts.

The official 1976 gas depletion plan is depicted in Figures 16.1 and 16.2.

16.4 ENERGY SUPPLY WORKSHEETS

This section includes worksheets in the following order (see Section 1.2.1 of Chapter 1 for the definitions of line items):

Energy supply summary

Crude oil

Natural gas (2)

WAES NATIONAL ENERGY SUPPLY STUDIES, 1972-2000
ENERGY SUPPLY SUMMARY

Country: The Netherlands

Scenario Assumptions	1972	1985					2000			
WAES CASE	Base Year	A	B	C	D	E	C-1	C-2	D-7	D-8
GNP 1976-85; 1985-2000		High	Low	High	Low	High	High	High	Low	Low
Oil/Energy Price*		11.50-17.25	11.50-17.25	11.50	11.50	11.50-7.66	11.50-17.25	11.50-17.25	11.50	11.50
National Policy Response		Vig	Vig	Vig	Res	Res	Vig	Vig	Vig	Vig
Principal Replacement Fuel							Coal	Nuclear	Coal	Nuclear
Energy Supply **Units**	10^{15} Joules									
Crude Oil	67			80	80		80	80	80	80
Natural Gas 1)	2058			3318	3318		760	760	760	760
Coal	82									
Nuclear 2)	4			172	103		688	1100	447	688
Hydro-electric										
Geothermal										
Solar Thermal							15	15	8	8
Solar Electric										
Total Energy Supply	2211			3570	3501		1543	1955	1295	1536

*1975 US dollars per barrel of Arabian light crude oil, fob Persian Gulf.
All Figures are primary production prior to any processing or conversion.

1) Onshore gas has been calculated as 1 $m^3 = 35.2 \times 10^6 J$
Offshore gas has been calculated as 1 $m^3 = 39.4 \times 10^6 J$

2) All reactors are LWR's with an operating time of 6,500 hours per year, and a steam cycle efficiency of 34 percent. Capacities represent the maximum potential under WAES assumptions, not necessarily the likely capacities.

INDIGENOUS SUPPLY ASSESSMENT WORKSHEET

CRUDE OIL

Country: The Netherlands

Crude Oil:
Offshore X
Onshore X

WAES SCENARIO CASE (1985): A, B, C, D, E

WAES SCENARIO CASE (2000): C1, C2, D7, D8

Total Resource in Place: unknown

Domestic Oil Price: ______

Technology Assumptions: 1976 technology

	(Units)	1972	1975	1980	1985	1990	2000	2010	2025
(1) Ultimate Recoverable Resource: 1)	10^9 kg		46						
(2) Discoveries:									
(3) Reserves Proved to Date:	10^9 kg		46						
(4) Potential Production:									
(5) Expected Annual Production 10^9 kg	heavy	1.64	1.5		1.8		1.8		
	light								
(6) Cumulative Production:	10^9kg		44		61		88		
(7) Remaining Proved Reserves:									
(8) Reserve/Production Ratio:									
(9) Host Government Take ($/unit of production):									
(10) Capital Cost ($/unit of daily capacity):									
(11) Operating Cost ($/unit of production):									

1) Presently, the oil reserves are not worth further analysis.

INDIGENOUS SUPPLY ASSESSMENT WORKSHEET

NATURAL GAS

Country: The Netherlands

Natural Gas:
Offshore X
Onshore

WAES SCENARIO CASE (1985): A, B, C, D, E

WAES SCENARIO CASE (2000): C1, C2, D7, D8

Total Resource in Place: 700 x $10^9 m^3$ at 0 °C

Domestic Gas Price:

Technology Assumptions: 1976 technology

	(Units)	1972	1975	1980	1985	1990	2000	2010	2025
(1) Ultimate Recoverable Resource: 1)	$10^9 m^3$ at 0 °C	0	400	500	500	500	500		
(2) Discoveries:									
(3) Reserves Proved to Date: 1)		0	94	120	180	250	330		
(4) Potential Production: 1)									
(5) Expected Annual Production: 2)		0	0.1	20	19	13	5		
(6) Cumulative Production:			0.1	50	148	228	318		
(7) Remaining Proved Reserves:			94	70	32	22	12		
(8) Reserve/Production Ratio:									
(9) Host Government Take ($/unit of production):									
(10) Capital Cost ($/unit of daily capacity):									
(11) Operating Cost ($/unit of production):									

1) Many speculative estimates have been published of the possible extent of the natural gas reserves of The Netherlands' continental shelf. This is one of them.

It is still too early to give anything resembling a realistic prognosis of reserves and potential production.

2) Expected annual production has been assumed to be roughly 20 percent of the 1976 gas depletion plan.

INDIGENOUS SUPPLY ASSESSMENT WORKSHEET

NATURAL GAS

Country: The Netherlands

Natural Gas:
Offshore ______
Onshore X

WAES SCENARIO CASE (1985): A, B, C, D, E

WAES SCENARIO CASE (2000): C1, C2, D7, D8

Total Resource in Place: $3500 \times 10^9 m^3$ at 0 °C

Domestic Gas Price: ______

Technology Assumptions: 1976 technology

	(Units)	1972	1975	1980	1985	1990	2000	2010	2025
(1) Ultimate Recoverable Resource:	$10^9 m^3$ at 0 °C	2000	2300	2300	2300	2400	2500		
(2) Discoveries:									
(3) Reserves Proved to Date:		1660	1660	1800	2000	2200	2400		
(4) Potential Production:	1)								
(5) Expected Annual Production:	2)	57.5	89	82	73	52	16		
(6) Cumulative Production:		186	473	900	1288	1600	1940		
(7) Remaining Proved Reserves:		1474	1187	900	712	600	460		
(8) Reserve/Production Ratio:									
(9) Host Government Take ($/unit of production):	3)								
(10) Capital Cost ($/unit of daily capacity):									
(11) Operating Cost ($/unit of production):									

1) Virtually all known Netherlands' onshore gas reserves are now committed against an official depletion plan. By 1979 the reserves will be at their peak production level.

2) Expected annual production has been assumed to be roughly 80% of the 1976 gas depletion plan.

3) Government take on natural gas is influenced by the State participating in the production. It also depends on the location of the reservoir. For the Groningen field, government take varies between 70 and 95 percent of revenues (minus production costs) dependent on the level of revenues. These levels are subject to price indexing.

17 Norway

17.1 SUMMARY

The indigenous supply estimates for year 2000 are about 5.1 million TJ (2.3 MBDOE) for scenarios C-1, C-3, and D-3 (high energy price). For scenarios D-7 and D-8 (medium energy price) the supply potential is about 4.1 million TJ (1.9 MBDOE). Oil and gas from the Norwegian sector of the North Sea provide about 75 percent of the supply, or approximately 4.0 million TJ (1.9 MBDOE) for the high growth case and approximately 3.0 billion TJ (1.4 MBDOE) for the low growth case. The ratio of oil to gas is assumed to be 2:1 for all cases. The remaining supply is basically electricity; of this around 80–90 percent is expected to be provided by hydro power, with the higher percentage in the low economy scenarios. Nuclear power plants are assumed to cover most of the additional electricity. Coal represents less than 1 percent of the total supply.

In 1985 the supply estimate is a total of around 3.7 million TJ (1.7 MBDOE) for scenario C as well as for scenario D. Oil and gas represent about 70 percent of the total or approximately 2.6 million TJ (1.2 MBDOE). The oil to gas ratio is 2:1. Electricity is produced from hydro power only. Coal provides about 1 percent of the total supply.

The GNP growth rate is expected to influence the level of oil and gas production only after 1985. Before that there is no difference between the high and low scenarios because production is a result of the commitments already made. With oil prices at or above the present level, this anticipated production will not be affected. Most of this production will be exported, that is, around two-thirds of the oil and around 90 percent of the gas. There is potential for higher production but it will most likely be limited due to its inflationary effect on the Norwegian economy as well as its social effects. So far the official maximum limit on oil and gas production is about 90 million TOE/year (1.8 MBDOE).

Hydro power will probably be fully developed sometime during the early 1990s, but the rate of development will depend somewhat on the rate of the economic growth. Hydro power, a renewable source, will be developed to its optimum, based on a comparison of the economic and environmental effects of it and other forms of electricity production. Nuclear power plants for electricity production are assumed to be introduced by the beginning of the 1990s. Uranium will probably have to be imported since Norway does not have any high grade nuclear resources. Because strong social resistance to nuclear is expected to continue, at least one oil- or gas-fired plant might be built in a transition period between hydro and nuclear power.

The report was prepared by Kai Killerurd and Paal Kristensen.

17.2 CONCLUSIONS

The supply picture for Norway is very satisfactory in year 2000 as well as in 1985. Most of the abundant supply of oil and gas in the North Sea will be exported. Hydro power for electricity production is projected to be sufficient up to the beginning of the 1990s under all scenarios. Nuclear power will probably be introduced for further expansion of electricity production, particularly under a severe global energy deficit, especially of oil. The potential for higher than assumed production of oil and gas in the North Sea exists, but the current official policy has limited this production to prevent the inflationary and other negative effects that higher production would have on the Norwegian economy.

17.3 COMMENTS ON EACH RESOURCE

17.3.1 Oil and Gas

The resources of oil and gas on the Norwegian Continental Shelf appear significant. To date, however, relatively few test drillings have been carried out, and these only south of 62°N. Even so, there are confirmed resources of more than 1 billion TOE about evenly distributed between oil and gas. It seems very likely that the potential resources south of 62°N will run over 4 billion TOE. On the shelf north of 62°N magnetic and seismological investigations have been conducted, but the first drilling will not be done before 1978. Given the volume of sediments, the total potential resources of oil and gas in this area could be about eight to ten times the potential resources south of 62° N. This estimate, however, is subject to great uncertainties, and exploration, production, and transportation in the north will be more expensive. Thus the potential resources might be much less.

To limit the possible inflationary and social effect of overproducing oil and gas, the Parliament of Norway has decided to limit production to a certain share of the GNP. This limit has been set at about 90 million TOE a year (1.8 MBDOE). The existing official production forecast for the Norwegian Continental Shelf based on existing fields reaches a maximum of 76 million TOE (1.5 MBDOE) in 1985 and then declines to 59 million TOE (1.2 MBDOE) in 1990.

For this analysis, the forecast has been revised to reflect probable new potential reserves and influence from the international economic activity. For the case of high global economic growth during the period 1990–2000 production has been estimated to have a growth rate equal to the government-forecasted medium national economic growth (3.4 percent per year). For the low global economic growth case a flat rate of growth (1.1 percent per year) is assumed for the whole period 1980–2000. This means high and low production rates of 95 and 73 million TOE/year (1.9 and 1.5 MBDOE) in 2000. These estimates are considered reasonably realistic under existing government policies.

17.3.2 Hydro Power

Given the yearly precipitation and topography of Norway, the theoretic energy potential is estimated at 1,800,000 TJ/year (500 TWh/year). However, only 30 percent of this (150 TWh/year) can be utilized. Of this, slightly more than half (80 TWh/year) has been developed so far.

Because of environmental concerns, about 17 TWh/year of the potential 150 TWh/year has been excluded from development, at least up to 1983. Further environmental restrictions are also likely to be introduced. How much of the remaining potential hydro power will be developed in the future depends on economic and environmental comparisons with electricity produced from oil, gas, and uranium. The latest official energy document estimates that about 415,000 TJ/year (115 TWh/year) can be developed at costs equivalent to nuclear power. In this analysis we have chosen a development ceiling of 102 TWh/year, which is identical to the government's most recent plans for 1985. We assume that development will proceed at a rate slightly higher than the demand rate for electricity and that this level will be reached somewhere between 1985 and the early 1990s.

17.3.3 Nuclear Energy

Nuclear power will be a potential energy resource for production of electricity in Norway in about fifteen years. The price of electricity produced from nuclear power will generally be substantially lower than that from thermal plants using oil and gas. It is reasonable to assume that in the longer term oil and gas will be reserved as feedstock for the chemical industry and fuel for transportation where the fuel choices are more limited. Nuclear power will be relatively easy to introduce in Norway's electricity system, which will be dominated by hydro for a long time. Norway does not have any high-grade resources of uranium that can be utilized for present-day reactors.

17.3.4 Coal

Potential reserves of coal at Svalbard (Spitsbergen) are around 30 million tons. The production today is approximately 0.5 million tons per year; this level is assumed to triple as new mines are developed. Most of the present production is being utilized in Norway. However, a coal-fired power plant of 600 MW would use 1 million tons a year, only enough to supply 3 percent of the total electricity production in year 2000. These reserves, therefore, are of minor significance for electricity production but will undoubtedly play an increasing role in the metallurgical industry.

17.3.5 Other Resources

Solar energy might be important after the period 1985 to 1990 for space heating, particularly within the residential sector. Research and development are needed to establish the importance this energy might have. We have assumed that solar energy will supply only about 0.5 percent of Norway's total energy demand in year 2000. Resources like wind, tidal and wave power, and geothermal energy have not been considered significant before 2000. Some R&D is, however, being done in Norway on the possibility of producing electricity from tidal water energy.

17.4 MAXIMUM POTENTIAL LEVEL OF DOMESTIC RESOURCES AVAILABLE FOR EXPORT

Norwegian oil and gas resources in the North Sea are also important internationally. A present political ceiling of about 1.8 MBDOE for future production exists. In the high growth scenarios, a production at about this level has been assumed for the year 2000. The political ceiling might be adjusted upward, to 3 MBDOE for example, if the global development in the energy field in the 1980s and early 1990s indicates that increased production of North Sea oil and gas is essential to the international supply of these energy sources. The supply module below is based on this reasoning. However, such an adjustment does not seem very probable today. Even so, it is highly probable that new reserves could be found to permit production of about 3 MBDOE for a reasonable time.

Description: Production of oil and gas is increased more rapidly than is now foreseen, and in year 2000 a production of 150 million TOE is assumed rather than the 90 million TOE that is today regarded as the present political ceiling.

Oil and gas production: Today's reserves are not sufficient to support this production. Thus new reserves have to be found. It is, however, highly probable that enough new reserves to support the assumed production could be found.

Key numbers for one module (0.1 MBDOE)
Size: 0.1 MBDOE
Total lead time: 5–10 years from find date to full production
Investment: \$0.7–\$1.2 billion/module
Energy cost: \$5–\$12/BOE

Maximum potential: 1.1 MBDOE (equivalent to 150 million tons per year minus the scenario production of 95 million tons per year). This is based on the assumption that the political ceiling of about 90 million tons/year could be changed over time because of the realization that increased production might be an essential contribution to the international supply of oil and gas.

17.5 ENERGY SUPPLY WORKSHEETS

This section includes the following worksheets (see Section 1.2.1 of Chapter 1 for the definitions of line items):

Energy supply summary

Crude oil (2)

Natural gas (2)

Coal

Uranium and nuclear

Hydroelectric

WAES NATIONAL ENERGY SUPPLY STUDIES, 1972-2000
ENERGY SUPPLY SUMMARY

Country: NORWAY

Scenario Assumptions		1972	1985					2000			
WAES CASE		Base Year	A	B	C	D	E	C-1	C-2	D-7	D-8
GNP 1976-85; 1985-2000			High	Low	High	Low	High	High	High	Low	Low
Oil/Energy Price*			11.50-17.25	11.50-17.25	11.50	11.50	11.50-7.66	11.50-17.25	11.50-17.25	11.50	11.50
National Policy Response			Vig	Vig	Vig	Res	Res	Vig	Vig	Vig	Vig
Principal Replacement Fuel								Coal	Nuclear	Coal	Nuclear
Energy Supply	**Units**										
Crude Oil	10^6bdoe	.03	.83	.83	.83	.83	.83	1.27	1.27	.98	.98
Natural Gas	10^6bdoe	.00	.42	.42	.42	.42	.42	.64	.64	.46	.46
Coal	10^6bdoe	.01	.02	.02	.02	.02	.02	.02	.02	.02	.02
Nuclear											
Hydro-electric ~~(primary)~~	10^6bdoe	.30	.46	.46	.46	.46	.46	.46	.46	.46	.46
Geothermal											
Solar Thermal											
Solar Electric											
Total Energy Supply		.34	1.73	1.73	1.73	1.73	1.73	2.39	2.39	1.92	1.92

*1975 US dollars per barrel of Arabian light crude oil, fob Persian Gulf.
All Figures are primary production prior to any processing or conversion.

INDIGENOUS SUPPLY ASSESSMENT WORKSHEET

CRUDE OIL

Country: NORWAY

Crude Oil:
Offshore x
Onshore ____

WAES SCENARIO CASE (1985): A=B=C=D=E

WAES SCENARIO CASE (2000): C-1, C-2, D-3

Total Resource in Place: ____

Domestic Oil Price: ____

Technology Assumptions: ____

	(Units)	1972	1975	1980	1985	1990	2000	2010	2025
(1) Ultimate Recoverable Resource:	10^9toe	36	36	36	36	36	36		
	10^9boe	266	266	266	266	266	266		
(2) Discoveries:									
(3) Reserves Proved to Date:	10^9toe	.30	.60	.76	1.08	1.28	2.09		
	10^9boe	2.2	4.4	5.6	7.1	9.4	12.4		
(4) Potential Production:	10^6toe/y	1.6	8.0	40	42	45	63		
	10^6bdoe	.032	.16	.80	.83	.90	1.27		
(5) Expected Annual Production	0% heavy								
	100% light			Same as Potential Production					
(6) Cumulative Production:	10^9toe	.002	.016	.17	.38	.60	1.14		
	10^9boe	.015	.12	1.26	2.7	4.4	8.4		
(7) Remaining Proved Reserves:	10^9toe	-	-	.59	.60	.68	.95		
	10^9boe	-	-	4.4	4.4	5.0	7.0		
(8) Reserve/Production Ratio:		-	-	15	15	15	15		
(9) Host Government Take ($/unit of production):				13% of production volume plus 50% income tax					
(10) Capital Cost* ($/unit of daily capacity)	$/toe/y	200	200	200	200	200	200		
	$10^3$$/bdoe	9.9	9.9	9.9	9.9	9.9	9.9		
(11) Operating Cost** ($/unit of production):	$/toe	5.6	5.6	5.6	5.6	5.6	5.6		
	$/boe	.75	.75	.75	.75	.75	.75		

* CAPEX + 30%

** OPEX

INDIGENOUS SUPPLY ASSESSMENT WORKSHEET

CRUDE OIL

Country: NORWAY

WAES SCENARIO CASE (1985): A=B=C=D=E

WAES SCENARIO CASE (2000): D-7, D-8

Total Resource in Place: ____

Domestic Oil Price: ____

Technology Assumptions: ____

Crude Oil:
Offshore X
Onshore ____

	(Units)	1972	1975	1980	1985	1990	2000	2010	2025
(1) Ultimate Recoverable Resource:	10^9 toe	36	36	36	36	36	36		
	10^9 boe	266	266	266	266	266	266		
(2) Discoveries:									
(3) Reserves Proved to Date:	10^9 toe	.30	.60	.76	1.08	1.24	1.79		
	10^9 boe	2.2	4.4	5.6	7.1	9.1	13.1		
(4) Potential Production:	10^6 toe/y	1.6	8.0	40	42	43	49		
	10^6 bdoe	.032	.16	.80	.83	.86	.98		
(5) Expected Annual Production	0% heavy / 100% light	Same as Potential Production							
(6) Cumulative Production:	10^9 toe	.002	.016	.17	.38	.59	1.05		
	10^9 boe	.015	.12	1.26	2.7	4.3	7.7		
(7) Remaining Proved Reserves:	10^9 toe	–	–	.59	.60	.65	.74		
	10^9 boe	–	–	4.4	4.4	4.8	5.4		
(8) Reserve/Production Ratio:		–	–	15	15	15	15		
(9) Host Government Take ($/unit of production):		13% of production volume plus 50% income tax							
(10) Capital Cost * ($/unit of daily capacity):	$/toe/y	200	200	200	200	200	200		
	10^3/bdoe	9.9	9.9	9.9	9.9	9.9	9.9		
(11) Operating Cost ** ($/unit of production):	$/toe	5.6	5.6	5.6	5.6	5.6	5.6		
	$/boe	.75	.75	.75	.75	.75	.75		

* CAPEX + 30%

** OPEX

INDIGENOUS SUPPLY ASSESSMENT WORKSHEET

NATURAL GAS

Country: NORWAY

Natural Gas:
Offshore X
Onshore ____

WAES SCENARIO CASE (1985): A=B=C=D=E

WAES SCENARIO CASE (2000): C-1,C-2,D-3

Total Resource in Place: ____

Domestic Gas Price: ____

Technology Assumptions: ____

	(Units)	1972	1975	1980	1985	1990	2000	2010	2025
(1) Ultimate Recoverable Resource:	10^9toe 10^9boe	24 178	24 178	24 178	24 178	24 178	24 178		
(2) Discoveries:									
(3) Reserves Proved to Date:	10^9toe 10^9boe	.20 1.5	.25 1.8	.33 2.5	.50 3.7	.64 4.7	1.05 7.7		
(4) Potential Production:	10^6toe/y 10^6bdoe	0 0	0 0	20 .40	21 .42	23 .45	32 .64		
(5) Expected Annual Production:				Same as Potential Production					
(6) Cumulative Production:	10^9toe 10^9boe	0 0	0 0	.08 .59	.18 1.3	.29 2.1	.57 4.2		
(7) Remaining Proved Reserves:	10^9toe 10^9boe	- -	- -	.25 1.9	.32 2.3	.35 2.6	.48 3.5		
(8) Reserve/Production Ratio:		-	-	6	15	15	15		
(9) Host Government Take ($/unit of production):				13% of production volume plus 50% income tax					
(10) Capital Cost * ($/unit of daily capacity):	$/toe/y $10^3$$/bdoe	200 9.9	200 9.9	200 9.9	200 9.9	200 9.9	200 9.9		
(11) Operating Cost ** ($/unit of production):	$/toe $/boe	5.6 .75	5.6 .75	5.6 .75	5.6 .75	5.6 .75	5.6 .75		

* CAPEX + 30%

** OPEX

INDIGENOUS SUPPLY ASSESSMENT WORKSHEET

NATURAL GAS

Country: NORWAY

Natural Gas:
Offshore X
Onshore

WAES SCENARIO CASE (1985): A=B=C=D=E

WAES SCENARIO CASE (2000): D-7,D-8

Total Resource in Place:

Domestic Gas Price:

Technology Assumptions:

	(Units)	1972	1975	1980	1985	1990	2000	2010	2025
(1) Ultimate Recoverable Resource:	10^9toe 10^9boe	24 178	24 178	24 178	24 178	24 178	24 178		
(2) Discoveries:									
(3) Reserves Proved to Date:	10^9toe 10^9boe	.20 1.5	.25 1.8	.33 2.5	.50 3.7	.63 4.6	.91 6.7		
(4) Potential Production:	10^6toe/y 10^6bdoe	0 0	0 0	20 .40	21 .42	22 .43	25 .46		
(5) Expected Annual Production:				Same as Potential Production					
(6) Cumulative Production:	10^9toe 10^9boe	0 0	0 0	.08 .59	.19 1.4	.30 2.2	.53 3.9		
(7) Remaining Proved Reserves:	10^9toe 10^9boe	- -	- -	.25 1.9	.31 4.7	.33 2.4	.38 2.8		
(8) Reserve/Production Ratio:		-	-	13	15	15	15		
(9) Host Government Take ($/unit of production):				13% of production volume plus 50% income tax					
(10) Capital Cost * ($/unit of daily capacity):	$/toe $10^3$$/bdoe	200 9.9	200 9.9	200 9.9	200 9.9	200 9.9	200 9.9		
(11) Operating Cost ** ($/unit of production):	$/toe $/boe	5.6 .75	5.6 .75	5.6 .75	5.6 .75	5.6 .75	5.6 .75		

* CAPEX + 30%

** OPEX

INDIGENOUS SUPPLY ASSESSMENT WORKSHEET

COAL

Country: NORWAY

Coal:
Surface Mined ______
Deep Mined X

WAES SCENARIO CASE (1985): A=B=C=D=E

WAES SCENARIO CASE (2000): C-1, C-2, D-7, D-8

Total Resource in Place: ______

Domestic Coal Price: ______

Technology Assumptions: ______

		1972	1975	1980	1985	1990	2000	2010	2025
(1) Ultimate Recoverable Resource:	10^6 tce		assumed 115						
(2) Reserves Proved to Date:	10^6 tce								
(3) Potential Production:	10^6 bdoe	.006	.006	.013	.020		.020		
(4) Expected Annual Production:	10^6 tce/y	.47	.49	1.00	1.50		1.50		
	10^6 bdoe	.006	.006	.013	.020		.020		
(5) Cumulative Production:									
(6) Remaining Proved Reserves:	10^6 tce		34	35					
(7) Reserve/Production Ratio:			70	35					
(8) Primary Transportation* —Capital Cost ($/unit of daily capacity)	$/t	2.5	2.5		2.5		2.5		
—Operating Cost ($/unit of daily capacity)	$/boe	.52	.52		.52		.52		
** (9) Capital Cost ($/unit of daily capacity):	$/tce/y	220	220		220		220		
	10^3 $/bdoe	17	17		17		17		
*** (10) Operating Cost ($/unit of production):	$/tce	17	17		17		17		
	$/boe	3.51	3.51		3.51		3.51		

*From mine mouth to first processing stage.

** CAPEX + 25%

*** OPEX

NONFOSSIL SOURCES AND CONVERSION PROCESSES WORKSHEET

URANIUM AND NUCLEAR

Country: NORWAY

WAES SCENARIO CASE (1985) A=B=C=D=E

WAES SCENARIO CASE (2000) C-1, C-2, D-7, D-8

Total Resource in Place: none

Domestic Uranium Price: ______

Technology Assumptions: ______

	(Units)	1972	1975	1980	1985	1990	1995	2000	2010
1. Ultimate Recoverable Uranium Resource: ***		0	0	0	0	0	?	?	
2. Reserve of Uranium Proved to Date:		0	0	0	0	0	?	?	
3. Potential Production of Uranium:		0	0	0	0	0	0	?	
4. Expected Production of Uranium:		0	0	0	0	0	0	0	
5. Reserve/Production Ratio:		–	–	–	–	–	–	–	
6. Maximum Installed Capacity (GWe)* L f.~0.7		0	0	0	0	1 LWR	2 LWR	4 LWR	
7. Capital Cost ($/KWe):			550	575	600	600	600	600	
8. Operating Cost (¢/Kwhr)**:			.80	.84	.88	.9	.9	.9	

*Indicate assumed load factor and LWR or HWR.

**Excluding capital charges.

*** No ore with a uranium content sufficient for economical production has been found so far, however, the possibility that such ore may be found can not be excluded.

NONFOSSIL SOURCES AND CONVERSION PROCESS WORKSHEET

Hydroelectric X
Coal Gasification
Coal to Syncrude
Gas Liquifaction
Regasification
Other* (specify):

Country: NORWAY

WAES SCENARIO CASE (1985): A=B=C=D=E

WAES SCENARIO CASE (2000): C-1,C-2,D-7,D-8

Total Resource in Place: ~150 TWh/year electrical

Technology Assumptions:

	(Units)	1972	1975	1985	1990	2000	2010	2025
(1) Yield: (fictive η):			.35	.35		.35		
(2) Maximum Potential Output (electrical)	TWh/y		.75	104		133		
	10^6 bdoe		.12	.17		.21		
(3) Capital Cost ** ($/unit of daily capacity):	$/MWh/y		139	158		235		
(electr. output)	10^3 $/bdoe		85.2	97		144		
(4) Cumulative Capital Investment:								
(5) Operating Cost *** ($/unit of production):	c/kWh		.05	.05		.05		
(electr. output)	$/boe		.83	.83		.83		

*Specify other relevant non-fossil energy source or conversion process, i.e., geothermal, solar, wind, etc.

** GEMM CAPEX: η + 35%

*** GEMM OPEX : η

Sweden

18.1 OVERVIEW AND SUMMARY

The supply projections to 2000 are based essentially on recent forecasts made by the Energy Forecast Commission published in 1974 and ongoing studies at the National Swedish Board for Energy Source Development. The results in the present report have been adjusted to conform to the WAES scenario cases.

The potential domestic energy production under the WAES scenario assumptions is summarized in Tables 18.1 and 18.2.

Until 1975 about 75 percent of the Swedish energy demand was satisfied by imports, mainly of crude oil and oil products. In absolute numbers oil imports through 1985 are not expected to change significantly, but the percentage of Swedish demand supplied by imported energy, including oil products, coal, and coke, is planned to decrease from about 75 percent to about 60 percent.

In 1978 Parliament will adopt a new energy policy for the 1980s. To that end an energy commission has been appointed to study the whole energy issue and develop the foundation for the new policy. This commission will present energy alternatives based on different assumptions about economic growth, energy demand, development of nuclear programs, and so on. The consequences of the alternatives will be expressed in terms of their economic, social, and environmental effects.

During the period 1975 to 2000 it is assumed that because of a vigorous policy response, domestic energy sources such as peat and organic fuels will start to replace oil in Sweden's energy system. The overall goal for any government in Sweden, then, is to use renewable energy sources as much and as quickly as possible. Solar energy for heating homes might help to reduce Sweden's relatively large oil consumption.

The supply projections in this report do not reflect the opinion of any official institution or government body; they are simply the results of technical and economic judgments. Since the Swedish contribution to the Third Technical Report of WAES includes a detailed analysis of energy policies for Sweden, the policy discussion in this volume is brief.

Table 18.1
Domestically produced primary energy, 1972 and 1985 (TWh)

	1972 Base Year	1985 scenarios				
		A	B	C	D	E
Nuclear		140	140	140	35	
Hydroelectric	154	190	190	190	190	190
Other[a]	31	40	39	42	40	38
Total	185	370	369	372	370	328

[a]Mainly internal energy in the pulp and paper industry, but also includes energy from wastes.

18.2 CONCLUSIONS AND DISCUSSION

Until now about 75 percent of the Swedish energy demand has been satisfied through imports, mainly in the form of crude oil and oil products. Although the absolute volume of oil imports is not expected to change greatly before 1985, imported energy, including oil products, coal, and coke, is planned to decrease to about 60 percent of Swedish energy consumption.

In 1975 the Swedish Parliament adopted an energy policy for the period up to 1985. The goal of this policy is to reduce the annual growth rate of energy consumption from the 4.5 percent that has prevailed since the postwar period to 2 percent per year for the period ending in 1985. Beyond 1985 the goal is to keep consumption constant from the beginning of the 1990s. A second goal is to meet the future heavy requirements of energy needed to raise universal living standards. These two goals will be met by

Table 18.2
Domestically produced primary energy, 2000 (TWh)

	2000 scenarios			
	C-1	C-2	D-7	D-8
Nuclear	140	340	140	240
Hydroelectric	243	214	243	214
Solar thermal	10	2	8	2
Other[a]	91	44	59	47
Total	484	600	450	503

[a]Mainly internal energy in the pulp and paper industry, but also includes contributions from wind power, especially in Cases C-1 and D-7. Other contributions may come from peat and methanol.

- deliberate conservation
- comprehensive research and development
- a guaranteed power supply
- an active oil policy
- international cooperation

The new government elected in September 1976 declared energy to be one of the main issues in its new policy. They strongly support energy conservation, especially the more efficient use of energy in heating and in industrial processes. Energy research will be aimed at developing techniques for better use of renewable energy sources such as solar energy and wind power.

The new declaration devoted special attention to nuclear power, particularly the reprocessing and storage of radioactive waste. The nuclear power program adopted by the former parliament in May 1975 has been slowed down recently, and the new government is now drawing up legislation prohibiting operation of a new nuclear power plant unless the utility can guarantee that the radioactive waste will be taken care of—including initial storage, transportation, and reprocessing—and then permanently and safely stored.

In 1978 Parliament will adopt a new energy policy for the 1980s. The government has appointed a special energy commission to develop a foundation for the new policy. This commission will study nuclear safety problems, present energy alternatives, devise a plan for ending the nuclear program if safety problems are not solved, and clarify the possibility of organizing a referendum on the nuclear issue if political opinion in 1978 is still strongly divided.

Parliament's May 1975 goal for nuclear power was to have an installed capacity of 10.4 GWe by 1985. The total annual production of nuclear electricity would then be 63 TWh—12 percent of the total net energy production, or about 40 percent of all electricity produced. At that level, nuclear power plants would produce almost as much electricity as hydro power plants. The new political situation in Sweden, however, has made the nuclear program uncertain, even though work on the reactors under construction during late 1976 and early 1977 is being continued.

The numbers given in the nuclear data worksheets at the end of this chapter are based on Parliament's program until 1975 and licenses given until that year to start constructing plants. Yet the lead time for constructing a big power plant is so long that decisions to construct stations to come on stream by 1985 must be made no later than 1977. Therefore, the 1985 figures represent maximum likely installed capacity that will be difficult to exceed if decisions to build new plants come too late. The figures may, of course, be lower if the government revises its plans based on the energy commission's findings.

In addition to nuclear energy, new energy resources will be studied between 1975 and 1985; these alternatives have received close attention. Alternative resources, however, will not make important contributions before the 1990s. Only energy conservation in the residential sector might yield a significant contribution by 1985.

As a result of the vigorous policy response assumed in the WAES scenarios beyond 1985, oil would be partly replaced by domestic fuels and raw materials such as peat, various kinds of waste, and organic fuels. Solar energy for heating homes—especially if long-term storage systems could be developed—could help reduce dependence on oil and thus relieve a serious burden on the balance of payments. The goal is to introduce renewable energy sources as quickly and as extensively as possible.

18.3 COMMENTS ON EACH RESOURCE

18.3.1 Hydroelectric Power

During the last few years government commissions have published reports on the possibility of further developing hydroelectric power in Sweden. The present total production per year is about 61 TWh. The potential that may be further economically developed is 34 TWh. This figure includes resources that have been excluded until now for environmental reasons. At present the extent to which these resources will be developed is uncertain because of the strong environmental concern for many of the unharnessed rivers.

The energy bill adopted by the previous parliament in May 1975 included an expansion to 5 TWh above the present level. Further expansion is likely to be postponed until the recently appointed energy commission has evaluated the material published to date. Hydroelectric power must then be discussed in conjunction with electrical power as a whole: how much will be generated by hydropower, how much by nuclear power, and how much by back-pressure power.

18.3.2 Nuclear Energy

Uranium is an abundant resource in Sweden. The total resource in place is about 1 million tons, of which about 30 percent is considered ultimately recoverable. Recent plans were to extract 1300 metric tons per year starting in the early 1980s, corresponding to the needs of the reactors that were planned according to the 1975 energy program. Since Sweden has no enrichment facilities, this uranium would be exported in exchange for enriched reactor fuel.

In view of the present political situation it is not clear if and when Sweden will start mining uranium, or if and when enrichment facilities will be in operation. Because these installations require long lead times, it is hard to imagine that domestic uranium will contribute to the Swedish energy system before 1985. The whole nuclear issue is being carefully examined by the new energy commission. Parliament will base the energy policy for the 1980s on the commission's 1978 reports. Thus their decision will probably come too late to permit any uranium mining and enrichment installations to be in operation by 1985. Moreover, if they decide not to increase the nuclear program during the 1980s and 1990s, it may be better not to mine the domestic uranium resources, provided enough uranium can be imported.

All the relevant nuclear problems—economic, environmental, social, dependence on energy imports—will be studied and considered in evaluating the level of nuclear energy Sweden will adopt.

These assessments of the potential for developing different parts of the nuclear fuel cycle are purely technical and economical judgments made within the WAES scenarios. They in no way reflect the opinion of any particular political party, official institution, or government body. Within the WAES scenarios the Swedish nuclear program is small, although the level of development on the national scale may change considerably.

18.3.3 Waste Heat

The recovery of waste heat is not properly a form of energy production. It is, however, an essential part of a fuel substitution–conservation program; as such, it is appropriate to discuss it along with energy resources.

A waste heat program uses surplus low-grade heat from power production or industrial processes mainly for heating buildings. On the one hand, this recovery should lead to a reduced demand for primary energy (oil, coal, nuclear energy). On the other hand, recovered waste heat will be used even where it is not directly replacing other forms of energy.

It was recently estimated that waste heat used to heat buildings could supply 30–40 percent of energy demand by the year 2000. A total of 50–70 TWh (thermal energy) waste heat per year could be used, corresponding to about 5–7 MTOE per year. This estimate, however, assumes that the nuclear program in Sweden will continue, because most of the usable waste heat will be generated when nuclear power is produced. If the nuclear program is stopped at the level decided by Parliament in 1975 (13 reactors), the waste heat potential will only be about half of this estimate.

18.3.4 Internal Fuel

The use of internal fuel, mainly consisting of spent liquors, bark, and wood waste from the pulp and paper industry, will not increase much from now until 2000. Instead, those industries are expected to conserve more energy as old production lines are replaced by new and more efficient ones. In particular, the pulp and paper industries will demand small quantities of imported fuels by the end of the century. The National Board of Industry recently showed that within only a few years the two main processing industries (pulp and paper, iron and steel) will be able to deliver waste heat for municipal power plants and district heating systems.

18.3.5 Peat

If Sweden maintained its consumption of fossil fuel at today's level and replaced oil with peat, the Swedish supply of peat would be exhausted in 100 to 200 years, depending on extraction techniques. About 15 percent of the Swedish landscape would then be changed. A more reasonable effect on the landscape, however, would take place if only 5 percent of the peat resources are extracted. Production could then be maintained at 2 MTOE per year (about 7 percent of today's oil import) for about 100 years.

Whether and to what extent a peat program is implemented will depend primarily on environmental and social factors, rather than on technical development. Efficient energy systems based on peat could be developed within ten years, a pilot plant of about 50 MW could be in operation by 1990, and a significant energy production could be established by the turn of the century. As a comparison, Finland plans to produce peat equivalent to about 2 MTOE per year by the end of the 1980s.

A 10 percent reduction in the price of peat would be sufficient to make peat competitive with oil at today's price. To achieve that reduction, however, production cost would have to be held within certain limits, and transport over distances of more than 60 km would have to be avoided. Conventional combustion systems based on peat could be used efficiently within a couple of years, whereas developing conversion techniques (to synthetic fuels) would take longer, probably fifteen years. Because of the limited areas in which peat bogs are found and the associated transport problems, the best choice is to develop relatively small power plants (about 100 MWe), about one-tenth the size of today's big power units for base-load production.

18.3.6 Methanol

At present methanol is produced on an industrial scale in several countries. The resource base is essentially natural gas. Large projects are underway in the United States, the Federal Republic of Germany, and South Africa to produce methanol from solid fuels such as coal.

To investigate the possibility of introducing methanol to Sweden, Volvo and the state-owned Berol Kemi AB have formed a company to study and develop different methods of using methanol in Otto engines. A high estimate is that 1 million metric tons of methanol could be used in 1985; that is 17 percent of the transport fuel consumption in 1972. Most of it would have to be imported, however, because it does not seem possible to start producing methanol from domestic resources before 1985. By the end of the 1980s methanol production in Sweden may be sufficient to meet the demand.

The present research program includes a study of the resource base used to produce methanol from domestic resources such as shale, peat, urban solid waste, and waste from forestry, and from imported resources such as rest oil from the refineries, coal, and natural gas.

18.3.7 Solar Energy

Before the year 2000 solar energy in Sweden will be used primarily to heat buildings. Direct (via solar cells) or indirect (via steam) production of electricity is not expected to be significant before the end of the century, even though intensive R&D may well yield some interesting results before then.

The main problem in using solar energy for heating homes is that of storage, but interesting developments are being reported, for example, storage of water heated by solar energy falling on "flat plate" collectors in well-insulated tanks or pools. Some systems seem to be close to commercialization, and solar energy will yield a considerable portion of the energy required for a standard house even in northern latitudes. It is important for houses to have well-insulated walls and three-pane windows and for heat to be recovered for use in the ventilation system.

Other systems under development are based on combinations of solar collectors and heat pumps.

Heating of buildings now accounts for about 40 percent of the total Swedish energy demand. Since oil supplies the major portion of Swedish energy, it is natural that all possible means of reducing oil consumption for heating purposes have received great attention lately. The use of solar energy for heating buildings may be one way to reduce oil consumption.

18.3.8 Wind Energy

Wind energy is not expected to contribute significantly to the Swedish energy supply before the end of the century. Much R&D is still necessary before wind power can be used efficiently. The Board of Energy Source Development recently estimated that by the end of the century, wind energy could provide 10–15 TWhe, equivalent to about 15 percent of today's electricity production in Sweden.

The cost of producing electricity from wind power is still uncertain. The potential of 10–15 TWhe is therefore essentially a technical limit, which may be reached if the cost of producing wind power units can be made sufficiently low.

18.3.9 Shale

The energy that can be extracted from Swedish shales using techniques plausible within the near future is of the order of 1000 MTOE. (The Swedish import of oil products during the 1970s has been about 30 MTOE per year.) The economic and environmental problems, however, are considerable, and the solutions to them will determine the practical possibilities of using this large quantity of energy.

The organic content of Swedish shales makes it less suitable for oil production than, for example, North American shales. However, uranium and other minerals attractive to the nuclear fuel industry can be extracted from it.

If and when shale becomes a useful energy source, a combined extraction process, with simultaneous production of uranium and other minerals and fuels, will probably be used. The technique, economy, and environmental impact of fuel production will therefore depend on how the combined process can be developed.

18.3.10 Other

The National Swedish Board for Energy Source Development is considering many kinds of domestic energy sources for the future. One not yet mentioned in the present report is biomass in different forms. R&D in Sweden will be performed in close cooperation with research organizations abroad, though there are certain reasons to start an intensive national program:

- Sweden is comparatively rich in water and land well suitable for this type of plantation.
- Choices of types of plants and plantation methods are locally or regionally dependent.
- Sweden is too dependent on energy imports.
- Further increases of imported energy may lead to severe consequences for the balance of trade.
- Preliminary cost estimates show that domestic biomass and peat—each or in combination—today are close to being economically feasible energy sources.

18.4 MUNICIPAL ENERGY PLANNING

A government commission recently published a report on municipal energy planning. The commission found that local authorities have had little opportunity to direct energy planning despite its importance to municipalities. The commission suggested that local energy planning be organized to give local authorities the major responsibility for energy planning. The purpose of this arrangement is

- To promote optimal solutions for local energy use, for instance through the selection of technical systems or methods of cooperation between municipalities and industry.
- To improve the basis for prognosis and investigations in the energy field in Sweden.
- To make local goals for energy production and distribution consistent with national goals.

Since imported fuel must be utilized as efficiently as possible, local energy planning will be especially important in a program for distributing waste heat from industry to local district heating systems.

18.5 ENERGY AND THE ENVIRONMENT

During the last few years it has become increasingly clear that rising energy consumption often has unfavorable side effects, including environmental problems such as emission of sulphur from the combustion of fossil fuel, local and possibly global impacts on the climate, and special safety problems in the case of nuclear power. An increasing concern for the preservation of certain areas and ecological systems makes some people hesitate to continue the expansion of hydroelectric power and the extraction of uranium and other minerals from shale.

A special problem related to energy and environment exists in certain industries. To improve working conditions—for example, to make the air inside a steel plant cleaner—better ventilation systems, which often require more energy, must be installed. The requirements for improved environment will in such cases work against energy conservation. Problems of this kind are being studied intensively in Sweden as well as in many other countries.

18.6 REFERENCES

1. Energiprognosutredningen, "Energi 1985-2000," SOU 1974:64-65, Ministry of Industry, 1974.

2. Energiprogramkommitten, "Energiforskning":
SOU 1974:72-76, 1974
DsI 1974:8, 1974
DsI 1975:1, "Behov av prototyper och demonstrationsanläggningar," 1975
DsI 1975:3, "Förslag till Plan för Allmänna energisystemstudier m m," 1975
Ministry of Industry.

3. Federation of Swedish Industries, "Energy Conservation in Swedish Industry," 1975. Report from an "Energy in Industry Day" on May 2, 1974. A similar symposium was held on May 18, 1976. Proceedings from the latter symposium are being published.

4. National Board of Industry, "Hushållning med energi inom industrin," report to the Swedish Government, June 1976.

5. "Energihushållning m m," the Swedish Government's proposal, 1975:30, to the Riksdag on energy management. A summary presentation in English titled "Energy Planning in Sweden" is available from the Ministry of Industry.

6. Ångpanneföreningen, "Verksamheten," 1971-1975, Stockholm.

7. Ångpanneföreningen, "Spillvärmeutnyttjande" (Use of Waste Heat), report to the National Board of Industry, 1976.

8. Ministry of Finance, "Långtidsutredningen 1975," SOU 1975:89, 1975.

9. National Board of Industry, "Energiförsörjningen 1975-1980," SOU 1975:96, Ministry of Finance, 1975.

10. AKA-utredningen, "Använt kärnbränsle och radioaktivt avfall," SOU 1976:30, Ministry of Industry, 1976. A summary of the report in English titled "Spent nuclear and radioactive waste," SOU 1976:32, 1976, is available from the Ministry of Industry.

11. Industrins Utredningsinstitut, "IUI:s långtidsbedömning 1976," 1976.

12. Ministry of Industry, "Petroindustrin i Sverige—petrokemisk industri," SOU 1976:59, 1976.

13. Ministry of Housing, "Vattenkraft och miljö 3," SOU 1976:28, 1976.

14. Ministry of Industry, "Kommunal energiplanering," SOU 1976:55, 1976.

15. Ministry of Trade, "Beredskapslagring av olja, kol och uran," SOU 1976:67, 1976.

16. The National Swedish Board for Energy Source Development, research program.

17. The Secretariat for Future Studies, *Energy and Society.*

18. The National Board of Consumers, "Hushållens energikonsumtion," KOV 1976:1, 1976.

19. Federation of Swedish Industries, conference on the utilization of solar energy, November 1, 1976. Proceedings being published.

20. Ministry of Industry, *Instructions to the Energy Commission,* December 22, 1976.

21. The Royal Academy of Engineering Sciences, *Study on the Swedish Energy Future,* 1977.

22. The Energy R&D Commission (affiliated with the Ministry of Industry), long-term plan of energy research.

23. Svensk författningssamling, "Lag om allmänna fjärrvärmeanläggningar," SFS 1976:838.

24. Transportforskningsdelegationen, "Drivmedelsförbrukningen i Sverige emot år 1990," Personbilar, 1976.

25. Statens Planverk, "Energisparmöjligheter i befintlig bebyggelse—preliminär rapport," B1719/76, 1976.

18.7 ENERGY SUPPLY WORKSHEETS

The following worksheets are included (see Section 1.2.1 of Chapter 1 for the definitions of line items):

Energy supply summary

Uranium and nuclear (4)

Hydroelectric (2)

Internal fuel

Wind (3)

Methanol

Shale oil

Peat

WAES NATIONAL ENERGY SUPPLY STUDIES, 1972-2000
ENERGY SUPPLY SUMMARY

Country: Sweden

Scenario Assumptions	1972	1985					2000			
WAES CASE	Base Year	A	B	C	D	E	C-1	C-2	D-7	D-8
GNP 1976-85; 1985-2000		High	Low	High	Low	High	High	High	Low	Low
Oil/Energy Price*		11.50-17.25	11.50-17.25	11.50	11.50	11.50-7.66	11.50-17.25	11.50-17.25	11.50	11.50
National Policy Response		Vig	Vig	Vig	Res	Res	Vig	Vig	Vig	Vig
Principal Replacement Fuel							Coal	Nuclear	Coal	Nuclear
Energy Supply **Units**					(TWh)					
Crude Oil										
Natural Gas										
Coal										
Nuclear		140	140	140	35		140	340	140	240
Hydro-electric	154	190	190	190	190	190	243	214	243	214
Geothermal										
Solar Thermal							10	2	8	2
Solar Electric										
Others	31	40	39	42	40	38	91	44	59	47
Total Energy Supply	185	370	369	372	370	328	484	600	450	503

*1975 US dollars per barrel of Arabian light crude oil, fob Persian Gulf.
All Figures are primary production prior to any processing or conversion.

NONFOSSIL SOURCES AND CONVERSION PROCESSES WORKSHEET

URANIUM AND NUCLEAR

Country: Sweden

WAES SCENARIO CASE (1985): ABC

WAES SCENARIO CASE (2000): C1

Total Resource in Place: 10^6 t

Domestic Uranium Price: ________

Technology Assumptions: ________

	(Units)	1972	1975	1980	1985	1990	1995	2000	2010
1. Ultimate Recoverable Uranium Resource:	(10^5t)	3	3		3				
2. Reserve of Uranium Proved to Date:	(10^5t)								
3. Potential Production of Uranium:	(t)				800			800	
4. Expected Production of Uranium:	(t)				800			800	
5. Reserve/Production Ratio:									
6. Maximum Installed Capacity (GWe)*:	LWR 70%	0.44	3.2		7.4			7.4	
7. Capital Cost ($/KWe):		500	700						
8. Operating Cost (¢/Kwhr)**:		0.2	0.6						

*Indicate assumed load factor and LWR or HWR.

**Excluding capital charges.

NONFOSSIL SOURCES AND CONVERSION PROCESSES WORKSHEET

URANIUM AND NUCLEAR

Country: Sweden

WAES SCENARIO CASE (1985): ABC

WAES SCENARIO CASE (2000): C2

Total Resource in Place: 10^6 t

Domestic Uranium Price: ____________

Technology Assumptions: ______ ______

	(Units)	1972	1975	1980	1985	1990	1995	2000	2010
1. Ultimate Recoverable Uranium Resource:	(10^5t)	3	3		3				
2. Reserve of Uranium Proved to Date:	(10^5t)								
3. Potential Production of Uranium:	(t)				800			2000	
4. Expected Production of Uranium:	(t)				800			2000	
5. Reserve/Production Ratio:									
6. Maximum Installed Capacity (GWe)*:	LWR 70%	0.44	3.2		7.4			20	
7. Capital Cost ($/KWe):		500	700						
8. Operating Cost (¢/Kwhr)**:		0.2	0.6						

*Indicate assumed load factor and LWR or HWR.

**Excluding capital charges.

NONFOSSIL SOURCES AND CONVERSION PROCESSES WORKSHEET

URANIUM AND NUCLEAR

Country: Sweden

WAES SCENARIO CASE (1985): D

WAES SCENARIO CASE (2000): D7

Total Resource in Place: 10^6 t

Domestic Uranium Price: ______

Technology Assumptions: ______

	(Units)	1972	1975	1980	1985	1990	1995	2000	2010
1. Ultimate Recoverable Uranium Resource:	(10^5t)								
2. Reserve of Uranium Proved to Date:	(10^5t)								
3. Potential Production of Uranium:	(t)				800			800	
4. Expected Production of Uranium:	(t)				200			800	
5. Reserve/Production Ratio:									
6. Maximum Installed Capacity (GWe)*:	LWR 70%	0.44	3.2		7.4			7.8	
7. Capital Cost ($/KWe):		500	700						
8. Operating Cost (¢/Kwhr)**:		0.2	0.6						

*Indicate assumed load factor and LWR or HWR.

**Excluding capital charges.

NONFOSSIL SOURCES AND CONVERSION PROCESSES WORKSHEET

URANIUM AND NUCLEAR

Country: Sweden

WAES SCENARIO CASE (1985): D

WAES SCENARIO CASE (2000): D8

Total Resource in Place: ______

Domestic Uranium Price: ______

Technology Assumptions: ______

	(Units)	1972	1975	1980	1985	1990	1995	2000	2010
1. Ultimate Recoverable Uranium Resource:	(10^5t)								
2. Reserve of Uranium Proved to Date:	(10^5t)								
3. Potential Production of Uranium:	(t)				800			2000	
4. Expected Production of Uranium:	(t)				200			1400	
5. Reserve/Production Ratio:									
6. Maximum Installed Capacity (GWe)*:	LWR 70%	0.44	3.2		7.4				
7. Capital Cost ($/KWe):		500	700						
8. Operating Cost (¢/Kwhr)**:		0.2	0.6						

*Indicate assumed load factor and LWR or HWR.

**Excluding capital charges.

NONFOSSIL SOURCES AND CONVERSION PROCESS WORKSHEET

Hydroelectric X
Coal Gasification ______
Coal to Syncrude ______
Gas Liquifaction ______
Regasification ______
Other* (specify): ______

Country: Sweden

WAES SCENARIO CASE (1985): ABCDE

WAES SCENARIO CASE (2000): Cl D7

Total Resource in Place: 160 TWhe

Technology Assumptions: ______

	(Units)	1972	1975	~~1980~~ 1985	1990	2000	2010	2025
(1) Yield:	(TWhe)	54	59	66	72	85		
(2) Maximum Potential Output	(TWhe)	95	95	95	95	95		
(3) Capital Cost ($/unit of daily capacity):	($/kWe)	600	750	800				
(4) Cumulative Capital Investment:	($ 10^9)			3.8		7		
(5) Operating Cost ($/unit of production):	(c/kWhe)	0.1	0.1	0.2		0.2		

*Specify other relevant non-fossil energy source or conversion process, i.e., geothermal, solar, wind, etc.

NONFOSSIL SOURCES AND CONVERSION PROCESS WORKSHEET

Hydroelectric X
Coal Gasification
Coal to Syncrude
Gas Liquifaction
Regasification
Other* (specify):

Country: Sweden

WAES SCENARIO CASE (1985): ABCDE

WAES SCENARIO CASE (2000): C2 D8

Total Resource in Place: 160 TWhe

Technology Assumptions:

	(Units)	1972	1975	1985 (1980 struck out)	1990	2000	2010	2025
(1) Yield:	(TWhe)	54	59	66	70	75		
(2) Maximum Potential Output	(TWhe)	95	95	95	95	95		
(3) Capital Cost ($/unit of daily capacity):	($/kWe)	600	750	800				
(4) Cumulative Capital Investment:	($ 10^9)			3.8		7		
(5) Operating Cost ($/unit of production):	(c/kWhe)	0.1	0.1	0.2	0.2	0.2		

*Specify other relevant non-fossil energy source or conversion process, i.e., geothermal, solar, wind, etc.

NONFOSSIL SOURCES AND CONVERSION PROCESS WORKSHEET

Hydroelectric ______
Coal Gasification ______
Coal to Syncrude ______
Gas Liquifaction ______
Regasification ______
Other* (specify): Internal Fuel*

Country: Sweden

WAES SCENARIO CASE (1985): ABCDE

WAES SCENARIO CASE (2000): 12 78

Total Resource in Place: ______

Technology Assumptions: ______

	(Units)	1972	1975	1985 ~~1980~~	1990	2000	2010	2025
(1) Yield:	(10^6toe)	2.8	2.9	3		3.7		
(2) Maximum Potential Output	(10^6toe)			3.5		4.0		
(3) Capital Cost ($/unit of daily capacity):	($/kW)		350					
(4) Cumulative Capital Investment:								
(5) Operating Cost ($/unit of production):	(c/kWh		0.2					

*Specify other relevant non-fossil energy source or conversion process, i.e., geothermal, solar, wind, etc.

Note: Internal fuel is mainly spent liquors, bark, and wood waste in the pulp and paper industries.

NONFOSSIL SOURCES AND CONVERSION PROCESS WORKSHEET

Hydroelectric ______
Coal Gasification ______
Coal to Syncrude ______
Gas Liquifaction ______
Regasification ______
Other* (specify): Wind Energy

Country: Sweden

WAES SCENARIO CASE (1985): ______

WAES SCENARIO CASE (2000): Cl

Total Resource in Place: ______

Technology Assumptions: ______

	(Units)	1972	1975	~~1980~~ 1985	1990	2000	2010	2025
(1) Yield:	(TWhe)			1		15		
(2) Maximum Potential Output	(TWhe)			2		20		
(3) Capital Cost ($/unit of daily capacity):								
(4) Cumulative Capital Investment:								
(5) Operating Cost ($/unit of production):	(c/kWhe)					0.6		

*Specify other relevant non-fossil energy source or conversion process, i.e., geothermal, solar, wind, etc.

NONFOSSIL SOURCES AND CONVERSION PROCESS WORKSHEET

Hydroelectric ______
Coal Gasification ______
Coal to Syncrude ______
Gas Liquifaction ______
Regasification ______
Other* (specify): Wind Energy

Country: Sweden

WAES SCENARIO CASE (1985): ______

WAES SCENARIO CASE (2000): D7

Total Resource in Place: ______

Technology Assumptions: ______

	(Units)	1972	1975	~~1980~~ 1985	1990	2000	2010	2025
(1) Yield:	(TWhe)			0.5		5		
(2) Maximum Potential Output	(TWhe)			2		15		
(3) Capital Cost ($/unit of daily capacity):								
(4) Cumulative Capital Investment:								
(5) Operating Cost ($/unit of production):	(c/kWhe)					0.6		

*Specify other relevant non-fossil energy source or conversion process, i.e., geothermal, solar, wind, etc.

NONFOSSIL SOURCES AND CONVERSION PROCESS WORKSHEET

Hydroelectric ____
Coal Gasification ____
Coal to Syncrude ____
Gas Liquifaction ____
Regasification ____
Other* (specify): Wind Energy

Country: Sweden

WAES SCENARIO CASE (1985): ____

WAES SCENARIO CASE (2000): D8

Total Resource in Place: ____

Technology Assumptions: ____

	(Units)	1972	1975	1985 (~~1980~~)	1990	2000	2010	2025
(1) Yield:	(TWhe)			0.25		1		
(2) Maximum Potential Output	(TWhe)			0.5		2		
(3) Capital Cost ($/unit of daily capacity):								
(4) Cumulative Capital Investment:								
(5) Operating Cost ($/unit of production):	(c/kWhe)					0.6		

*Specify other relevant non-fossil energy source or conversion process, i.e., geothermal, solar, wind, etc.

NONFOSSIL SOURCES AND CONVERSION PROCESS WORKSHEET

Hydroelectric ____
Coal Gasification ____
Coal to Syncrude ____
Gas Liquifaction ____
Regasification ____
Other* (specify): Methanol

Country: Sweden

WAES SCENARIO CASE (1985): ____

WAES SCENARIO CASE (2000): C2 D8

Total Resource in Place: ____

Technology Assumptions: ____

	(Units)	1972	1975	1980	1990	2000	2010	2025
(1) Yield:	(10^6toe)				0.5	1.5		
(2) Maximum Potential Output	(10^6toe)				1	2		
(3) Capital Cost ($/unit of daily capacity):								
(4) Cumulative Capital Investment:								
(5) Operating Cost ($/unit of production):								

*Specify other relevant non-fossil energy source or conversion process, i.e., geothermal, solar, wind, etc.

NONFOSSIL SOURCES AND CONVERSION PROCESS WORKSHEET

Hydroelectric ______
Coal Gasification ______
Coal to Syncrude ______
Gas Liquifaction ______
Regasification ______
Other* (specify): Shale Oil

Country: Sweden

WAES SCENARIO CASE (1985): ______

WAES SCENARIO CASE (2000): 12 78

Total Resource in Place: 20×10^9 boe

Technology Assumptions: ______

	(Units)	1972	1975	1980	1990	2000	2010	2025
(1) Yield:	(10^6 boe)				3	6		
(2) Maximum Potential Output	(10^6 boe)				5	10		
(3) Capital Cost ($/unit of daily capacity):								
(4) Cumulative Capital Investment:								
(5) Operating Cost ($/unit of production):								

*Specify other relevant non-fossil energy source or conversion process, i.e., geothermal, solar, wind, etc.

NONFOSSIL SOURCES AND CONVERSION PROCESS WORKSHEET

Hydroelectric ______
Coal Gasification ______
Coal to Syncrude ______
Gas Liquifaction ______
Regasification ______
Other* (specify): Peat

Country: Sweden

WAES SCENARIO CASE (1985): ______

WAES SCENARIO CASE (2000): 12 78

Total Resource in Place: ______

Technology Assumptions: ______

	(Units)	1972	1975	1980	1990	2000	2010	2025
(1) Yield:	(10^6toe)				0.5	1.5		
(2) Maximum Potential Output	(10^6toe)				1.0	2.0		
(3) Capital Cost ($/unit of daily capacity):								
(4) Cumulative Capital Investment:								
(5) Operating Cost ($/unit of production):								

*Specify other relevant non-fossil energy source or conversion process, i.e., geothermal, solar, wind, etc.

19.1 INTRODUCTION

The United Kingdom has large deposits of coal, oil, and gas. The size of oil and coal reserves, however, is offset to some extent by the cost of extraction—most of the coal must be deep mined, while the North Sea is one of the most difficult environments in which oil engineers operate. Nevertheless, at the energy prices envisaged in the WAES scenarios, the U.K. has significant economically recoverable reserves of all types of fossil fuels. However, the rate of production may well depend on technical problems, policy decisions, and social forces, which cannot easily be case in economic terms or realistically modelled with available techniques. To project future production rates, we have estimated the technical constraints and then modified the technically possible rates with other factors relevant to the U.K.

19.2 COAL

An estimate of U.K. coal reserves and resources is shown in Table 19.1. Total "physically workable" reserves (reserves in seams greater than 2 feet thick and at depths less than 4000 feet) could be between 90×10^9 and 160×10^9 tonnes of coal. Of this total about 6.5×10^9 tonnes are in existing mines or prospective new developments. This is equivalent to approximately fifty years of production at current (1976) rates and is large by comparison with the working reserves of the oil and gas industries. However, it is probably conservative since recent survey work by the National Coal Board has confirmed the existence of several large potentially workable new coal fields. In view of this, coal reserves are not likely to be the main limitation on the production rate to the end of the century.

Table 19.1
Coal reserves and resources

Category of reserves	Amount (10^9 tonnes)
1. Classified reserves (all coal accessible from existing collieries and currently economically extractable)	3.5
2. Unclassified reserves (all coal accessible from existing collieries that could become economically extractable with further proving or with beneficial changes in mining methods and/or marketing requirements)	2.5
3. Potential classified reserves (classified reserves at new mine prospects)	0.5
4. All other physically workable coal (at least 2 feet thick, no more than 4000 feet deep) in proved coalfields	92[a]

[a]In addition to the cumulative total of about 98×10^9 tonnes above, it is thought that additional resources in improved coalfields could make the cumulative total of physically workable coal up to about 160×10^9 tonnes.

The coal industry is the oldest fossil fuel industry in the U.K., and its current state is to a large extent a legacy of its past. Over the last two decades, while oil was relatively cheap, there was little investment in new mines, and as a result the average age of U.K. mines is about eighty years. Many of the existing mines will approach the end of their useful life in the period to 2000.

Because of the rise in oil prices in 1973–1974, the National Coal Board (the nationalized industry which produces virtually the entire U.K. coal output) and the U.K. government reassessed their policy towards coal production up to 1985. The "Plan for Coal" announced in July 1974 is the basis for a policy to arrest the decline in production that took place in the two decades up to 1973 and if possible to increase production from 120 million tonnes in 1973 to about 130 million tonnes in 1985.

This policy was based mainly on investment in existing mines to open up new coal faces, but it also envisages the sinking of new mines before 1985. We have assumed that the aims of the Plan for Coal will be achieved and that production in all the 1985 scenarios will therefore be 130 million tonnes.

The production rate between 1985 and 2000 will depend on the investment policy to be decided between now and the early 1980s. Investment to extend the life of existing mines can only have a limited effect, and in the longer term an increasing amount of investment will be required for sinking new mines which take about ten years to bring into full production. Other factors that will determine the level of production in the long term are manpower availability and national policy on areas of natural beauty under which many of the unworked coal reserves lie.

19.3 OIL

The U.K. has produced a small volume of oil for many years, but the prospects have been transformed by the discoveries of large reserves in the North Sea over the past decade. Important discoveries were made in the early 1970s, and production from these fields commenced at the end of 1975. As more fields are brought into operation production will rapidly increase into the early 1980s. Further development of proven fields and future discoveries should allow for production to continue to increase into the mid 1980s.

An estimate of the reserves of oil at the end of 1975 is shown in Table 19.2. Reserves of 2 billion tonnes seem reasonably assured and this might be increased to 3 billion tonnes by possible discoveries in areas of the North Sea already under investigation. This total could be raised further by discoveries in other parts of the North Sea and in offshore areas to the West and South of the U.K. Here there is much less certainty, and the estimate of 1 billion tonnes for such further additions may be optimistic. However, we have based our estimates of future production on a total reserves figure of 4 billion barrels per year.

The rate of production from these reserves will depend on the rate at which they are discovered and developed. Theoretically it would be possible to produce more than 200 million tonnes per year if reserves of 4 billion tonnes were discovered and developed within the next few years. However, it is more likely that the rate of discovery and development will continue to be more gradual. This together with possible policy restraints on the rate of development or production from new fields will result in a much lower maximum production rate.

Table 19.2
Petroleum reserves (Department of Energy end-1975 estimates)

Category of Reserves[a]	Amount (10^9 tonnes)
1. Proven (technically and economically producible)	1.35
2. Probable (greater than 50% chance of being technically and economically producible)	0.96
3. Possible (less than 50% chance of being technically and economically producible)	0.88

[a]These estimates refer to all areas of the U.K. Continental Shelf that have been licensed. The possible cumulative total of 3.19×10^9 tonnes could be raised to 4.5×10^9 tonnes when unlicensed, but already designated, areas are taken into account.

Production from fields under development at the end of 1976 could reach 100 million tonnes per year in 1980 but is likely to peak and begin to decline in the early 1980s. Further increases in output will require production from proven fields not yet under development and from new discoveries. We have therefore assumed that the rate of development will slow and reach a level of around 160 million tonnes per year (7×10^{18} J/yr) by 1985. We assume that production will remain around this level until technical factors force a decline. The rate of production in 2000 is taken to be around 75 million tonnes/year (3.4×10^{18} J/yr)—based on a reserves-to-production ratio of 10 to 1 in that year.

This production profile has been assumed for all the scenarios for both 1985 and 2000. Differences in economic factors such as oil price or economic growth might have some effect on the production rate, but the effect is likely to be small. For example, the higher energy price in 2000 in scenarios C-1 and C-2 compared with scenarios D-7 and D-8 might make some additional fields commercial. However, their contribution is likely to be small and is not expected to make a significant difference in the overall production profile.

19.4 NATURAL GAS

Natural gas was first discovered in the southern North Sea in the early 1960s, and its development enabled the U.K. gas industry to switch progressively from gas manufactured from oil to natural gas and at the same time to expand rapidly the use of gas in the domestic and commercial and industry sectors. The production from these fields is now around the maximum feasible; further increases in supply will have to come from the reserves of gas in the northern North Sea—to a large extent in association with oil.

Future prospects for oil depend on exploration success and on the depletion policy adopted. Table 19.3 shows an estimate of the remaining reserves of gas in discoveries made before the end of

Table 19.3
Gas reserves (Department of Energy estimates for gas remaining in known discoveries at 31 December 1975)

Category of Reserves[a]	Amount (10^{12} ft^3)	Amount (10^9 tonnes of oil equivalent
Proven	28.7	0.69
Probable	11.5	0.28
Possible	10.3	0.25
Total	50.5	1.22

[a]These figures exclude any gas associated with undiscovered oil and also include only the U.K. sector of the Frigg Field.

1975. It excludes any gas associated with undiscovered oil and also excludes gas in the Norwegian sector of the Frigg field (which straddles the boundary between the U.K. and Norwegian sectors of the North Sea and from which the entire production will be consumed in the U.K.).

These reserves figures are much more uncertain than those for oil, partly because of the uncertainties in the quantities of gas that might be associated with oil. In this analysis we have assumed a total gas reserve figure of 1.44×10^9 tonnes oil equivalent (of which 0.14×10^9 tonnes oil equivalent have been produced).

Although these assumptions on reserves are consistent with the latest figures, they may be rather conservative since they exclude reserves of gas that might be found associated with oil that has yet to be discovered. Allowing for this additional associated gas might put the remaining reserves more optimistically in the range 1.3 to 1.9×10^9 tonnes of oil equivalent. Our supply estimates are based on the lower end of this range and may therefore underestimate gas supply potential, especially in the year 2000.

In the projection of supply to 1985 we have assumed that U.K. production will meet demand after taking into account imports from the Norwegian sector of the Frigg field of about 9 million tonnes of oil equivalent per year. This would require total production of between 35 and 39 million tonnes of oil equivalent per year. There are small variations between scenarios due to variations in natural gas demand.

In estimating production in 2000 it has been assumed that production will meet U.K. demand for as long as the reserves-to-production ratio exceeds 10 to 1. Based on ultimate reserves of 1.44 billion tonnes oil equivalent, this results in production of 22 million tonnes oil equivalent in scenarios C-1 and C-2 and 25 million tonnes oil equivalent in scenarios D-7 and D-8. If a higher reserves figure of 1.6 billion tonnes oil equivalent had been assumed, the annual gas production in the year 2000 could have been about 10 million tonnes oil equivalent higher than the figures used in these scenarios.

19.5 NUCLEAR AND ELECTRICITY

Future electricity production is determined by the electricity demand projected under the WAES scenario assumptions. One important factor in determining how the electricity is generated will be the policy towards and the consequent development of nuclear power.

The first nuclear power station in the U.K. was a Magnox reactor commissioned in 1956, but the development since then has been slow. About 5 GW of Magnox stations were installed between 1956 and 1970 but their effective output capacity is only 3.55 GWe. The second phase of nuclear construction was based on Advanced Gas-Cooled Reactors (AGR) of which about 6.5 GW is now under construction or in operation. All these stations are expected to be in operation by 1980.

The third phase of the nuclear program, announced in 1974, is to be based on the Steam-Generating Heavy Water Reactor (SGHWR). In meeting supply estimates for 1985 we have assumed that 4×0.66 GWe of SGHWR plants will be commissioned by 1985. In addition, we have assumed that two further nuclear power stations of 1 GWe each will be installed. These are not yet planned and we therefore do not specify the type.

These estimates were made in late 1975 and early 1976, and at the time of writing (January 1977) these forecasts for nuclear power in 1985 appear optimistic. The SGHWR program has been subject to delay and no firm contracts for their construction have been placed. The large overcapacity in the electricity supply industry which is expected to last for several years is also likely to slow the development of nuclear supply.

We have further assumed that the nuclear stations will operate at a 60 percent load factor. Table 19.4 shows how this compares with the estimated load factor for other types of power stations.

Table 19.4
Estimated output capacity and load factors for electric power in 1985

Source	Estimated 1985 Output Capacity (GWe)	Load Factor (%)
Coal- and Dual-Fired[a]	52	40
Oil	19	20–40[b]
Nuclear	15	60
Hydro[c]	1.3	—
Other	3.3	—
Total	90	—

[a]Dual-fired stations can use both coal and oil.
[b]Depends on scenario; U.K. policy is to make maximum possible use of coal-fired power stations that consume indigenous coal. Therefore the variations in electricity demand between scenarios will be reflected in changes in the supply of electricity from oil-fired stations.
[c]Excluding pumped storage.

To estimate the nuclear capacity in 2000, we have started from the C-2 scenario (high nuclear) and attempted to outline a possible plan for ordering nuclear stations. We assume that 1 GWe is ordered annually until 1983, when the first SGHWR was originally planned to be commissioned, followed by annual orders of 2 GWe to 1986, when, with commercial operation fully proved, the ordering rate is increased to 3, then 4, then 5 GWe, the last increase being delayed until 1991 by the assumption that capital requirements for nuclear power station building would be a constraint.

Assuming six years from ordering to commission, this gives 58 GWe in 2000 for scenario C-2. In this case, electricity demand is such that additional coal-fired power stations are also required. This is also true, but to greater extent, in scenario C-1, a high coal scenario, in which we assume 33 GWe of nuclear power. For the low economic growth cases D-7 and D-8, the installed nuclear capacities assumed are 29 GWe and 50 GWe, the last figure being derived by an argument similar to that applied in scenario C-2.

19.6 SUMMARY

The estimated U.K. supply of fuels is summarized in the worksheets at the end of this chapter, while the assumptions used for fossil fuel reserves are set out in Table 19.5. With the supplies indicated, the U.K. could become a net exporter of energy during the 1980s and early 1990s, as shown in the demand study (WAES First Technical Report, Chapter 14) and the supply-demand integration (WAES Third Technical Report). However, under all the scenarios considered, it is expected that by 2000 the U.K. will once more require imports of energy to satisfy domestic demand. The extent of the import requirement is uncertain, not only because of the doubts about the growth of U.K. energy demand reflected in our scenarios, but also because of uncertainty in the supply prospects.

Table 19.5
Assumptions about effective reserves

Fuel	Assumptions
Coal	No effective limit due to reserves. Production is limited by investment policy, manpower estimates, and lead times for new development.
Oil	For the projections, reserves of 4×10^9 tonnes are assumed to have been proved by the late 1980s. This assumption is considered to be somewhat optimistic.
Gas	For the projections, we assume a reserves figure of 1.3×10^9 tonnes of oil equivalent. This assumption is now considered to be somewhat pessimistic.

19.7 CONVERSION FACTORS AND CONVENTIONS

The conversion factors used in this report are shown in Table 19.6. It should be noted that the thermal content of U.K. coal is lower than the conventional figure of 7 million kilocalories (28.8 GJ) per tonne that is often used internationally.

Table 19.6
Conversion factors

Energy	10^{18} joules	$= 10^9$ GJ
		$= 947 \times 10^6$ Btu
		$= 278$ TWh $= 278 \times 10^9$ kwh
		$= 22.7 \times 10^6$ tonnes of oil equivalent
		$= 39.2 \times 10^6$ tonnes of coal equivalent
		$= 25.0 \times 10^9$ m^3 of natural gas
Power	10^{18} joules per year	$= 32$ GW $= 32 \times 10^9$ Watts
		$= 0.45 \times 10^6$ barrels per day oil equivalent
Calorific values	1 tonne oil equivalent = 44.0 GJ	
	1 tonne coal equivalent = 25.5 GJ	
	10^3 standard m^3 natural gas (at 0°C 760 mm dry)	
		$= 37.9 \times 10^3$ standard ft^3 (at 60°F 30″ wet)
		$= 40.0$ GJ
	10^6 BDOE	$= 2.2 \times 10^9$ GJ per year
		$= 2.2 \times 10^{18}$ J per year
Electricity	A 1000 MW (= 1 GW) power station at 60 percent load factor and at 35 percent conversion efficiency (assumed for 1985) requires annually a primary energy input of 54×10^{15} joules or 1.23 million tonnes of oil equivalent or 2.12 tonnes of coal equivalent.	
	In 2000 it has been assumed that the efficiency will be 37.5 percent. A 1000 MW (= 1 GW) power station at 60 percent load factor and a 37.59 efficiency requires annually an input of 50.5×10^{15} joules or 1.15 million tonnes of oil equivalent or 1.98 million tonnes of coal equivalent.	

19.8 ENERGY SUPPLY WORKSHEETS

The following worksheets are included in this section (see Section 1.2.1 of Chapter 1 for the definitions of line items):

Crude oil

Natural gas (5)

Coal (4)

Uranium and nuclear (4)

Hydroelectric

INDIGENOUS SUPPLY ASSESSMENT WORKSHEET

CRUDE OIL

Country: UNITED KINGDOM

WAES SCENARIO CASE (1985): A,B,C,D,E

WAES SCENARIO CASE (2000): C1,C2,D7,D8

Total Resource in Place: -

Domestic Oil Price: -

Technology Assumptions: -

Crude Oil:
Offshore X
Onshore

	(Units)	1972	1975	1980	1985	1990	2000	2010	2025
(1) Ultimate Recoverable Resource:	10^9 Tonnes		4.00		4.00		4.00		
(2) Discoveries:									
(3) Reserves Proved to Date:	10^9 Tonnes		1.35						
(4) Potential Production:									
(5) Expected Annual Production	10^6 Tonnes	0.1	1.1	100	159.1		77.1		
(6) Cumulative Production:									
(7) Remaining Proved Reserves:									
(8) Reserve/Production Ratio:	Years						∿ 10		
(9) Host Government Take ($/unit of production):									
(10) Capital Cost ($/unit of daily capacity):									
(11) Operating Cost ($/unit of production):									

INDIGENOUS SUPPLY ASSESSMENT WORKSHEET

NATURAL GAS

Country: UNITED KINGDOM

WAES SCENARIO CASE (1985): A

WAES SCENARIO CASE (2000): –

Total Resource in Place: ________

Domestic Gas Price: ________

Technology Assumptions: ________

Natural Gas:
Offshore X
Onshore ____

All in 10^6 Tonnes Oil Equiv	1972	1975	1980	1985	1990	2000	2010	2025
(1) Ultimate Recoverable Resource:	1440	1440		1440		1440		
(2) Discoveries:								
(3) Reserves Proved to Date:		830						
(4) Potential Production:								
(5) Expected Annual Production:	26.2	31.6		38.2				
(6) Cumulative Production:	54.5	141						
(7) Remaining Proved Reserves:		690						
(8) Reserve/Production Ratio:								
(9) Host Government Take ($/unit of production):								
(10) Capital Cost ($/unit of daily capacity):								
(11) Operating Cost ($/unit of production):								

INDIGENOUS SUPPLY ASSESSMENT WORKSHEET

NATURAL GAS

Country: UNITED KINGDOM

Natural Gas:
Offshore X
Onshore

WAES SCENARIO CASE (1985): B

WAES SCENARIO CASE (2000): -

Total Resource in Place:

Domestic Gas Price:

Technology Assumptions:

All in 10^6 Tonnes Oil Equiv	1972	1975	1980	1985	1990	2000	2010	2025
(1) Ultimate Recoverable Resource:	1440	1440		1440		1440		
(2) Discoveries:								
(3) Reserves Proved to Date:		830						
(4) Potential Production:								
(5) Expected Annual Production:	26.2	31.6		35.2				
(6) Cumulative Production:	54.5	141						
(7) Remaining Proved Reserves:		690						
(8) Reserve/Production Ratio:								
(9) Host Government Take ($/unit of production):								
(10) Capital Cost ($/unit of daily capacity):								
(11) Operating Cost ($/unit of production):								

INDIGENOUS SUPPLY ASSESSMENT WORKSHEET

NATURAL GAS

Country: UNITED KINGDOM

Natural Gas:
Offshore X
Onshore

WAES SCENARIO CASE (1985): C

WAES SCENARIO CASE (2000): C1,C2

Total Resource in Place:

Domestic Gas Price:

Technology Assumptions:

All in 10^6 Tonnes Oil Equiv	1972	1975	1980	1985	1990	2000	2010	2025
(1) Ultimate Recoverable Resource:	1440	1440		1440		1440		
(2) Discoveries:								
(3) Reserves Proved to Date:		830						
(4) Potential Production:								
(5) Expected Annual Production:	26.2	31.6		38.6		22.0		
(6) Cumulative Production:	54.5	141						
(7) Remaining Proved Reserves:		690						
(8) Reserve/Production Ratio:								
(9) Host Government Take ($/unit of production):								
(10) Capital Cost ($/unit of daily capacity):								
(11) Operating Cost ($/unit of production):								

INDIGENOUS SUPPLY ASSESSMENT WORKSHEET

NATURAL GAS

Country: UNITED KINGDOM

WAES SCENARIO CASE (1985): D

WAES SCENARIO CASE (2000): D7, D8

Total Resource in Place: ______

Domestic Gas Price: ______

Technology Assumptions: ______

Natural Gas:
Offshore X
Onshore ______

All in 10^6 Tonnes Oil Equiv	1972	1975	1980	1985	1990	2000	2010	2025
(1) Ultimate Recoverable Resource:	1440	1440		1440		1440		
(2) Discoveries:								
(3) Reserves Proved to Date:		830						
(4) Potential Production:								
(5) Expected Annual Production:	26.2	31.6		35.1		25.0		
(6) Cumulative Production:	54.5	141						
(7) Remaining Proved Reserves:		690						
(8) Reserve/Production Ratio:								
(9) Host Government Take ($/unit of production):								
(10) Capital Cost ($/unit of daily capacity):								
(11) Operating Cost ($/unit of production):								

INDIGENOUS SUPPLY ASSESSMENT WORKSHEET

NATURAL GAS

Country: UNITED KINGDOM

Natural Gas:
Offshore X
Onshore ____

WAES SCENARIO CASE (1985): E

WAES SCENARIO CASE (2000): -

Total Resource in Place: ____

Domestic Gas Price: ____

Technology Assumptions: ____

All in 10^6 Tonnes Oil Equiv	1972	1975	1980	1985	1990	2000	2010	2025
(1) Ultimate Recoverable Resource:	1440	1440		1440		1440		
(2) Discoveries:								
(3) Reserves Proved to Date:		830						
(4) Potential Production:								
(5) Expected Annual Production:	26.2	31.6		39.0				
(6) Cumulative Production:	54.5	141						
(7) Remaining Proved Reserves:		690						
(8) Reserve/Production Ratio:								
(9) Host Government Take ($/unit of production):								
(10) Capital Cost ($/unit of daily capacity):								
(11) Operating Cost ($/unit of production):								

INDIGENOUS SUPPLY ASSESSMENT WORKSHEET

COAL

Country: UNITED KINGDOM

WAES SCENARIO CASE (1985): A,B,C,D,E

WAES SCENARIO CASE (2000): C1

Total Resource in Place: 160×10^9 Tonnes

Domestic Coal Price: ________

Technology Assumptions: ________

Coal:
Surface Mined X
Deep Mined X
} TOTAL

10^6 Tonnes of Coal Equivalent

	1972	1975	1980	1985	1990	2000	2010	2025
(1) Ultimate Recoverable Resource:								
(2) Reserves Proved to Date:								
(3) Potential Production:								
(4) Expected Annual Production:	119	128		130		140		
(5) Cumulative Production:								
(6) Remaining Proved Reserves:		4000						
(7) Reserve/Production Ratio:								
(8) Primary Transportation* —Capital Cost ($/unit of daily capacity): —Operating Cost ($/unit of daily capacity):								
(9) Capital Cost ($/unit of daily capacity):								
(10) Operating Cost ($/unit of production):								

*From mine mouth to first processing stage.

INDIGENOUS SUPPLY ASSESSMENT WORKSHEET

COAL

Country: UNITED KINGDOM

WAES SCENARIO CASE (1985): A,B,C,D,E

WAES SCENARIO CASE (2000): C2

Total Resource in Place: 160×10^9 Tonnes

Domestic Coal Price: ______

Technology Assumptions: ______

Coal:
Surface Mined X
Deep Mined X
} TOTAL

10^6 Tonnes Coal Equivalent

	1972	1975	1980	1985	1990	2000	2010	2025
(1) Ultimate Recoverable Resource:								
(2) Reserves Proved to Date:								
(3) Potential Production:								
(4) Expected Annual Production:	119	128		130		90		
(5) Cumulative Production:								
(6) Remaining Proved Reserves:		4000						
(7) Reserve/Production Ratio:								
(8) Primary Transportation* —Capital Cost ($/unit of daily capacity): —Operating Cost ($/unit of daily capacity):								
(9) Capital Cost ($/unit of daily capacity):								
(10) Operating Cost ($/unit of production):								

*From mine mouth to first processing stage.

INDIGENOUS SUPPLY ASSESSMENT WORKSHEET

COAL

Country: UNITED KINGDOM

WAES SCENARIO CASE (1985): A,B,C,D,E

WAES SCENARIO CASE (2000): D7

Total Resource in Place: 160 x 10^9 Tonnes

Domestic Coal Price: ______

Technology Assumptions: ______

Coal:
Surface Mined X
Deep Mined X
} TOTAL

10^6 Tonnes Coal Equivalent

	1972	1975	1980	1985	1990	2000	2010	2025
(1) Ultimate Recoverable Resource:								
(2) Reserves Proved to Date:								
(3) Potential Production:								
(4) Expected Annual Production:	119	128		130		123		
(5) Cumulative Production:								
(6) Remaining Proved Reserves:		4000						
(7) Reserve/Production Ratio:								
(8) Primary Transportation* —Capital Cost ($/unit of daily capacity): —Operating Cost ($/unit of daily capacity):								
(9) Capital Cost ($/unit of daily capacity):								
(10) Operating Cost ($/unit of production):								

*From mine mouth to first processing stage.

INDIGENOUS SUPPLY ASSESSMENT WORKSHEET

COAL

Country: UNITED KINGDOM

Coal:
Surface Mined X
Deep Mined X
} TOTAL

WAES SCENARIO CASE (1985): A,B,C,D,E

WAES SCENARIO CASE (2000): D8

Total Resource in Place: 160×10^9 Tonnes

Domestic Coal Price:

Technology Assumptions:

10^6 Tonnes Coal Equivalent

	1972	1975	1980	1985	1990	2000	2010	2025
(1) Ultimate Recoverable Resource:								
(2) Reserves Proved to Date:								
(3) Potential Production:								
(4) Expected Annual Production:	119	128		130		82		
(5) Cumulative Production:								
(6) Remaining Proved Reserves:		4000						
(7) Reserve/Production Ratio:								
(8) Primary Transportation* —Capital Cost ($/unit of daily capacity): —Operating Cost ($/unit of daily capacity):								
(9) Capital Cost ($/unit of daily capacity):								
(10) Operating Cost ($/unit of production):								

*From mine mouth to first processing stage.

NONFOSSIL SOURCES AND CONVERSION PROCESSES WORKSHEET

URANIUM AND NUCLEAR

Country: UNITED KINGDOM

WAES SCENARIO CASE (1985): A,B,C,D,E

WAES SCENARIO CASE (2000): C2

Total Resource in Place:

Domestic Uranium Price:

Technology Assumptions: Programme based on SGHWR

	(Units)	1972	1975	1980	1985	1990	1995	2000	2010
1. Ultimate Recoverable Uranium Resource:									
2. Reserve of Uranium Proved to Date:									
3. Potential Production of Uranium:									
4. Expected Production of Uranium:									
5. Reserve/Production Ratio:									
6. Maximum Installed Capacity (GWe)*:		4.98[a] 3.55[b]	5.18[a] 3.76[b]		15			58	
7. Capital Cost ($/KWe):									
8. Operating Cost (¢/Kwhr)**:									

*Indicate assumed load factor and LWR or HWR.

**Excluding capital charges.

a. Installed capacity
b. Output capacity

NONFOSSIL SOURCES AND CONVERSION PROCESSES WORKSHEET

URANIUM AND NUCLEAR

Country: UNITED KINGDOM

WAES SCENARIO CASE (1985): A,B,C,D,E

WAES SCENARIO CASE (2000): C1

Total Resource in Place: ________

Domestic Uranium Price: ________

Technology Assumptions: Programme based on SGHWR

	(Units)	1972	1975	1980	1985	1990	1995	2000	2010
1. Ultimate Recoverable Uranium Resource:									
2. Reserve of Uranium Proved to Date:									
3. Potential Production of Uranium:									
4. Expected Production of Uranium:									
5. Reserve/Production Ratio:									
6. Maximum Installed Capacity (GWe)*:		4.98[a] 3.55[b]	5.18[a] 3.76[b]		15			33	
7. Capital Cost ($/KWe):									
8. Operating Cost (¢/Kwhr)**:									

*Indicate assumed load factor and LWR or HWR.

**Excluding capital charges.

a. Installed capacity
b. Output capacity

NONFOSSIL SOURCES AND CONVERSION PROCESSES WORKSHEET

URANIUM AND NUCLEAR

Country: UNITED KINGDOM

WAES SCENARIO CASE (1985): A,B,C,D,E

WAES SCENARIO CASE (2000): D7

Total Resource in Place: ______

Domestic Uranium Price: ______

Technology Assumptions: Programme based on SGHWR

	(Units)	1972	1975	1980	1985	1990	1995	2000	2010
1. Ultimate Recoverable Uranium Resource:									
2. Reserve of Uranium Proved to Date:									
3. Potential Production of Uranium:									
4. Expected Production of Uranium:									
5. Reserve/Production Ratio:									
6. Maximum Installed Capacity (GWe)*:		4.98[a] 3.55[b]	5.18[a] 3.76[b]		15			29	
7. Capital Cost ($/KWe):									
8. Operating Cost (¢/Kwhr)**:									

*Indicate assumed load factor and LWR or HWR.

**Excluding capital charges.

a. Installed capacity
b. Output capacity

NONFOSSIL SOURCES AND CONVERSION PROCESSES WORKSHEET

URANIUM AND NUCLEAR

Country: UNITED KINGDOM

WAES SCENARIO CASE (1985): A,B,C,D,E

WAES SCENARIO CASE (2000): D8

Total Resource in Place: ________

Domestic Uranium Price: ________

Technology Assumptions: Programme based on SGHWR

	(Units)	1972	1975	1980	1985	1990	1995	2000	2010
1. Ultimate Recoverable Uranium Resource:									
2. Reserve of Uranium Proved to Date:									
3. Potential Production of Uranium:									
4. Expected Production of Uranium:									
5. Reserve/Production Ratio:									
6. Maximum Installed Capacity (GWe)*:		4.98[a] 3.55[b]	5.18[a] 3.76[b]		15			50	
7. Capital Cost ($/KWe):									
8. Operating Cost (¢/Kwhr)**:									

*Indicate assumed load factor and LWR or HWR.

**Excluding capital charges.

a. Installed capacity
b. Output capacity

NONFOSSIL SOURCES AND CONVERSION PROCESS WORKSHEET

Hydroelectric X
Coal Gasification ______
Coal to Syncrude ______
Gas Liquifaction ______
Regasification ______
Other* (specify): ______

Country: UNITED KINGDOM

WAES SCENARIO CASE (1985): A,B,C,D,E

WAES SCENARIO CASE (2000): C1,C2,D7,D8

Total Resource in Place: ______

Technology Assumptions: ______

	(Units)	1972	1975	1985	1990	2000	2010	2025
(1) Yield:	10^{15}J(e)	13	15.5	14		15		
(2) Maximum Potential Output								
(3) Capital Cost ($/unit of daily capacity):								
(4) Cumulative Capital Investment:								
(5) Operating Cost ($/unit of production):								

*Specify other relevant non-fossil energy source or conversion process, i.e., geothermal, solar, wind, etc.

20.1 INTRODUCTION

Although the United States possesses abundant energy resources, the future course of development of those resources to meet growing energy demand is not clear. The WAES scenarios to 1985 and 2000 have led to an estimated range of domestic energy supply availability that falls short of likely domestic needs for oil and gas, while providing a large potential surplus of coal. This report does not explicitly deal with energy development in response to specific energy demands or with questions of import and export availability. Rather, it is an analysis of the maximum potential from each domestic energy resource, given the assumptions of the various scenarios.

Potential domestic energy production under the WAES scenario assumptions is summarized in Figures 20.1 and 20.2. In 1985 the estimated production ranges from 61 to 86 Quads (1 Quad = 10^{15} Btu) compared to 62 Quads in 1975. In the year 2000 the range is from 93 to 122. The supply study indicates that the key determinant of the level of U.S. energy production in 1985 is national policy response, with domestic energy prices (influenced strongly by national policy response) playing a lesser but still important role.

The importance of national policy response to the development of domestic energy resources requires a serious national debate on energy policy issues. The relatively high levels of energy production in the "vigorous" response scenarios can be achieved only if the assumed policy decisions are taken and implemented in a timely manner. Given lead times ranging from five to ten or more years for most energy development projects, continued delay may foreclose some of the supply options

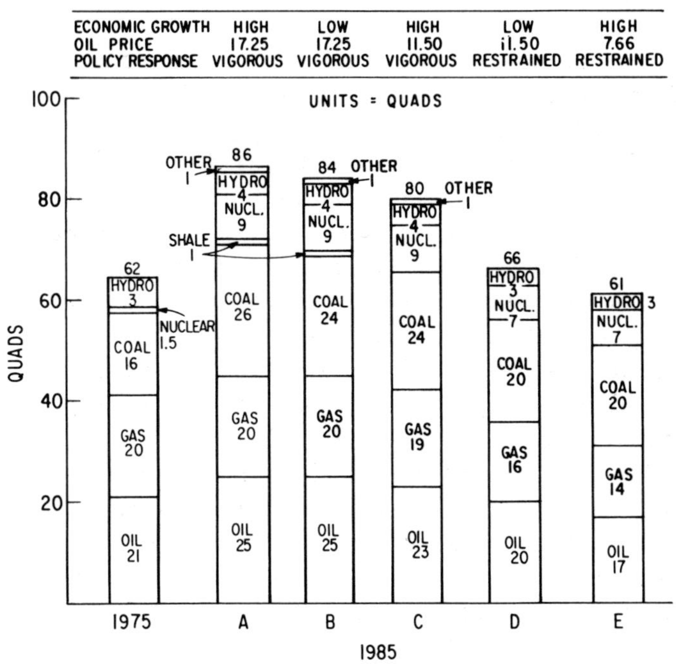

Figure 20.1
How much primary energy can the United States produce domestically in 1985?

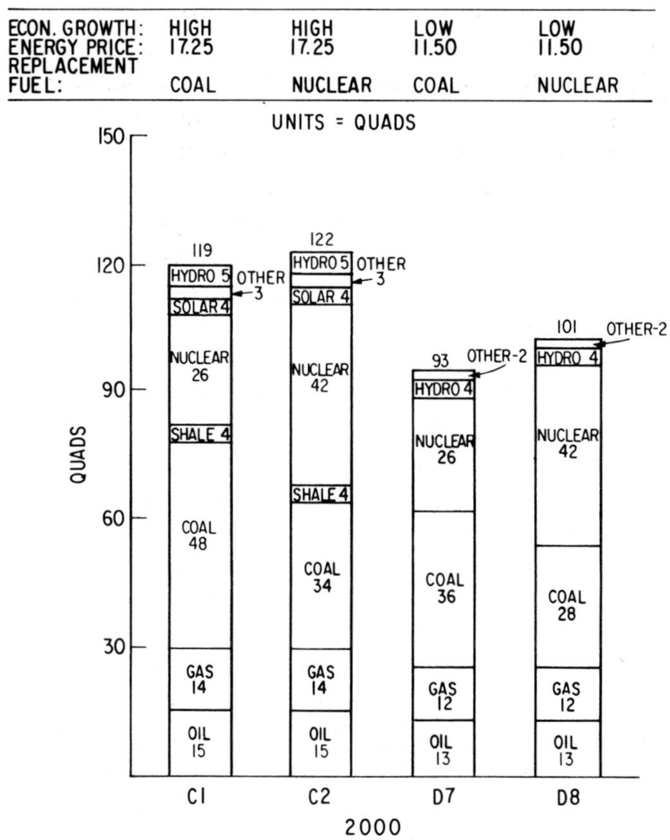

Figure 20.2
How much primary energy can the United States produce domestically in 2000?

This study is the work of the WAES Associates from the United States and was coordinated by Edward Griffith with the assistance of Walter Allaire and John Gardner. We are especially indebted to Mr. Gary Mader and to Dr. Michael Gallagher of Bechtel Corporation for their analysis of the resources required to develop coal and synthetic fuels, and to ERDA/APAE, currently supporting ongoing work in this area, for making possible the dissemination of the information in this study.

included in this study from significant contributions during the twentieth century.

In the WAES scenarios for 2000 vigorous policy response is assumed in all cases beyond 1985, with the policy variable being a choice of principal replacement fuel, either coal or nuclear. The differences in total energy supply between cases is primarily a function of energy prices. The choice between coal and nuclear is roughly a trade-off in terms of primary energy between cases with the same price assumption. In all cases oil and gas decline to 30 percent or less of primary energy supply compared to 66 percent in 1975. The big increases are in coal and nuclear. Solar energy begins to make a significant contribution by 2000 in higher energy price cases, at about 4 Quads or 3 percent of primary energy. While solar will become a major energy source in the twenty-first century, its development is not likely to exceed the levels assumed in the WAES scenarios until after the year 2000.

The massive expansions of coal and nuclear capacity inherent in all of the 2000 scenarios will also require massive infrastructure expansions for energy production and transportation. The question of long lead times and timely policy decisions is again crucial to the realization of the indicated levels of production. The levels of output contemplated in the year 2000 result from a continuous expansion of capacity over a long period. Failure to achieve the necessary increases in the earlier years will make later expansions more difficult, if not impossible.

The WAES study does not include an explicit scenario with long delays or a nuclear moratorium, nor does it consider a long postponement of the development of federal coal lands in the west, although the "restrained" response scenarios to 1985 assume some continuing delays in both nuclear and coal development. The scenarios to 2000 assume a policy commitment to either coal or nuclear, as specified.

The development of synthetic fuels and renewable energy resources is not large in the WAES scenarios, but is believed to represent credible estimates that are consistent with the scenario assumptions. Large-scale development of these energy sources would require either unforeseen technological developments or energy prices higher than the WAES case assumptions.

This report includes projections of potential supply by fuel type for each of the five cases to 1985 and for each of the four cases to 2000.

Fossil fuel resource estimates were developed for each scenario, and a resulting production potential was calculated using the fossil fuel worksheet taking account of the impacts of energy price and national policy response on the development rates of the estimated potential reserves. Consideration was also given to infrastructure development lead times and to environmental constraints. The results were then compared with the range of published estimates and reviewed in detail with knowledgeable individuals.

The production estimates for nonfossil fuels and conversion processes were developed in relation to available published estimates and their related assumptions; these have also been reviewed with knowledgeable individuals.

A brief description of key scenario variables relating to U.S. energy supply is included for each of the scenarios and each energy source.

The capital cost data in the worksheets are taken from a number of sources, including "The Energy Supply Planning Model" (Energy Systems Group, Bechtel Corporation, August 1975). The costs are based on standard facilities as defined in the Bechtel report and represent the cost of building a unit of daily capacity. Operating cost data are being developed in conjunction with the National Center for Analysis of Energy Systems at the Brookhaven National Laboratory. Unless otherwise noted, all energy prices and costs in this report are stated in constant 1975 U.S. dollars.

The data in this study provide one view of potential energy supply for each scenario. The range of estimates in the scenarios are meant to indicate the likely range of energy development potential and should not be taken as forecasts.

20.2 WAES SCENARIO ASSUMPTIONS AND IMPACT ON U.S. ENERGY SUPPLY

WAES scenarios are characterized by three sets of assumptions: real economic growth rate, real oil or energy prices, and national policy response or principal replacement fuel. The specific WAES scenarios to 1985 and 2000, and the associated assumptions, are shown in Table 20.1.

Table 20.1
Couplings of 1985 scenarios to 1985–2000 scenarios

1972–1985 scenarios	A	B	C	D	E
World economic growth[a]	high	low	high	low	high
Oil price[b]	high	high	medium	medium	low
National policy response	vigorous	vigorous	vigorous	restrained	restrained

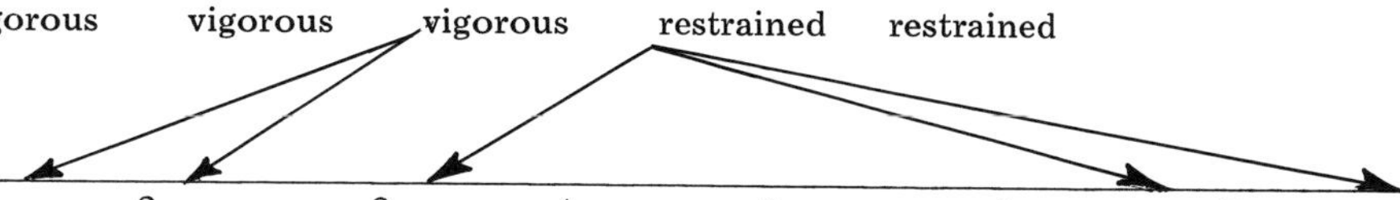

1985–2000 scenarios	1	2	3	4	5	6	7	8
World economic growth[c]	high	high	low	low	high	high	low	low
Energy price[d]	high	high	high	high	medium	medium	medium	medium
Principal replacement fuel	coal	nuclear	coal	nuclear	coal	nuclear	coal	nuclear

[a]High, 6 percent; low, 3.5 percent.

[b]High, $11.50 increasing to $17.25; medium, $11.50 constant; low, $11.50 decreasing to $7.66 (per barrel of light Arabian crude FOB Persian Gulf).

[c]High, 5 percent; low, 3 percent.

[d]High, $17.25; medium, $11.50 (per barrel of oil equivalent).

Table 20.2
WAES U.S. supply policy assumptions 1972–1985

Fuel	Cases A, B, and C Vigorous Policy Response	Cases D and E Restrained Policy Respose
Oil	Rapid outer continental shelf leasing	Delayed outer continental shelf leasing
	Decontrol of crude oil prices in 1979.	Oil price controls on crude oil not removed by 1985.
	Tax incentives to promote secondary and tertiary recovery of crude oil.	No recovery incentives for crude oil.
Gas	Deregulation of new natural gas prices.	Natural gas price regulation continued.
Coal	Surface mining laws eliminate uncertainties about reclamation standards.	Reclamation uncertainties are not resolved soon.
	Federal coal lands leasing program is initiated immediately.	Delays continue in leasing of federal coal lands.
	Government provides assistance to railroads for coal transportation and grants rights to coal slurry pipelines.	Government does not assist expansion of transportation capacity.
Nuclear	Nuclear development is not seriously impacted by additional delays. Environmental and safety concerns about both reactors and the fuel cycle are resolved.	Environmental and safety concerns continue to lead to delays and deferrals in nuclear capacity expansion.
Other	Government research, development, and demonstration programs including financial assistance for the commercialization of emerging energy technologies, such as synthetic fuels, solar, and geothermal.	No additional federal assistance for massive development of alternative sources.

Table 20.3
Estimated U.S. energy supply (physical units)

		1985 scenarios					2000 scenarios			
Fuel source	Units	A	B	C	D	E	C-1	C-2	D-7	D-8
Crude oil (including NGL)	MBD	12.0	12.0	11.0	9.5	8.0	7.0	7.0	6.0	6.0
Natural gas	TCF	20.0	20.0	19.0	16.0	14.0	14.0	14.0	11.5	11.5
Coal	10^6 tons	1040	950	950	800	800	2212	1600	1660	1220
Shale oil	MBD	0.4	0.4	0.2	0	0	2.0	2.0	0	0
Nuclear[a]	GWe	166	166	166	127	127	380	620	380	620
Hydroelectric[a]	GWe	90	90	90	67	67	100	100	80	80
Geothermal[a]	GWe	10	10	10	4	4	50	50	20	20
Solar electric	10^{15} Btu						1.5	1.5	0.5	0.5
Solar heating and cooling	10^{15} Btu	0.2	0.2	0.2	0.1	0.1	2.4	2.4	1.2	1.2
Biomass	10^{15} Btu						1.0	1.0	0	0
Synthetic oil	MBD	0.4	0.4	0.2	0	0	2.0	2.0	0	0
Synthetic gas	TCF	1.4	1.4	0.6	0.2	0.2	6.0	6.0	4.0	0

[a]Installed nameplate capacity.

Table 20.4
Estimated U.S. energy supply (MBDOE)

	1985 scenarios					2000 scenarios			
Fuel source	A	B	C	D	E	C-1	C-2	D-7	D-8
Crude oil (including NGL)	12.0	12.0	11.0	9.5	8.0	7.0	7.0	6.0	6.0
Natural gas	9.4	9.4	9.0	7.6	6.6	6.6	6.6	5.4	5.4
Coal	12.4	11.3	11.3	9.5	9.5	22.7	16.2	16.9	13.1
Shale oil	0.4	0.4	0.2	0	0	2.0	2.0	0	0
Nuclear[a]	4.0	4.0	4.0	3.1	3.1	12.3	20.0	12.3	20.0
Hydroelectric[a]	2.0	2.0	2.0	1.5	1.5	2.2	2.2	1.7	1.7
Geothermal[a]	0.2	0.2	0.2	0.1	0.1	1.4	1.4	0.6	0.6
Solar electric[a]						0.7	0.7	0.2	0.2
Solar heating and cooling	0.1	0.1	0.1	0	0	1.1	1.1	0.6	0.6
Biomass						0.5	0.5	0	0
Total energy supply	40.5	39.4	37.8	31.3	28.8	56.5	57.7	43.7	47.6
Synthetic oil availability						2.0	2.0	0	0
Synthetic gas availability	0.7	0.7	0.3	0.1	0.1	2.8	2.8	1.9	0

[a]Primary energy input prior to generating losses.

In the United States, energy prices and policy response assumptions are the key determinants of potential supplies of oil, gas, and nuclear energy. Synthetic fuels and renewables are also highly constrained by the rate of technology development and long lead times in the WAES time frame to 1985 and 2000. Coal production is constrained primarily by demand, which is expected to remain below potential production. However, sharp increases in coal demand could outstrip the country's ability to rapidly expand coal production and transportation infrastructure. For these reasons possible lead-time constraints and the pace of potential supply increases were carefully considered in our work.

For the United States the most critical and difficult assumptions are those related to national policy response. Some examples of the most critical policy response assumptions to 1985 are shown in Table 20.2. Tables 20.3 and 20.4 recap the results of our analysis for the WAES 1985 and 2000 scenarios in physical units and MBDOE, respectively.

20.3 OIL AND GAS SUPPLY

The United States has historically been one of the world's largest producers of oil and gas, and until recent years it has satisfied most of its own needs from domestic resources. In the early 1970s, however, both oil and gas production peaked and began to decline while demand continued to grow. Imports began to play an increasing role in the U.S. energy supply, and the potential consequences of importing oil to satisfy a significant share of energy demand were suddenly dramatized by the oil embargo in the winter of 1973–1974.

One result of the oil crisis was renewed interest in the potential for increasing domestic production of oil and gas to reduce the nation's dependence on imports. This has led to sharp debates on the potential size of undiscovered oil and gas resources. While any estimate of undiscovered resources must be highly uncertain, recent estimates of the United States Geological Survey are of the same order of magnitude as many other published estimates, and they were adopted for this assessment. We believe that our results are plausible given the scenario assumptions. Our methodology is straightforward and easily replicated. Those who wish to test our results against different assumptions can readily do so.

20.3.1 Methodology

Estimates of future oil and gas production were prepared, consistent with the various scenario assumptions, using the WAES supply worksheets designed expressly for that purpose. The basic approach was to estimate total potential reserves yet to be found, to estimate likely annual additions to reserves consistent with scenario energy price and national policy response assumptions, and to calculate production based on an assumed minimum reserve-to-production (R/P) ratio. The key variable is annual additions to reserves, a judgmental estimate based on both policy assumptions and energy prices. A description of the key policies affecting oil and gas supplies follows for each 1985 scenario. The issues will be similar in year 2000 cases, and vigorous policy response is assumed for all scenarios, with coal or nuclear as the principal replacement fuel.

20.3.2 Scenario Assumptions

Scenarios A and B: 1985

• 1985 crude oil price: $17.25/bbl FOB Persian Gulf

• National policy resposne: Vigorous

In this scenario, high energy prices are a strong incentive to the development of conventional and synthetic energy resources. Vigorous government response leads to extensive domestic energy development by the private sector in response to both market incentives and strong policy signals, including:

1. Adherence to current (1975) outer continental shelf leasing schedules.

2. New crude oil price allowed to rise to landed cost of imports by 1979, followed by full decontrol.

3. Deregulation of new natural gas prices with old prices gradually rising to parity.

4. Full production of Naval Petroleum Reserves 1, 2, and 3 and development of NPR4.

5. Tax incentives to promote secondary and tertiary recovery of crude oil.

6. Government research, development, and demonstration programs, including financial support for the commercialization of emerging energy technologies.

Scenario C: 1985; Scenarios C-1 and C-2: 2000

- 1985 crude oil price: $11.50/bbl FOB Persian Gulf
- National policy response: Vigorous
- 2000 energy price: $17.25/BOE

In this scenario, flat energy prices provide continued incentives to develop most conventional sources of energy supply. Some synthetic sources come onstream in response to government R&D and financial support programs. A vigorous government response is adopted to encourage the development of domestic energy resources and reduce import dependence. Specific policy actions include:

1. Adherence to current (1975) outer continental shelf leasing schedules.

2. New crude oil price allowed to rise to landed cost of imported oil by 1979, followed by full decontrol.

3. Deregulation of new natural gas prices with old prices gradually rising to parity.

4. Full production of Naval Petroleum Reserves 1, 2, and 3 and development of NPR4.

5. Tax incentives to promote secondary and tertiary recovery of crude oil.

6. Government research, development, and demonstration programs, including financial support for the commercialization of emerging energy technologies.

Scenario D: 1985; Scenarios D-7 and D-8: 2000

- 1985 crude oil price: $11.50/bbl FOB Persian Gulf
- National policy response: Restrained
- 2000 energy price: $11.50/BOE

Flat energy prices provide incentive for the continued development of most conventional fossil fuel resources, with the exception of particularly remote or hostile environments. There is insufficient incentive for massive development of alternative sources, although some come onstream in response to government research, development, and demonstration programs. Enhanced recovery techniques that are economically attractive at prevailing prices are implemented. Restrained government policy is characterized by:

1. Full leasing of outer continental shelf not completed by 1985.

2. Price controls on crude oil not removed.

3. Natural gas price regulation continued with modest price escalation on new interstate gas.

4. Partial production from Naval Petroleum Reserves 1, 2, and 3. NPR4 explored, but shut-in.

5. No enhanced recovery incentives for crude oil.

Scenario E: 1985

- 1985 crude oil price: $7.66/bbl FOB Persian Gulf
- National policy response: Restrained

Falling energy prices provide modest incentive for the development of conventional fossil fuel resources, but restricted to large fields and excluding new arctic areas. There is not sufficient incentive for development of alternative sources or for extensive implementation of secondary and tertiary recovery techniques. Restrained government policy is characterized by:

1. Full leasing of outer continental shelf not completed by 1985.

2. Price controls on crude oil not removed.

3. Natural gas price regulation continued with modest price escalation on new interstate gas.

4. Partial production from Naval Petroleum Reserves 1, 2, and 3. NPR4 explored, but shut-in.

5. No enhanced recovery incentives for crude oil.

20.3.3 Liquid Hydrocarbons: Crude Oil and Natural Gas Liquids

Our estimates of potential U.S. crude oil and natural gas liquids production are shown in Figure 20.3. Proved reserves of crude oil and natural gas liquids amounted to 40 billion barrels in 1974 following cumulative prior production of 122

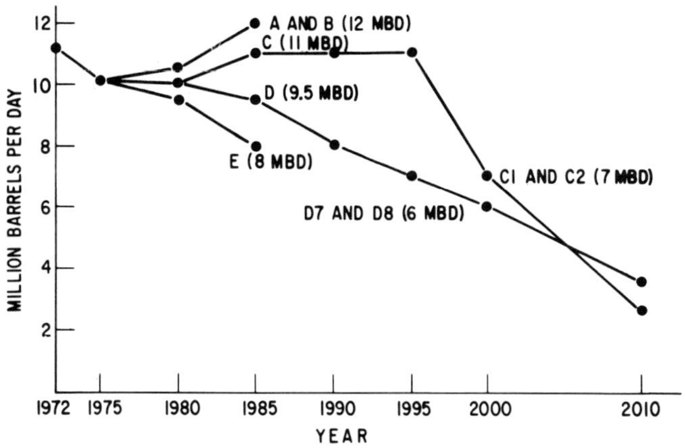

Figure 20.3
U.S. crude oil and NGL production

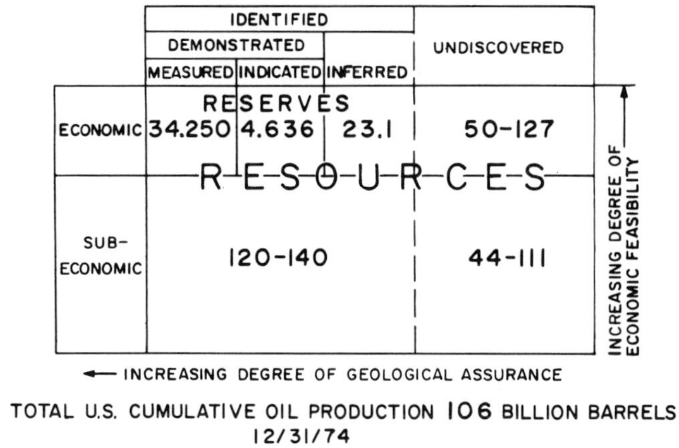

Figure 20.4
Crude oil resources of the United States (10^9 bbl).
Source: U.S. Geological Survey.

billion barrels. The recent U.S. Geological Survey estimates of additional undiscovered recoverable resources give a range from 61 to 149 billion barrels with a mean value of 98 billion barrels. Because this is well within the range of other recent estimates, WAES adopted the USGS value for our analysis. Figure 20.4 shows the USGS estimates of total crude oil resources (without NGL).

Future production rates do not depend strictly on total ultimately recoverable resources but more precisely on the quantity of resources that can be found and developed within the relevant time period. Thus, based on the assumptions of each scenario, we have estimated the percentage of the total remaining resource base that could be developed and put into production by 1985 and 2000. As shown in Tables 20.5 and 20.6, we estimate from 20 to 40 percent of remaining resources by 1985, and from 60 to 80 percent by the year 2000. The resulting reserve additions are then added to existing proved reserves, and an annual maximum production is calculated by dividing by the assumed minimum R/P ratio of 8. (This is about the current ratio in the United States. It is lower than the minimum ratio in many other regions and is consistent with the R/P ratio of 15 assumed in Chapter 3.) For each successive year, the assumed annual production is subtracted from reserves, the potential discoveries added, and a new production rate calculated using the fossil fuel worksheets described earlier in this chapter. This procedure leads to a range of potential crude oil and NGL production of 8 to 12 MBD in 1985 and 6 to 7 MBD in 2000 for the WAES scenarios.

20.3.4 Natural Gas

Our estimates for potential natural gas production are shown in Figure 20.5. These estimates were developed by the same procedure described earlier for crude oil and natural gas liquids. Proved reserves of natural gas at the end of 1974 amounted to 237 trillion cubic feet (TCF) after cumulative production of 481 TCF. The U.S. Geological Survey estimates undiscovered recoverable resources of 322–655 TCF, with a mean value of 484 TCF; these were again used as the basis of the WAES analysis. Published estimates range between 361 and 530 TCF. USGS resource estimates are shown in Figure 20.6.

Our analysis is based primarily on the portion of undiscovered resources that can be found and developed in the WAES time frame to 1985 and 2000, and our assumed minimum R/P ratio of 11 to 1 for natural gas. As shown in Tables 20.7 and 20.8, we estimate that 25–40 percent of the potential resources will be developed by 1985, and 60–80 percent by the year 2000, as a result of our specific scenario assumptions. Calculation on the WAES fossil fuel worksheet leads to an estimated production potential in the range 14–20 TCF in 1985 and 11.5–14 TCF in 2000.

20.4 COAL

Under the assumed WAES scenario conditions, it is apparent that coal will be required to supply an increasing share of the energy burden. A principal

Table 20.5
Scenario variables influencing 1985 U.S. crude oil and NGL production

Variable	1985 scenario A	B	C	D	E
World crude oil price 1985 (1975 $/bbl)	17.25	17.25	11.50	11.50	7.66
Crude oil price control:					
New oil price rises to import cost by 1979, then decontrol	X	X	X		
Price controls on crude oil not removed				X	X
Domestic crude oil price 1985 (1975 $/bbl)	18.25	18.25	12.50	11.50	8.66
Potential discoverable reserves (10^9 bbl)	98	98	98	98	98
Portion leased and proved by 1985 (%)	40	40	35	25	20
Reserve additions through 1985 (10^9 bbl)	39	39	34	25	20
Minimum R/P ratio	8	8	8	8	8
Expected annual production (MBD)	12.0	12.0	11.0	9.5	8.0

Table 20.6
Scenario variables influencing year 2000 U.S. crude oil and NGL production

Variable	2000 scenario C-1	C-2	D-7	D-8
World crude oil price 2000 (1974 $/bbl)	17.25	17.25	11.50	11.50
Domestic crude oil price 2000 (1974 $/bbl)	18.25	18.25	12.50	12.50
Potential discoverable reserves (10^9 bbl)	98	98	98	98
Portion leased and proved by 2000 (%)	80	80	60	60
Reserve additions through 2000 (10^9 bbl)	78	78	59	59
Minimum R/P ratio	8	8	8	8
Expected annual production (MBD)	7.0	7.0	6.0	6.0

Table 20.7
Scenario variables influencing 1985 U.S. natural gas production

Variable	1985 scenario A	B	C	D	E
World crude oil price 2000 (1975 $/bbl)					
Domestic crude oil price 2000 (1975 $/bbl)	X	X	X		
Continued regulation, gradual escalation in ceiling price				X	X
Potential discoverable reserves (TCF)	484	484	484	484	484
Portion leased and proved by 1985 (%)	40	40	35	30	25
Reserve addition through 1985 (TCF)	194	194	169	145	121
Minimum R/P ratio	11	11	11	11	11
Incremental associated gas production related to increased oil production (TCF)	0.6	0.6	0.5	0.3	—
Expected annual production (TCF)	20.0	20.0	19.0	16.0	14.0

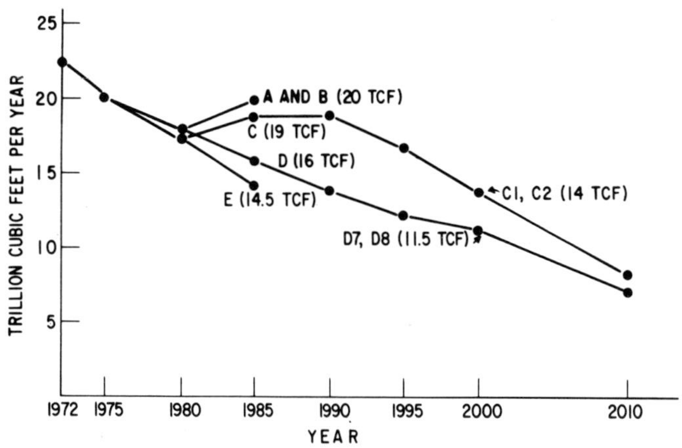

Figure 20.5
U.S. natural gas production.

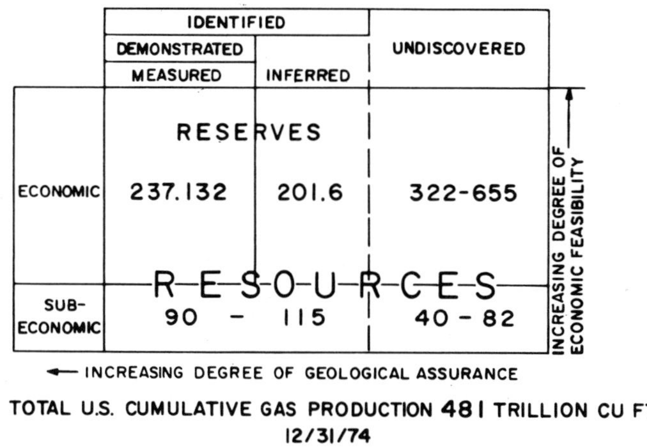

Figure 20.6
Natural gas resources of the United States (10^{12} cubic feet). Source: U.S. Geological Survey.

contributor to the needed world coal requirements will be the vast coal reserves of the United States, which have lain largely undeveloped in modern times.

Coal production in the United States actually declined between 1947 and 1961, from 630 to 402 million tons. The oil embargo renewed interest in U.S. coal, resulting in the production of 640 million tons in 1975. To further develop U.S. coal to the level required to meet future energy demands will require positive and accommodative national policy actions as well as manpower, money, and time.

In this study we have assessed the likely levels of maximum coal production that might be achieved under the assumptions of each scenario. The results are shown in Figure 20.7.

20.4 Policy Issues in United States Coal Development

Producing U.S. coal at the maximum levels consistent with the scenario assumptions would require massive expansion of western surface mining and sizable increases in eastern underground mining. Such large-scale development would require early resolution of current uncertainties regarding surface mining regulations and air quality standards.

Widespread surface mining in previously undisturbed areas is opposed by environmental groups and local agricultural interests. The two most important objections to western surface mining are (1) the potential despoiling of large tracts of land and (2) potentially undesirable social and economic impacts in a primarily rural, sparsely populated, agriculturally oriented society of small towns and isolated ranches. Ranchers are concerned that some of the ranch land currently used to graze livestock may be ruined for future agricultural use and that water requirements for mining and reclamation may conflict with the needs of farmers.

New mines are now being developed in the Powder River Basin of the Northern Great Plains. Soon

Table 20.8
Scenario variables influencing year 2000 U.S. natural gas production

Variable	2000 scenario C-1	C-2	D-7	D-8
Domestic natural gas price (1974 $/BOE)	18.25	18.25	12.50	12.50
Potential discoverable reserves (TCF)	484	484	484	484
Portion leased and proved by 2000 (%)	80	80	60	60
Reserve addition through 2000 (TCF)	387	387	290	290
Minimum R/P ratio	11	11	11	11
Expected annual production (TCF)	14.0	14.0	11.5	11.5

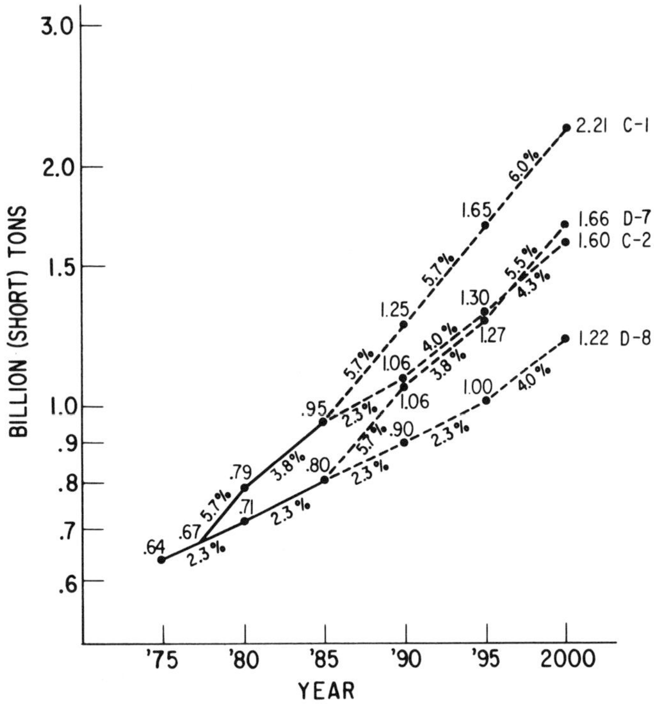

Figure 20.7
The outlook for U.S. coal production to 2000.

some of these mines will be operating, and the mine developers will be in a position to demonstrate whether they can conduct mining operations in an acceptable manner and effectively rehabilitate disturbed lands. The mining companies are confident that this can be accomplished at a reasonable cost in relation to the value of the coal produced. If the acceptability of these rehabilitation programs is demonstrated, as has been done in Europe, environmental issues are not likely to be a major constraint on future coal mine development.

Recent federal legislation has provided assistance to local governments in planning for and meeting the cost of expanded social and community services. Most of the states involved have also adopted strict surface mining regulations that will allow new mine development with concurrent reclamation to prevent the type of abuse that has occurred in the mountainous coal mining areas of the eastern United States.

On the other hand, Congress strongly supports federal legislation to regulate surface mining activities at the federal level and to limit mining in certain areas. This possible legislation and other federal/state conflicts—over regulatory jurisdiction, royalty sharing, land use planning, and local government assistance—may slow the expansion of western surface mining activities. The significance of these issues and the related policy choices is demonstrated by the difference of 600 million short tons in estimated U.S. annual production potential between the rising energy price/coal scenario (C-1) and the rising energy price/nuclear scenario (C-2).

The scenarios with higher estimates also include substantial increases in eastern underground mining in the United States. Development of this high-cost coal is consistent with the rising energy price scenarios. The relative costs of underground coal may continue to rise because of increasingly stringent mine safety regulations, associated declines in underground mining productivity, and the gradual depletion of readily accessible reserves. Increasing productivity due to technological advancements may lead to improved productivity and reduce the problem of potential manpower shortages in underground coal mining. Development of eastern underground coal would stimulate employment and economic activity in a region of the country that has had chronic economic development and employment problems for many years.

Resolution of the current debate over air quality standards is necessary to allow the anticipated increases in demand for coal to materialize. The outcome may increase the cost to consumers of burning coal, and thus affect its competitiveness with other fuels, because of the added cost of controlling sulfur dioxide and other pollution.

Increased coal production will also require expanded transportation facilities. This may require policy action allowing the development of coal slurry pipelines and expanded port facilities for coal exports.

Table 20.9 sets out the basic assumptions on coal policies that underlie our projections for 1985. The "vigorous" policies have been assumed in the year 2000 scenarios with coal as the principal replacement fuel.

20.4.2 Coal Resource Base

Recent estimates of U.S. coal resources are set out in Table 20.10. Recoverable reserves in reach of current technology and economics are on the order of 273 billion tons, over 400 times current annual production levels.

Table 20.9
Definitions of national policy responses for U.S. coal scenarios

Policy area	Vigorous response	Restrained response
1. Surface mining laws	Debate over surface mining laws is resolved to eliminate the uncertainties about reclamation standards	Reclamation uncertainties are not resolved soon
2. Coal lands leasing programs	An effective coal lands leasing program is initiated by 1977	Delays continue in the establishment of an effective program to lease federal coal lands
3. Environmental impact statements	Government shortens the processing time of environmental impact statements to hasten the expansion stimulated by other government policies	Processing time of environmental impact statements does not improve
4. Air quality standards	Ambient air quality standards permit the burning of western coal without stack scrubbers	Stack scrubbers required for western coal
5. Miner/manpower training programs	Government supports a miner training program to meet industry's manpower requirements	Government does not support miner/manpower training programs
6. Labor productivity	Government assists in increasing miner productivity	Government does not assist in increasing labor productivity
7. Equipment productivity	Tax incentives are immediately created to induce industry to purchase more productive mining equipment	Tax incentives for increased equipment productivity are not instituted
8. R&D programs for mining techniques	Government supports R&D programs to improve mining techniques	No government-supported R&D programs for mining techniques established
9. Slurry pipelines	Governments grant rights of eminent domain to coal slurry pipelines to support coal transportation needs	Coal slurry pipelines continue to have right-of-way problems
10. Railroad transportation	Government provides assistance to railroads for their expansion and maintenance of capacity to ensure increased coal transportation capabilities	Government does not assist expansion of railroad's coal transportation capacity
11. Federal siting laws	Government streamlines authorization procedures for new plants and facilities	Present siting authorization procedures and delays continue

Table 20.10
U.S. coal resources

Description	Estimated resources (10^9 tons)
Total to a depth of 6000 feet in beds at least 14 inches thick for anthracite and bituminous coal and 2.5 feet for subbituminous and lignite	3200
To a depth of 3000 feet	1600
In reach of current technology and economics	437
Recoverable with 50% recovery underground and 90% by surface mining	273

The distribution of coal resources between the eastern and western regions and between surface minable and deeper reserves is shown in Figure 20.8.

20.4.3 General Methodology (1975 to 1985)

1985 projections for surface and deep mined coal were developed by

1. Projecting to 1985 total U.S. coal production for each scenario based on the appropriate economic growth, oil price, and national policy response assumptions.

2. Comparing these coal projections with production estimates from other sources.

3. Assuming for each year through 1985 the proportion of total coal produced from surface mines and deep mines.

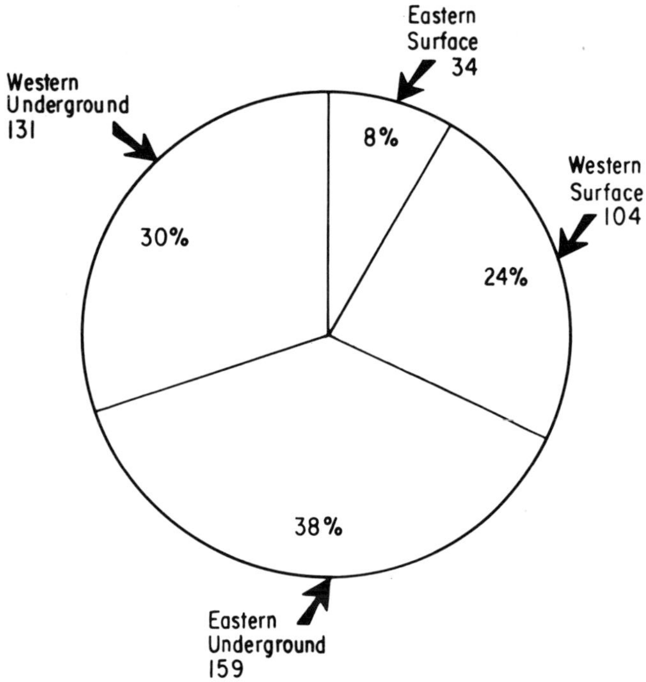

Figure 20.8
U.S. demonstrated coal reserve base, January 1974.

4. Calculating 1985 projections for surface and deep mined coal.

For each scenario, growth rates of total coal production were estimated for the periods 1975 to 1977, 1977 to 1980, and 1980 to 1985, as shown in the Table 20.11.

Growth in coal production is a function of all three scenario variables: national policy response, economic growth rate, and oil price. Although a quantitative relationship between each of these variables and coal growth is difficult to specify, it is felt that national policy response at the present and over the period 1975 to 1985 will be more important to the rate of growth in coal production than will the other two variables. Apart from policy, high economic growth rates or high oil prices would of course encourage higher coal production levels than would low economic growth rates or low oil prices.

Our specific estimates for the 1985 scenarios follow.

Scenario E (High Economic Growth, Low Oil Price, Restrained National Policy Response)
The environment characterized by scenario E would not cause coal growth rates to exceed the historical experience. While low oil prices might induce a slower than usual coal growth, the high economic growth rate would stimulate higher demand. An average annual growth rate of 2.3 percent, close to the historical rate, was selected for scenario E.

Table 20.11
Projected annual growth rates in U.S. total coal production

Scenario	1975–1977	1977–1980	1980–1985
A	2.3%	8.2%	4.2%
B and C	2.3%	5.7%	3.8%
D and E	2.3%	2.3%	2.3%

Scenario D (Low Economic Growth, Medium Oil Price, Restrained National Policy Response)
The coal production of scenario D is not significantly different from that of scenario E. Both scenarios assume the same restrained government response. While the somewhat higher oil prices of scenario D would tend to increase coal production, a low economic growth rate could nullify the effect of price. The historical growth rate of 2.3 percent was also selected for scenario D.

Scenario C (High Economic Growth, Medium Oil Price, Vigorous National Policy Response)
The higher oil price of this scenario provides a definite stimulus to coal production. High economic growth reinforces the effect of price, and vigorous government policy response becomes apparent in coal production levels by 1978.

This scenario and the remaining ones also require consideration of the U.S. political environment during the next few years. Inaction by government and industry may occur during this period as both groups await a resolution of the uncertainties in the national energy policy. This inaction would result in continuation of the low historical growth of 2.3 percent until the end of 1977 for all remaining scenarios.

In scenario C coal production growth rates are stimulated by all three scenario variables; the national policy response, instead of impeding expansion as in scenarios D and E, will encourage it. With vigorous government actions, scenario C growth rates would be more than two percentage points (the apparent effect of increased oil prices between 1973 and 1975) greater than the growth rate that would be expected with restrained policy response.

The 1977–1980 growth rate was estimated to be 5.7 percent.

After 1980 the initial impetus of vigorous government action would begin to weaken. Government may perceive its job as done after successfully establishing a higher production base from which future higher growth rates would continue. Thus the growth rate after 1980 for scenario C is projected to fall from its 1977–1980 level to 3.8 percent, reflecting the diminishing impact of vigorous government action.

Scenario B (Low Economic Growth, High Oil Price, Vigorous National Policy Response)
Coal production growth rates under scenario B are projected to be the same as those used for scenario C. Both scenarios are stimulated by the same vigorous government action. The effect of reducing coal production growth in scenario B relative to scenario C is assumed to be offset by the relatively higher oil price in scenario B.

Scenario A (High Economic Growth, High Oil Price, Vigorous National Policy Response)
The coal production growth in scenario A, as in scenario C, is stimulated by all three scenario variables, and the oil price in scenario A is 50 percent greater than that in scenario C by 1985. Our historical perspective on coal production indicates that large increases in oil prices (such as those between 1973 and 1975) may be followed by higher growth rates in coal production. For scenario A a 1977–1980 growth rate of 8.2 percent was selected; this rate is 2.5 percentage points greater than the 1977–1980 rate for scenario C.

As in scenario C a decrease in the growth rate is projected after 1980. But with oil prices higher in scenario A than in scenario C, we might also expect a higher long-term growth rate. For 1980 to 1985, a 4.2 percent growth rate was assumed.

20.4.4 General Methodology (1985–2000)

For the 1985 to 2000 time period WAES has defined four standard scenarios:

- C-1 High growth, high energy price, high coal
- C-2 High growth, high energy price, high nuclear
- D-7 Low growth, medium energy price, high coal
- D-8 Low growth, medium energy price, high nuclear

The WAES methodology specifies that for the 1985 to 2000 period, the coal projections are primarily functions of only two WAES scenario descriptors—energy prices and government action (high coal or high nuclear). Therefore, for supply projections, the descriptions of the four standard scenarios can be simplified to:

- C-1 High energy price, high coal
- C-2 High energy price, high nuclear
- D-7 Medium energy price, high coal
- D-8 Medium energy price, high nuclear

A detailed explanation of the growth rates for coal production in each of the four scenarios for year 2000 follows.

Scenario D-8
With energy prices unchanged at $11.50 (in 1975$) and no increased governmental emphasis on coal supply, growth in production is projected to stay at the historical 2.3 percent rate through 1995. Nuclear energy will be assuming an ever greater market share in the electric utility industry prior to 1995. After 1995 the growth rate of coal production is projected to be 4 percent.

Scenario D-7
In the D-7 scenario, the impacts of the government's high coal policies are realized after 1985.

A rapid increase in coal's share of the electric utility market requires coal supply to grow at a rate of 5.7 percent per year between 1985 and 1990, the same rate of growth projected under scenario C conditions between 1977 and 1980. In 1990, however, coal's growth rate in scenario D-7 will decrease because coal use will approach saturation in the utility industry; between 1990 and 1995 coal's growth rate will be 3.8 percent, sufficient to maintain coal's share of the utility industry.

After 1995 coal supply for gasification will increase in response to policy support for synthetic gas. The increase is consistent with the policy assumption that coal will be the principal replacement fuel. The growth rate will increase 1.7 points to 5.5 percent.

Scenario C-2

In 1985 nuclear energy begins to increase significantly the share of energy used by the electric utility industry. Because coal is not emphasized, production will not rise above the 2.3 percent historical growth rate. The high energy prices assumed for scenario C-2 and policy support for synthetic fuels will lead to some coal gasification and liquefaction around 1990. The production of synthetic fuels from coal will increase the growth rate of coal production between 1990 and 1995 to 4.0 percent. Between 1995 and 2000 production growth is projected to increase to 4.3 percent, which is 2 percentage points above the 1985–1990 rate. This additional increase is due to increased use of coal for gasification resulting from the dwindling supplies of domestic natural gas.

Scenario C-1

In 1985 coal is significantly emphasized. Coal production growth is projected to increase to 5.7 percent as coal increases its market share in the utility industry. By 1990 we assume coal will have saturated the utility industry, and its growth rate for this use will diminish to about 4.0 percent. In 1990, however, as in scenario C-2, high energy prices and national policy support lead to demonstration of coal gasification technology. The growth in coal production is projected to increase about 1.7 percentage points to 5.7 percent. Between 1995 and 2000 interest in coal gasification continues as supplies of domestic natural gas decrease and as synthetic liquid production from coal begins in response to policy support for synthetic fuels. Because of these uses, the coal production growth rate will rise to 6.0 percent. The resulting estimates of annual coal production for selected years from 1975 to 2000 are shown by region and type of mining in Table 20.12.

20.4.5 United States Coal Infrastructure Requirements—A Test Case

This study indicates a coal production potential of up to 2.2 billion short tons per year by 2000 in Case C-1, which assumes rising energy prices, high economic growth, and coal as the major replacement fuel. The manpower, equipment, material, transportation, and capital required to achieve this level of coal production were assessed using Bechtel Corporation's Energy Supply Planning Model (ESPM).

Description of the Model

The model is structurally straightforward, although dimensionally quite large. An application of the model follows this procedure:

1. The user chooses a future energy mix.

2. A computer program then calculates, by year, the number of new energy facilities of various types that must be brought onstream and operated to supply the specified mix.

3. The program next calculates separately the annual requirements of capital, manpower, and material for each energy and transportation facility from initial commitment to plant startup and for operation throughout its economic life. (These requirements are expressed without regard to the availability of such resources.)

Some questions the model answers explicitly are the following:

- What energy facilities must be constructed to implement a specific national energy development program?
- When will they need to be online? committed?
- What direct capital, manpower, and materials are needed? When?
- What resources are required to operate these facilities?
- What are the requirements of a specific energy industry? (e.g., what level of manpower will the coal industry require in 1983?)
- What are the requirements in a specific region of the country?
- What transportation does a given national energy program require?
- How different are the requirements of different energy programs? What are the regional implications of the differences?

The model can be helpful in answering questions such as these:

- For which capital, manpower, and material resource categories are future shortages likely to occur unless remedial actions are taken?
- What are the indirect and total requirements of a given energy program?

Table 20.12
Distribution of U.S. coal production 1975-2000 (millions of short tons per year)

	1975	1980	1985	1990	1995	2000
Scenario C-1						
Eastern underground	287	326	348	424	515	651
Eastern surface	245	245	245	245	245	245
Western surface	95	205	344	568	877	1303
Western underground	13	13	13	13	13	13
Total	640	789	950	1250	1650	2212
Scenario C-2						
Eastern underground	287	326	348	359	401	467
Eastern surface	245	245	245	245	245	245
Western surface	95	205	344	446	635	875
Western underground	13	13	13	13	13	13
Total	640	789	950	1064	1295	1600
Scenario D-7						
Eastern underground	287	294	291	356	394	486
Eastern surface	245	245	245	245	245	245
Western surface	95	162	251	441	619	916
Western underground	13	13	13	13	13	13
Total	640	714	800	1055	1271	1660
Scenario D-8						
Eastern underground	287	294	291	301	308	353
Eastern surface	245	245	245	245	245	245
Western surface	95	162	251	337	438	609
Western underground	13	13	13	13	13	13
Total	640	714	800	896	1004	1220

The model does not answer questions such as the following:

- What should our national energy program be?
- What should our national resource program be?
- What should our national energy imports program be?

The model's data base was developed to calculate direct resource requirements in categories that may be potential constraints to energy development programs. Hence, major resource categories included in the data base, as well as the level of disaggregation in each category, were selected by identifying the categories of requirements that would yield information significant to analyzing the feasibility of proposed future energy policies.

The resulting data base consists of resources directly used in the engineering, construction, operation, and maintenance of energy facilities by establishments normally engaged in these activities, for each of the 91 energy-related facilities included in the model.

Capital resource requirements for the construction of energy facilities were developed in two-digit Bureau of Economic Analysis (U.S. Department of Commerce) categories (e.g., primary iron and steel production is BEA 37) to allow direct use in econometric and input-output models—that is, for planning at both the industry and the economywide levels—and to allow comparison with availability studies using this system of accounts.

Table 20.13
Schedule of annual facilities additions required for the coal production test case

Type of facility	Annual additions (number of facilities)[a]														
	1976	77	78	79	80	81	82	83	84	85	86	87	88	89	90
Mining facilities															
Underground eastern coal mine (2×10^6 tons/year)	11	11	11	10	10	10	9	8	9	10	12	14	15	15	15
Surface eastern coal mine (4×10^6 tons/year)	3	3	3	3	3	3	3	3	3	3	3	3	3	3	3
Surface western coal mine (6×10^6 tons/year)	4	4	4	4	4	4	5	5	5	6	7	7	8	8	9
Underground western coal mine (2×10^6 tons/year)	0	0	1	0	0	1	0	0	1	0	0	1	0	0	1
Transportation facilities															
Rail line (40 miles, single track)	105	101	106	106	121	56	101	91	101	79	133	131	161	161	183
Coal unit train (10,500 ton)	29	32	33	38	25	16	22	18	28	25	36	32	36	68	67
Conventional train (10,500 ton)	113	114	113	114	114	115	115	116	115	116	136	137	53	127	134
Slurry pipeline (25×10^6 tons/year)	0	0	0	0	0	0	0	1	0	1	0	1	0	1	0
Coal barge (21,000 ton)	14	15	14	15	14	12	13	12	12	12	18	18	17	18	18
Coal truck (25 ton)	390	391	390	391	390	411	410	409	411	410	455	466	211	503	443

[a]Annual schedules are derived from cumulative requirements for five-year intervals. No special significance should be attached to

91	92	93	94	95	96	97	98	99	2000	01	02	03	04	05	Total added 1976–2000
15	15	17	20	22	23	23	24	24	24	24	24	24	24	24	377
3	3	3	3	3	3	3	3	3	3	3	3	3	3	3	75
9	10	11	13	15	16	17	18	19	20	20	19	18	18	18	232
0	0	1	0	0	1	0	0	1	0	0	1	0	0	1	8
209	211	231	251	223	271	309	301	331	294	301	301	301	301	301	4367
74	82	83	93	81	80	77	93	103	94	107	107	107	107	107	1365
137	138	136	138	138	153	154	154	154	155	152	152	152	152	152	3189
1	0	1	0	1	0	1	0	1	0	1	0	1	0	1	9
20	21	20	30	28	37	38	38	43	32	36	36	36	36	36	529
389	386	382	369	365	341	310	259	241	261	280	280	280	280	280	9384

year-to-year variations within intervals.

Resource requirement data in selected categories of major construction materials and equipment items were also developed. The data base contains selected major items deemed essential to the construction of energy facilities (structural steel, steam turbine generators) and is significant in identifying possible constraints stemming from shortages in industrial production capacity for major commodities.

Selected raw materials (aluminum, carbon steel, and alloy steel) are also included, in which case total construction requirements are included, whether directly purchased or contained within purchased equipment.

Finally, resource requirements have been included for construction manpower in the major engineering disciplines (mechanical engineers, civil engineers), designers and draftsmen, supervisors and managers, and major skilled craftsmen (pipefitters, boilermakers).

Construction resources are expended at various rates during the construction period. To account for this effect, time-phased data on the percentage of the total resources required during each year of the construction project were developed.

Similar resource data on the direct manpower, materials, equipment, and utilities required to operate and maintain energy supply and transportation facilities were also developed. These data were not designed to establish the full operating cost for the energy-related facilities, but rather to establish the industrially significant quantities for each facility for the purpose of identifying potential constraints.

The sources for the construction data were the various operating divisions of Bechtel, relying on past construction experience, literature, and industry contacts. Bechtel was assisted by the Stanford Research Institute (SRI) in developing most of the operating data. Bechtel developed operations data for the remaining facilities and adjusted the SRI-supplied data.

Description of the Test Case

The test case simulates an expansion of the U.S. coal supply sector to a level of 2.2 billion short tons in the year 2000, starting from a capacity of 640 million tons in 1975. The largest capacity increase is specified to occur in western surface production, which increases more than 13-fold to twice the total current national capacity. An additional increment is accomplished by more than doubling eastern underground capacity. Production from eastern surface and western underground sources is held constant.

Mining Requirements

Coal mining facility requirements have been derived directly from scenario C-1 and the specification of a twenty-year coal mine lifetime and a 5 percent per year retirement rate for the existing mine capacity. In addition, to account for the impact of construction lead times on resource requirements through the year 2000, fuel requirements beyond the year 2000 must be considered; it was assumed that the development of coal mines reaches a plateau during the 1996 to 2005 period.

The complete schedule of additional mining facilities required for the coal production test case is given in Table 20.13. To implement the production level assumed in the test case would require opening the following numbers of new mines in the 1976 to 2000 period:

- 377 new eastern underground coal mines at 2 million tons/year
- 75 new eastern surface coal mines at 4 million tons/year
- 232 new western surface coal mines at 6 million tons/year
- 8 new western underground coal mines at 2 million tons/year

Thus, to achieve the production level specified in scenario C-1 the coal mine development program would have to add 2.5 billion tons/year new capacity. The total new capacity by the year 2000 would then have to be greater than the total year 2000 production forecast, since all existing capacity is retired before the turn of the century and all new capacity added in the 1976 to 1980 period is replaced between 1996 and 2000 at the end of its assumed twenty-year lifetime.

Transportation Requirements

Coal transportation facilities needed to support these production levels have also been derived for coal unit trains, conventional trains, rail lines (consisting of track and bed), coal barges, coal trucks, and coal slurry pipelines.

The transportation requirements are subject to

Table 20.14
Transportation facilities required for coal production of 2.2 billion short tons

Facility	Modal Split (Fraction) 1975	2000	Average Haul Length (Miles) 1975	2000	Lifetime (Years)
Rail line[a]	0.59	0.62	418	850	50
Unit train	0.20	0.48	550	1000	15
Conventional train	0.39	0.14	350		15
Coal barge	0.25	0.23	500		20
Coal truck	0.15	0.04	50		15
Slurry pipeline	0.01	0.11	1000		50

[a]Rail line modal split is the sum of unit train and conventional train fractions.

much greater uncertainty than the coal mine requirements because average transport haul lengths, fraction of the coal production carried by the competing modes, and other variables must be assumed. The resulting requirements, combined with assumptions regarding retirements and lifetimes, determine annual facility additions needed.

WAES specified explicitly that an additional thousand-mile coal slurry pipeline carrying 25 million tons per year was to begin operating every other year starting in 1983. Thus, nine slurry pipelines are operating by the year 2000 to carry 17 percent of the specified western coal production to markets.

Principal assumptions used to compute requirements for the other types of transportation facilities are given in Table 20.14.

The particular values used are approximations based on the current state of knowledge. They represent (with the exception of lifetimes) values generated by a previous application of a regional version of the Energy Supply Planning Model using a similar coal production scenario.

Note that the average haul length for unit trains increases because the increasing proportion of western coal is expected to be transported over greater distances than eastern coal.

A complete list of resulting capacity additions of coal transportation facilities appears in Table 20.13. The following new facilities are needed in the 1976 to 2000 period to support the coal production level:

- 9 new slurry pipelines at 25 million tons/year
- Approximately 1400 new coal unit trains at 10,500 tons
- Approximately 3200 new conventional trains at 10,500 tons
- Approximately 4400 new lengths of rail line at 40 miles
- Approximately 500 new coal barges at 21,000 tons
- Approximately 9400 new coal trucks at 25 tons

Capital, Manpower, Materials, and Equipment Requirements

Application of the ESPM provides schedules of capital, manpower, materials, and equipment directly required to design, construct, start up, operate, and maintain these mining and transportation facilities for the period 1976 to 2000. The following results are highlights of the extensive results produced.

- Cumulative constant dollar construction capital costs for the 1976 to 2000 period amount to $118 billion: mining requires $32 billion, and transportation $86 billion. (All capital costs are expressed in constant third-quarter 1974 dollars, unmodified by escalation factors. No owner's costs, such as interest during construction, land leasing costs, or the cost of environmental impact analyses, are included.)
- An annual inflation rate of 5 percent would increase the cumulative current dollars needed to $277 billion.
- Annual constant dollar capital requirements peak at $7.8 billion in 1998, while current dollar requirements peak at $26 billion.

Requirements for selected other resources are given in Table 20.15. Annual requirements typically increase over the time period by a factor of two to three for both construction and operation requirements.

As with any analysis, understanding the study's limitations is essential to the interpretation and use of the results. For this analysis, the most limiting consideration is perhaps that data describing the complex network of the energy transportation system in general and coal transportation in particular are inadequate for forecasting requirements for capacity expansion. As an example, determining

Table 20.15
Selected resource requirements for the coal production test case

Resource	Units	1975 requirements: Coal mining[a]	Coal transportation[a]	Total coal supply[a]	Total energy industry[b]	2000 requirements: Coal mining	Coal transportation	Total coal supply
Construction resources								
Total technical manpower	man years/ year	1600	5600	7200	76,000	5000	17,000	22,000
Total manual manpower	man-years/ year	5000	9000	14,000	300,000	12,000	30,000	42,000
Total carbon steel	million tons	0.2	1.9	2.1	9.7	0.5	4.7	5.2
Draglines	cubic yards	1500	0	1500	1700	2800	0	2800
Operations resources								
Total technical manpower	man-years/ year	16,000	15,000	31,000	140,000	40,000	40,000	80,000
Underground miners	man-years/ year	50,000	0	50,000	50,000	110,000	0	110,000
Total manual manpower	man-years/ year	110,000	60,000	170,000	620,000	270,000	150,000	420,000

[a]Values for 1975 coal sector requirements are model-derived estimates for the WAES coal production test case.
[b]Values for 1975 total energy industry requirèments are model-derived estimates from previous analyses.

requirements for additional rail line capacity, which accounts for two-thirds of the transportation capital requirements, is complicated by the number of ways in which capacity can be increased—spur lines may be built, poor sections of track upgraded, bypasses constructed around bottlenecks, main trunk lines added, and so on. WAES also internally prepared a special analysis of transportation facilities required for rapid expansion of U.S. coal production to 1985. This study concluded that the transportation capacity required to handle a doubling of production can be provided by either rail or slurry pipeline or probably some combination of the two. Rail transport is most suitable for local movements and low-to-moderate quantities, or where maximum flexibility is required. Slurry pipelines are most economical for high and consistent volume and long-distance transport. At the present time, the development of slurry pieplines in the United States faces severe political obstacles that would have to be resolved. The study also concluded that barges would probably not play a significant role in the movement of western coal because the barge traffic on the Missouri/Mississippi River system is already heavy.

Although the United States is the world's leading coal exporter (56 million metric tons in 1974), port facilities are generally deficient for any substantial expansion of coal export. The United States has very limited capacity to accommodate the larger bulk carriers that would be most economical for large coal shipments. In addition, coal facilities are presently concentrated in East Coast and Great Lakes ports far from the abundant western reserves. Clearly, if significant quantities of coal are to be exported, seaport expansion would be required.

The infrastructure requirements are large but not insurmountable given the time period involved and the 6 percent maximum annual rate for U.S. coal production in the WAES scenarios. But any delays in the early stages of development will make it more difficult, if not impossible, to achieve these levels of coal production by the end of the century.

20.5 SYNTHETIC FUELS

The development of synthetic fuels is estimated to reach moderate levels in the WAES scenarios. This is consistent with the scenario assumptions, since large-scale development of these energy sources would generally require new technological developments or higher energy prices than those assumed in the WAES cases.

Of the four types of synthetics considered—oil from shale, synthetic crude from coal, synthetic gas from coal, and biomass conversion—only synthetic gas and a small amount of shale oil are

Table 20.16
Potential synthetic fuel production in the year 2000

	2000 scenario			
	C-1	C-2	D-7	D-8
Shale oil (MBD)	2.0	2.0	0	0
Synthetic crude from coal (MBD)	2.0	2.0	0	0
Synthetic gas from coal (TCF)	6.0	6.0	4.0	0
Biomass conversion (TCF)	1.0	1.0	0	0

Table 20.17
Estimated shale oil production cost

Technique	Investment cost ($/daily bbl)	Operating cost ($/bbl)	Production cost ($/bbl; 15% DCF, 100% equity)
Modified in situ	5000–7000	3.50–5.00	8.00–11.00
Surface retorting	14,000–23,000	4.00–5.00	16.00–25.00

Source: *ERDA Weekly Announcement,* December 24, 1976.

Note: The oil produced by these methods may be prerefined to a quality comparable with Arabian Crudes for an additional $1 to $4 per barrel.

assumed to be available by 1985. Because of long lead times, coal liquefaction and biomass conversion are not considered likely before 1985 even under assumptions of high oil prices and a vigorous policy that includes some government-sponsored demonstration programs and financial assistance. Quantities potentially available by the year 2000 are shown in Table 20.16.

20.5.1 Oil Shale

Oil shale could become a large energy resource in the United States, which has the world's largest known resources.

Mining and in situ recovery techniques are two methods of extracting oil from oil shales, and substantial efforts are now being made to develop in situ production. A combination of the two methods is a likely pattern in the future. Underground mining is by room and pillar methods in seams 100 feet thick, but about half the shale is left in the pillars that support the roof. The oil in the shale, an organic material known as kerogen, is extracted from the shale by retorting the crushed shale at high temperature and treating it with hydrogen to produce a semirefined fuel oil. Part of the kerogen decomposes into gas, which can be burned on-site to furnish power or converted into pipeline quality gas. Consolidating the spent shale from which the oil has been extracted takes two to three barrels of water per barrel of oil recovered. This process may be largely avoided by in situ retorting in the mined cavern.

U.S. oil shale deposits are estimated to contain some 2000 billion barrels of oil. However, only about 6 percent of the deposits are accessible in seams thirty or more feet thick containing more than thirty gallons of oil per ton of rock. Only such deposits are of commercial interest at this time.

Intermittent federal research and development and private efforts to develop oil from shale have been underway since 1948, but there has been no commercial production to date. Table 20.17 gives current estimates of shale oil production costs in the United States.

Inflation has increased the estimated cost of commercial plants beyond the range of usual industrial funding. Environmental factors including disposal of spent shale and water requirements present major obstacles to developing a large oil shale industry in the United States. In situ technologies that might diminish environmental impacts are not yet commercially feasible. Uncertainty about future oil prices increases the investment risks.

Federal subsidies for at least the early stages of oil shale development appear necessary to demonstrate its technical, economical, and environmental feasibility. Such support implies policy decisions to move forward vigorously on shale oil development.

In the rising energy price scenarios in 2000 (Cases C-1 and C-2), production of oil from shale is assumed to be 2 MBDOE; in the constant energy price cases none is assumed. By 1985 shale oil production is estimated as potentially achieving 0.2 MBD in Case C (two or three demonstration plants in response to policy action) and 0.4 MBD in Cases A and B, in which higher prices for oil lead to more vigorous policy support.

20.5.2 Synthetic Gas

The production estimates for synthetic gas under WAES cases include 0.2 trillion cubic feet from liquids in 1985 and none from liquids in 2000. The remainder of the synthetic gas comes from coal and consists of high-Btu gas. For purposes of estimation, a standard facility yielding 2.5 million cubic feet per day is assumed. Such a gasification plant would require 27,250 tons per day of 8300 Btu/lb coal and would cost about $1 billion.

For Cases A and B (1985), which assume high energy prices and government assistance, the production of synthetic gas from coal is limited because of the long lead times required for facilities. Our estimates of potential gas production are shown in Table 20.18.

In the 2000 cases, the energy price is inadequate to bring on supply in the constant price cases (D-7, D-8) and only marginally adequate in rising price cases (C-1, C-2). Synthetic gas is assumed to be available only if there is a strong policy emphasis on coal and the demonstration or subsidy of alternative energy sources.

20.5.3 Synthetic Crude from Coal

Coal liquefaction is considered to be a supply potential only in the year 2000 Cases C-1 and C-2. These cases involve the higher price assumptions along with policy actions that encourage demonstration programs and an emphasis on coal. The policy action assumption leads to potential synthetic crude from coal production of 2.0 MBDOE by the year 2000.

The technology assumed in this estimate of coal liquefaction is conventional technology of a 24 thousand barrel per day plant requiring 10,000 tons per day of 11,000-Btu/lb coal.

20.5.4 Biomass Conversion

Another possible contribution to U.S. energy supply by 2000 is fuel from biomass conversion such as the conversion of municipal or agricultural wastes into methane. Under assumptions of high energy prices and vigorous policy support, it is estimated that an oil equivalent of about 0.5 million barrels per day would be available in Cases C-1 and C-2.

20.5.5 United States Synthetic Fuels—A Test Case

The WAES supply scenarios to the year 2000 provide for only modest development of synthetic fuels, given the scenario price assumptions of $17.25 and $11.50 per barrel. Because the resource base for greater development exists, WAES decided to explore the capital, manpower, equipment, and material resources needed to increase synthetic fuels significantly beyond the level assumed in the WAES scenarios. As in the coal test case, Bechtel's Energy Supply Planning Model was used to assess these requirements.

Description of the Test Case

The test case was constructed to simulate the development of a U.S. synthetic fuels industry from zero capacity in 1985 to a level of 3.5 MBDOE in the year 2000. Of this total, 3 MBDOE of synthetic oil and gas comes from coal and 0.5 MBDOE comes from shale oil—half by conventional retorting and half by in situ recovery (see Table 20.19). The facilities would start from zero level in 1985 and reach the target level in the year 2000.

Furthermore, to account for the impact of facility lead times on resource requirements through the year 2000, it is necessary to consider fuel requirements beyond the year 2000. For this case, it has been assumed that capacity identical to that installed in the years 1996 to 2000 is installed also in 2001 to 2005, bringing total capacity to a level of 5.25 MBDOE in 2005. This assumption simulates a situation in which synthetic fuel construction activity reaches a plateau for the 1996 to 2005 period.

Figure 20.9 is a graphical representation of the synthetic fuel scenario.

Facility Requirements

The Energy Supply Planning Model was developed to accept fuel mix inputs through the year 1995 and to produce resource requirements through the year 1985. To apply to this test case, the model had

Table 20.18
Synthetic gas estimates

	1985 scenarios					2000 scenarios			
	A	B	C	D	E	C-1	C-2	D-7	D-8
Physical units (10^{12} ft^3)	1.4	1.4	0.6	0.2	0.2	6.0	6.0	4.0	0
Oil equivalent (MBD)	0.7	0.7	0.3	0.1	0.1	2.8	2.8	1.9	0

to be modified to produce resource requirements through the year 2000.

One other modification was to directly input annual energy facility startup schedules rather than fuel mix variables. This modified procedure was used because fewer computer programs had to be modified. In addition, the number of facilities in the test case is a manageable one for manual calculation of facility schedules.

Facilities included in the test case were restricted to five synthetic fuel conversion facilities and four resource extraction supporting facilities. In addition, a power plant for utilization of the produced low-Btu gas is included. To simplify the test case, energy transportation, transmission, and distribution facilities, as well as oil refineries, have been excluded.

The complete schedule of facility additions required for implementation of the synthetic fuels test case and developed for model input is given in Table 20.20. As can be seen, implementation of the synthetic fuels program would require the construction of the following facilities in the 1986 to 2000 period:

- 42 nominal coal gasifaction facilities
- 47 nominal coal liquefaction facilities
- 6 nominal oil shale processing facilities

In addition, a coal mining capacity of more than 600 million tons per year would need to be added to support the year 2000 requirements. A corresponding schedule of facility commitment requirements, obtained by time-shifting the startup requirements by the engineering and construction lead times, is given in Table 20.21.

Table 20.19
Synthetic gas production in the test case (MBDOE)

	1985	1990	1995	2000
Coal gas[a]				
High-Btu gas	0	0.2	0.5	1.0
Low-and medium-Btu gas	0	0.2	0.5	1.0
Subtotal	0	0.4	1.0	2.0
Coal liquids[a]	0	0.2	0.5	1.0
Oil shale				
Mined	0	0.05	0.125	0.25
In situ	0	0.05	0.125	0.25
Subtotal	0	0.1	0.25	0.50
Total synthetic fuels	0	0.7	1.75	3.50

[a]Coal sources are specified to be 90 percent from western surface mines, 10 percent from western underground mines.

Direct Capital, Manpower, Materials, and Equipment Requirements

The outputs of the national computer run include the annual schedules of direct capital, manpower, materials, and equipment needed to design and construct the required facilities. Operating resource requirements were not developed for the test case. All requirements are expressed without regard to the availability of such resources. All capital costs are expressed in constant third-quarter 1974 dollars, unmodified by escalation factors. No owner's costs (interest during construction, cost of environmental impact statements) are included.

Total construction capital costs for the 1981 to 2000 period amount to $71.7 billion, summarized by sector as follows:

- Coal gasification: $34.1 billion
- Coal liquefaction: $19.1 billion
- Oil shale (including resource extraction): $4.7 billion
- Coal mines: $5.4 billion
- Low-Btu gas-fired power plants: $8.4 billion

Gross owner's cost factors would extend these results by 25 to 40 percent.

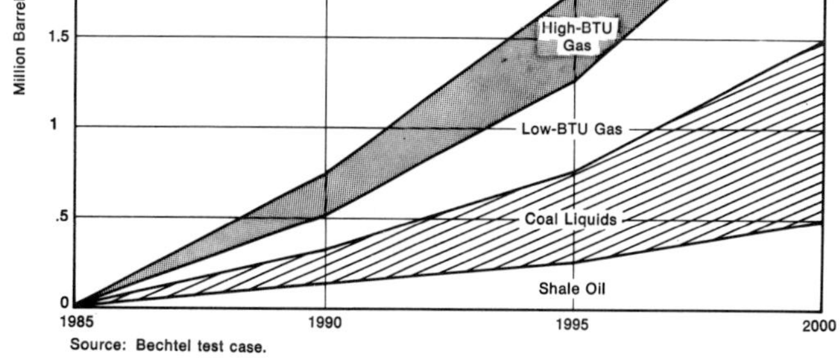

Figure 20.9
Synthetic fuel test case profile.

Table 20.20
Schedule of annual energy facilities additions required for WAES synthetic fuels test case

Type of facility	Annual additions (number of facilities)												
	1985	86	87	88	89	90	91	92	93	94	95	96	97
Processing facilities													
Coal gasification — high Btu (250×10^6 ft^3/day)	0	1	0	1	1	1	1	2	1	2	1	2	2
Coal gasification — low and medium Btu (1440 $\times 10^6$ ft^3/day)	0	0	1	1	1	1	1	1	2	1	1	2	2
Coal liquefaction — heavy fuel oil (21.7 MBD)	0	0	1	2	3	3	2	3	2	3	4	4	5
Oil shale retorting and upgrading (90 MBD)	0	0	1	0	0	0	0	1	0	0	0	0	1
Shale oil upgrading (90 MBD)	0	0	0	0	1	0	0	0	0	1	0	0	0
Mining facilities													
Western surface coal mine (6×10^6 tons/year)	0	2	3	5	5	5	5	6	5	6	7	8	9
Western underground coal mine (2×10^6 tons/year)	0	1	0	1	1	1	1	2	1	2	1	2	2
Oil shale mine (150,000 tons/day)	0	0	1	0	0	0	0	1	0	0	0	0	1
In situ shale oil recovery (37 MBD)	0	0	0	0	1	0	0	1	0	1	0	0	1
Other													
Low- to intermediate-Btu gas-fired power plant (800 MWe)	0	0	2	3	3	3	3	3	4	3	3	5	5

98	99	2000	01	02	03	04	05	Total added 1986–1990	1991–1995	1996–2000	2001–2005	1986–2000	1986–2005
3	2	2	2	2	3	2	2	4	7	11	11	22	33
2	2	2	2	2	2	2	2	4	6	10	10	20	30
5	5	5	4	5	5	5	5	9	14	24	24	47	71
0	0	0	0	1	0	0	0	1	1	1	1	3	4
0	1	0	0	0	0	1	0	1	1	1	1	3	4
11	11	10	8	9	11	11	10	20	29	49	49	98	147
3	2	2	2	2	3	2	2	4	7	11	11	22	33
0	0	0	0	1	0	0	0	1	1	1	1	3	4
0	1	1	0	1	0	1	1	1	2	3	3	6	9
6	6	5	5	5	6	6	5	11	16	27	27	54	81

Table 20.21
Schedule of annual energy facilities commitments required for WAES synthetic fuels test case

Type of facility	Lead time (years)	Annual commitments (number of facilities)											
		1981	82	83	84	85	86	87	88	89	90	91	92
Processing facilities													
Coal gasification — high Btu (250×10^6 ft^3/day)	4	0	1	0	1	1	1	1	2	1	2	1	2
Coal gasification — low and medium Btu (1440×10^6 ft^3/day)	4	0	0	1	1	1	1	1	1	2	1	1	2
Coal liquefaction — heavy fuel oil (21.7 MBD)	5	0	1	2	3	3	2	3	2	3	4	4	5
Oil shale retorting and upgrading (90 MBD)	5	0	1	0	0	0	0	1	0	0	0	0	1
Shale oil upgrading (90 MBD)	5	0	0	0	1	0	0	0	0	1	0	0	0
Mining facilities													
Western surface coal mine (6×10^6 tons/year)	2	0	0	0	2	3	5	5	5	5	6	5	6
Western underground coal mine (2×10^6 tons/year)	3	0	0	1	0	1	1	1	1	2	1	2	1
Oil shale mine (150,000 tons/day)	3	0	0	0	1	0	0	0	0	1	0	0	0
In situ shale oil recovery (37 MBD)	3	0	0	0	0	0	1	0	0	1	0	1	0
Other													
Low- to intermediate-Btu gas-fired power plant (800 MWe)	4	0	0	2	3	3	3	3	3	4	3	3	5

								Total committed				
93	94	95	96	97	98	99	2000	1981–1985	1986–1990	1991–1995	1996–2000	1981–2000
2	3	2	2	2	2	3	2	3	7	10	11	31
2	2	2	2	2	2	2	2	3	6	9	10	28
5	5	5	4	5	5	5	5	9	14	24	24	71
0	0	0	0	1	0	0	0	1	1	1	1	4
0	1	0	0	0	0	1	0	1	1	1	1	4
7	8	9	11	11	10	8	9	5	26	35	49	115
2	2	3	2	2	2	2	3	2	6	10	11	29
0	1	0	0	0	0	1	0	1	1	1	1	4
0	1	0	1	1	0	1	0	0	2	2	3	7
5	6	6	5	5	5	6	6	8	16	25	27	76

The peak annual requirement for engineering personnel is 8200 man-years/year, with an average annual requirement of 4800 man-years/year.

The peak annual requirement for manual manpower is 64,000 man-years/year, with an average annual requirement of 34,000 man-years/year. Largest requirements for building trades categories accrue to pipefitters (13,000 man-years/year average), electricians (3600 man-years/year average), and boilermakers (2600 man-years/year average).

Selected materials and equipment results indicate that:

- Peak annual requirement for carbon steel is 1.1 million tons.
- Peak annual requirement for nonnuclear pressure vessels (greater than 4-inch plate) is 33,000 tons.
- Peak annual requirement for draglines is 980 cubic yards.

Annual requirements of capital, engineering personnel, and manual manpower are indicated in Figure 20.10. Manpower, equipment, and material resources required are large. The lead times for constructing specific facilities range from three to five years. The total lead time—including policy decisions, investment commitments, design, and engineering—stretches from six to ten years. Therefore, policy decisions must be made in the next few years if facilities are to come on-stream by 1985 and if there is to be significant production by the year 2000. Otherwise, synthetic fuels from coal are not likely to contribute significantly to the overall national energy supply in this century.

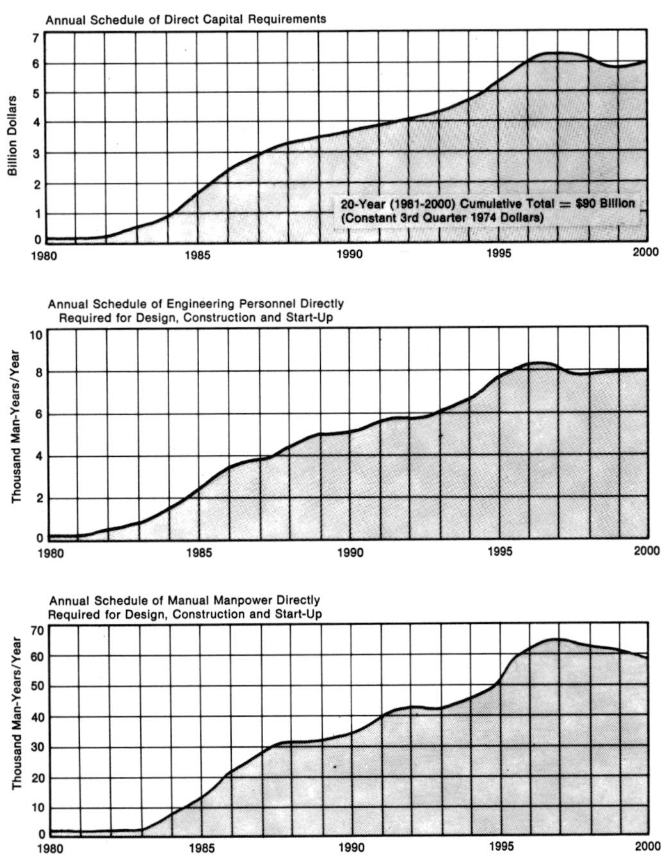

Source: Bechtel Corporation.

Figure 20.10
Synthetic fuel test case requirements.

20.6 NUCLEAR

The development of nuclear power as a replacement for oil and natural gas in the generation of electricity in the United States has become significant since the late 1960s and has been regarded by some to be the principal solution to the problem of satisfying future growth requirements in electricity generation. Since the first commercial generation of nuclear power in a prototype in 1957, installed commercial nuclear capacity had reached 15 GWe in 1972 and 40 GWe by 1975. Questions concerning safety and potential risks related to the nuclear fuel cycle have raised doubts about future growth and have slowed the rate of capacity expansions; as a result recent growth projections have had to be modified downward. In view of the character of the nuclear issues,

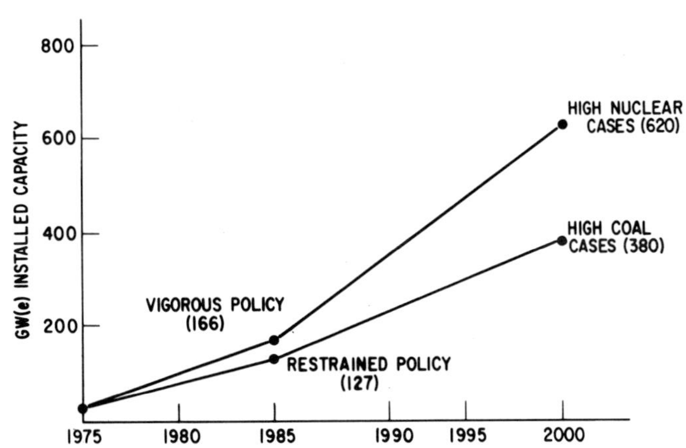

Figure 20.11
The outlook for U.S. nuclear development to 2000.

many of which are not technical but political, projections are very difficult to make.

Our estimates of nuclear electric generating capacity, shown in Figure 20.11, are based largely on ERDA estimates from October 1976. The projections to 1985 are largely based on current U.S. national estimates. The levels of nuclear growth assumed in our scenarios to 2000 are subject to a great degree of uncertainty. It is likely that these projections will change as the nuclear debate continues and the relevant issues are resolved to either permit the continued development of nuclear power or severely restrict its future application.

20.7 HYDROELECTRIC AND GEOTHERMAL

Hydroelectric and geothermal energy have recently attracted renewed interest because they are both clean and renewable. The long-term potential of geothermal energy is particularly interesting because it is based on the immense heat trapped deep within the earth. However, our assessments assume rather limited increases from these two sources within this century, as outlined in Table 20.22.

Table 20.22
Potential hydroelectric and geothermal capacity, 2000 (GWe installed)

	2000 scenario			
	C-1	C-2	D-7	D-8
Hydroelectric	100	100	80	80
Geothermal	50	50	20	20

20.7.1 Hydroelectric Generation

U.S. hydroelectric capacity in 1975 was about 64 GWe. Our estimates, which are similar to other published estimates, show a moderate growth of installed hydroelectric power to the year 2000. Future growth of hydroelectricity is somewhat constrained because the most attractive sites have largely been developed and growing environmental concerns may discourage the development of some remaining sites.

Our estimates for hydro capacity in 1985 range from 67 GWe to 90 GWe. The higher estimates assume that environmental issues do not delay development of additional hydroelectric capacity, while the lower estimates assume some delays. For the year 2000 the range is from 80 GWe to 100 GWe, with the lower estimates reflecting the time lags introduced from 1975 to 1985.

20.7.2 Geothermal Energy

In 1974 the electrical capacity produced from geothermal energy in the Geysers field in California was 0.4 GWe. Such dry steam fields require a combination of three geologic factors: (1) a natural underground source of water; (2) an impermeable layer above that traps the water and permits the formation of steam; and (3) a mass of hot rock near the natural water system. The occurrence of these three factors in the same area is relatively rare, but they are the only source of geothermal energy that has been commercially exploited to date.

It has been estimated that exploitation of geothermal energy from hot dry rocks and from high temperature hot water sources could create a supply of geothermal energy many times greater than that of geothermal fields now exploited.

Some experiments in the United States have involved fracturing hot rocks, injecting water into the lower part of the fractured area, and recovering steam or superheated water from the upper area, to use to generate electricity. However, this process is still subject to several problems: (1) the fracture system may leak water, preventing a build-up of pressure; (2) recovering heat and drilling in hot igneous rocks is difficult and costly.

Development of hot water geothermal resources in such areas as the Imperial Valley of California is also being tested. These sources are not currently feasible because of problems in controlling corrosion and scaling in the producing facilities and the need to dispose of the large volumes of brine produced.

All our estimates are for the conversion of geothermal energy to electricity. The direct application of geothermal energy for heat may also be possible, but it is limited in the United States by the remote location of most potential resources. Even if research, development, and demonstration prove geothermal commercially successful by 1985, electricity production from hot rocks and hot water by the year 2000 will still be in its initial stages because of the time involved in exploration, drilling, and development of geothermal fields and the subsequent integration of the developed fields into electricity generation and distribution systems.

Our 1985 scenario geothermal installed capacity estimates range from 4 to 10 GWe, growing to a range of 20 to 50 GWe by 2000.

20.8 SOLAR ENERGY

The technical feasibility of solar energy has been long established. The primary obstacle to large-scale implementation is cost. Other obstacles include reliability, architectural acceptability, energy storage and back-up techniques, problems of integration into national energy systems, dependence upon local insolation, and large front-end capital costs. The estimates in the WAES scenarios assume a vigorous R&D program with careful attention to credible demonstrations of commercial prototype systems. Estimates of the total solar energy impact in 1985 and 2000 range from 0 up to 1.8 MBDOE. Thus, solar energy plays a negligible role in the WAES time frame, but it is expected to be a more significant factor in the post-2000 time frame as technology is developed and relative costs decline.

In Table 20.23 judgmental dates are presented for each major solar technology for

- Start-up time, defined as the year when installation of commercial prototype units begins. Systems constructed may not themselves be economical, but there is sound engineering data to show that systems will be economical under mass production conditions.

- Take-off time, defined as the year when full-scale commercial implementation begins. By the time take-off is achieved, system reliability has been assured through field testing in the interval since start-up time and system costs are convincingly known to manufacturers and are competitive with competing energy forms (due to either intrinsically low costs or price guarantees, subsidies, and so on). The amount of system installed at start-up time is at least 1 GWe = 0.03 Quad/year for electricity-producing solar systems and 0.1 Quad/year for other systems (0.1 Quad/year corresponds to about 10^9 ft^2 of collector, or about 10^6 individual homes).

The stated start-up times are estimated to be achievable with a vigorous R&D program. The take-off times are the times for critical decision making, and national action may be required to ensure the projected levels.

Solar energy's potential is diverse. Some of the solar technologies being developed may fail, and others will take forms quite different from those presently envisaged. For long-range planning purposes such shifts have little importance, and we have attempted to assess only overall impacts.

Our estimates assume that intensive R&D will ensure successful economic competition with alternative energy forms such as synthetic oil and gas, nuclear, and coal-produced electricity.

20.9 REFERENCES

ADL Impact Services (Bruce Old), "An Outlook for Electric Power to 1985," February 1976.

Atlantic Richfield Co., "Development of Western Coal—A Producer's View," April 1975.

Bankers Trust Co., "Capital Resources for Energy through the Year 1990," 1976.

Bureau of Mines, U.S. Department of the Interior, *1973 Minerals Yearbook,* Vol. 1: *Metals, Minerals, and Fuels* (1975).

Continental Oil Co., "Outlook: 1975 and Beyond," A presentation to security analysts, May 13, 1975.

Lionel D. Edie & Co., "Coal, a New Growth Industry," May 1975.

"Exchange of Remarks between the President and Frank Zarb, Director of the Federal Energy Administration, in Reviewing the Federal Energy Administration's *National Energy Outlook* Report," Office of the White House Press Secretary, March 4, 1976.

Exxon Co., "World Energy Outlook," December 1975.

Federal Energy Administration, *Project Independence Report* (November 1974).

Federal Energy Administration, "Initial Report on Oil

Table 20.23
Solar energy impact analysis

	Year of start-up	Year of take-off	Impact in the year 2000 (Quads) C-1	C-2	D-7	D-8
Solar heating and cooling	1980	1985	2.4	2.4	1.2	1.2
Solar electric	1990	2000	1.5	1.5	0.5	0.5
Total			3.9	3.9	1.7	1.7

and Gas Resources, Reserves, and Productive Capacities," June 1975.

Federal Energy Administration, *National Energy Outlook* (1976).

Stewart W. Herman and James S. Cannon, "Energy Futures, Industry and the New Technologies," Inform, Inc., 1976.

Roger Legassie, ERDA, Testimony before the U.S. Congress Joint Committee on Atomic Energy.

National Academy of Sciences, *Mineral Resources and the Environment* (1975).

Petroleum Industry Research Foundation, Inc, "Electric Power and U.S. Oil Import Dependency," May 1975.

U.S. Department of the Interior, *Energy Perspectives* (February 1975).

U.S. Geological Survey, "Geological Estimate of Undiscovered Recoverable Oil and Gas Resources in the United States," Circular 725, 1975.

University of Oklahoma Science and Public Policy Program, "Energy Alternatives, A Comparative Analysis," May 1975.

Westinghouse Electric Corp., "Projections of Trends in Economics, Loads and Generation," May 1975.

Coal and Synthetics Test Cases

"Analysis of a Synthetic Fuels Test Case Using the Bechtel Energy Supply Planning Model," J. Michael Gallagher, Ralph G.Zimmerman, Ronald Barany, and Paul F. Paskert; prepared in support of the Workshop on Alternative Energy Strategies; August 30, 1976.

The Energy Supply Planning Model, Volume I: *Model Structure and Use;* Volume II: *User's Manual and Appendices;* M. Carasso J.M. Gallagher, K.J. Sharma, J.R. Gayle, and R. Barany; Bechtel Corporation; NTIS No. PB-245 382, PB-245 383; August 1975.

"Regional Requirements of Capital, Manpower, Materials and Equipment for Selected Energy Future," J. Michael Gallagher and Ralph G. Zimmerman, Bechtel Corporation, October 1976.

"Availability of Engineering Personnel for U.S. Energy Development Programs," Bechtel Corporation, October 1976.

"Availability of Skilled Craftsmen for U.S. Energy Development Programs," Bechtel Corporation, October 1976.

"Availability of Selected Materials and Equipment Commodities for U.S. Energy Development Programs," Bechtel Corporation, October 1976.

"Direct Requirements of Capital, Manpower, Materials and Equipment for Selected Energy Future," J. Michael Gallagher, Meir Carasso, Ronald Barany, and Ralph G. Zimmermann; Report to ERDA by Bechtel Corporation under Contract No. E(49-1)-3794; April 1976.

"The Energy Supply Planning Model—A Working Tool for Regional Analysis of Alternative National Energy Policies," Meir Carasso and J. Michael Gallagher; Bechtel Corporation; presented at the Seventh Annual Pittsburgh Conference on Modeling and Simulation, Pittsburgh, Pennsylvania, April 26–27, 1976.

20.10 ENERGY SUPPLY WORKSHEETS

The following worksheets are included in this section (see Section 1.2.1 of Chapter 1 for the definitions of line items):

Crude oil (4)

Natural gas (4)

Coal (10)

Shale oil (2)

Coal to syncrude (2)

Coal gasification (4)

Synthetic gas from liquids (1)

Uranium and nuclear (4)

Hydroelectric (2)

Geothermal (2)

INDIGENOUS SUPPLY ASSESSMENT WORKSHEET

CRUDE OIL

Country: U.S.A.

Crude Oil:
Offshore X)
Onshore X) TOTAL

WAES SCENARIO CASE (1985): A and B

WAES SCENARIO CASE (2000): ________

Total Resource in Place: 421-596 Bil. Bbls.

Domestic Oil Price: $18.25

Technology Assumptions: Current Technology

	(Units)	1972	1975	1980	1985	1990	2000	2010	2025
(1) Ultimate Recoverable Resource:	Bil. Bbls	260	260	260	260				
(2) Discoveries:	Bil. Bbls	-	4	16	39				
(3) Reserves Proved to Date:	Bil. Bbls	158	166	178	201				
(4) Potential Production:	MMBD	11.2	10.1	10.5	12.0				
(5) Expected Annual Production MMBD	heavy								
	light								
(6) Cumulative Production:	Bil. Bbls	115	126	145	166				
(7) Remaining Proved Reserves:	Bil. Bbls	43	40	33	35				
(8) Reserve/Production Ratio:		10	11	9	8				
(9) Host Government Take ($/unit of production):									
(10) Capital Cost ($/unit of daily capacity):									
(11) Operating Cost ($/unit of production):									

INDIGENOUS SUPPLY ASSESSMENT WORKSHEET

CRUDE OIL

Country: U.S.A.

Crude Oil:
Offshore X)
Onshore X) TOTAL

WAES SCENARIO CASE (1985): C

WAES SCENARIO CASE (2000): 1 and 2

Total Resource in Place: 421-596 Bil. Bbls.

Domestic Oil Price: $12.50(1985), $18.25(2000)

Technology Assumptions: 1975 Technology

	(Units)	1972	1975	1980	1985	1990	1995	2000	2010
(1) Ultimate Recoverable Resource:	Bil. Bbls	260	260	260	260	260	260	260	260
(2) Discoveries:	Bil. Bbls	-	4	14	34	54	74	78	82
(3) Reserves Proved to Date:	Bil. Bbls	158	166	176	196	216	236	240	244
(4) Potential Production:	MMBD	11.2	10.1	10.0	11.0	11.0	11.0	7.0	2.6
(5) Expected Annual Production* MMBD	heavy								
	light								
(6) Cumulative Production:	Bil. Bbls	115	126	144	163	183	203	219	236
(7) Remaining Proved Reserves:	Bil. Bbls	43	40	32	33	33	33	21	8
(8) Reserve/Production Ratio:		10	11	9	8	8	8	8	8
(9) Host Government Take ($/unit of production):									
(10) Capital Cost ($/unit of daily capacity):									
(11) Operating Cost ($/unit of production):									

INDIGENOUS SUPPLY ASSESSMENT WORKSHEET

CRUDE OIL

Country: U.S.A.

Crude Oil:)
Offshore X) TOTAL
Onshore X)

WAES SCENARIO CASE (1985): D

WAES SCENARIO CASE (2000): 7 and 8

Total Resource in Place: 421-596 Bil. Bbls.

Domestic Oil Price: $11.50(1985), $12.50(2000)

Technology Assumptions: 1975 Technology

	(Units)	1972	1975	1980	1985	1990	1995	2000	2010
(1) Ultimate Recoverable Resource:	Bil. Bbls	260	260	260	260	260	260	260	260
(2) Discoveries:	Bil. Bbls	-	4	10	25	38	50	59	68
(3) Reserves Proved to Date:	Bil. Bbls	158	166	172	187	200	212	221	230
(4) Potential Production:	MMBD	11.2	10.1	10.0	9.5	8.0	7.0	6.0	3.5
(5) Expected Annual Production *: MMBD	heavy light								
(6) Cumulative Production:	Bil. Bbls	115	126	144	161	178	192	203	220
(7) Remaining Proved Reserves:	Bil. Bbls	43	40	28	26	22	20	18	10
(8) Reserve/Production Ratio:		10	11	8	8	8	8	8	8
(9) Host Government Take ($/unit of production):									
(10) Capital Cost ($/unit of daily capacity):									
(11) Operating Cost ($/unit of production):									

INDIGENOUS SUPPLY ASSESSMENT WORKSHEET

CRUDE OIL

Country: U.S.A.

WAES SCENARIO CASE (1985): E

WAES SCENARIO CASE (2000): ________

Total Resource in Place: 421-595 Bil. Bbls.

Domestic Oil Price: $8.66

Technology Assumptions: 1975 Technology

Crude Oil:)
Offshore X) TOTAL
Onshore X)

	(Units)	1972	1975	1980	1985	1990	2000	2010	2025
(1) Ultimate Recoverable Resource:	Bil. Bbls	260	260	260	260				
(2) Discoveries:	Bil. Bbls	-	4	10	20				
(3) Reserves Proved to Date:	Bil. Bbls	158	166	172	182				
(4) Potential Production:	MMBD	11.2	10.1	9.5	8.0				
(5) Expected Annual Production*: MMBD	heavy				3.0				
	light				5.0				
(6) Cumulative Production:	Bil. Bbls	115	126	144	160				
(7) Remaining Proved Reserves:	Bil. Bbls	43	40	28	22				
(8) Reserve/Production Ratio:		10	11	8	8				
(9) Host Government Take ($/unit of production):									
(10) Capital Cost ($/unit of daily capacity):									
(11) Operating Cost ($/unit of production):									

* Indicate portion that is light crude and heavy crude in each year.

INDIGENOUS SUPPLY ASSESSMENT WORKSHEET

NATURAL GAS

Country: U.S.A.

WAES SCENARIO CASE (1985): A and B

WAES SCENARIO CASE (2000): ______

Total Resource in Place: 1372-1772 TCF

Domestic Gas Price: New gas deregulated

Technology Assumptions: 1975 Technology

Natural Gas:)
Offshore X) TOTAL
Onshore X)

	(Units)	1972	1975	1980	1985	1990	2000	2010	2025
(1) Ultimate Recoverable Resource:	TCF	1202	1202	1202	1202				
(2) Discoveries:	TCF	–	8	70	194				
(3) Reserves Proved to Date:	TCF	670	726	788	912				
(4) Potential Production:	TCF	22.5	20.1	18.0	20.0				
(5) Expected Annual Production:	TCF	22.5	20.1	18.0	20.0				
(6) Cumulative Production:	TCF	414	501	596	691				
(7) Remaining Proved Reserves:	TCF	266	225	192	221				
(8) Reserve/Production Ratio:		12	11	11	11				
(9) Host Government Take ($/unit of production):									
(10) Capital Cost ($/unit of daily capacity):									
(11) Operating Cost ($/unit of production):									

INDIGENOUS SUPPLY ASSESSMENT WORKSHEET

NATURAL GAS

Country: U.S.A.

Natural Gas:
Offshore X)
Onshore X) Total

WAES SCENARIO CASE (1985): C

WAES SCENARIO CASE (2000): C-1 and C-2

Total Resource in Place: New gas deregulated

Domestic Gas Price: 1975 Technology

Technology Assumptions:

	(Units)	1972	1975	1980	1985	1990	1995	2000	2010
(1) Ultimate Recoverable Resource:	TCF	1202	1202	1202	1202	1202	1202	1202	1202
(2) Discoveries:	TCF	–	8	68	169	270	337	387	437
(3) Reserves Proved to Date:	TCF	670	726	788	887	988	1055	1105	1155
(4) Potential Production:	TCF	22.5	20.1	17.5	19.0	19.0	17.0	14.0	8.5
(5) Expected Annual Production:	TCF	22.5	20.1	17.5	19.0	19.0	17.0	14.0	8.5
(6) Cumulative Production:	TCF	414	501	595	686	781	871	948	1060
(7) Remaining Proved Reserves:	TCF	266	225	193	201	207	184	157	95
(8) Reserve/Production Ratio:		12	11	11	11	11	11	11	11
(9) Host Government Take ($/unit of production):									
(10) Capital Cost ($/unit of daily capacity):									
(11) Operating Cost ($/unit of production):									

INDIGENOUS SUPPLY ASSESSMENT WORKSHEET

NATURAL GAS

Country: U.S.A.

Natural Gas:
Offshore X)
Onshore X) Total

WAES SCENARIO CASE (1985): D

WAES SCENARIO CASE (2000): D-7 and D-8

Total Resource in Place: 1372-1772

Domestic Gas Price: Continued regulation

Technology Assumptions: 1975 Technology

	(Units)	1972	1975	1980	1985	1990	1995	2000	2010
(1) Ultimate Recoverable Resource:	TCF	1202	1202	1202	1202	1202	1202	1202	1202
(2) Discoveries:	TCF	-	8	76	145	194	242	290	338
(3) Reserves Proved to Date:	TCF	680	726	794	863	912	960	1008	1056
(4) Potential Production:	TCF	22.5	20.1	18.0	16.0	14.0	12.5	11.5	7.5
(5) Expected Annual Production:	TCF	22.5	20.1	18.0	16.0	16.0	12.5	11.5	7.5
(6) Cumulative Production:	TCF	414	501	596	681	756	822	882	977
(7) Remaining Proved Reserves:	TCF	266	225	198	182	156	138	126	79
(8) Reserve/Production Ratio:		12	11	11	11	11	11	11	11
(9) Host Government Take ($/unit of production):									
(10) Capital Cost ($/unit of daily capacity):									
(11) Operating Cost ($/unit of production):									

INDIGENOUS SUPPLY ASSESSMENT WORKSHEET

NATURAL GAS

Country: U.S.A.

Natural Gas:
Offshore X)
Onshore X) Total

WAES SCENARIO CASE (1985): E

WAES SCENARIO CASE (2000): ______

Total Resource in Place: 1372–1772

Domestic Gas Price: Continued regulation

Technology Assumptions: 1975 Technology

	(Units)	1972	1975	1980	1985	1990	2000	2010	2025
(1) Ultimate Recoverable Resource:	TCF	1202	1202	1202	1202				
(2) Discoveries:	TCF	–	8	65	121				
(3) Reserves Proved to Date:	TCF	680	726	783	839				
(4) Potential Production:	TCF	22.5	20.1	17.5	14.5				
(5) Expected Annual Production:	TCF	22.5	20.1	17.5	14.5				
(6) Cumulative Production:	TCF	414	501	595	675				
(7) Remaining Proved Reserves:	TCF	266	225	188	164				
(8) Reserve/Production Ratio:		12	11	11	11				
(9) Host Government Take ($/unit of production):									
(10) Capital Cost ($/unit of daily capacity):									
(11) Operating Cost ($/unit of production):									

INDIGENOUS SUPPLY ASSESSMENT WORKSHEET

COAL

Country: U.S.A.

WAES SCENARIO CASE (1985): A

WAES SCENARIO CASE (2000): ______

Total Resource in Place: ______

Domestic Coal Price: ______

Technology Assumptions: ______

Coal:
Surface Mined X
Deep Mined ______

(Bil Tons)	1972	1975	1980	1985	1990	2000	2010	2025
(1) Ultimate Recoverable Resource:		137	137	137				
(2) Reserves Proved to Date:		123	123	123				
(3) Potential Production:		.340	.482	.645				
(4) Expected Annual Production:	.291	.340	.482	.645				
(5) Cumulative Production: From 1975		.340	2.366	5.249				
(6) Remaining Proved Reserves:		123	121	118				
(7) Reserve/Production Ratio:		362	51	22				
(8) Primary Transportation* —Capital Cost ($/unit of daily capacity): —Operating Cost ($/unit of daily capacity):								
(9) Capital Cost ($/unit of daily capacity):								
(10) Operating Cost ($/unit of production):								

*From mine mouth to first processing stage.

INDIGENOUS SUPPLY ASSESSMENT WORKSHEET

COAL

Country: U.S.A.

WAES SCENARIO CASE (1985): B

WAES SCENARIO CASE (2000):

Total Resource in Place:

Domestic Coal Price:

Technology Assumptions:

Coal:
Surface Mined X
Deep Mined

(Bil Tons)	1972	1975	1980	1985	1990	2000	2010	2025
(1) Ultimate Recoverable Resource:		137	137	137				
(2) Reserves Proved to Date:		123	123	123				
(3) Potential Production:		.340	.450	.589				
(4) Expected Annual Production:	.291	.340	.450	.589				
(5) Cumulative Production: From 1975		.340	2.304	4.960				
(6) Remaining Proved Reserves:		123	121	118				
(7) Reserve/Production Ratio:		362	53	24				
(8) Primary Transportation* —Capital Cost ($/unit of daily capacity): —Operating Cost ($/unit of daily capacity):								
(9) Capital Cost ($/unit of daily capacity):								
(10) Operating Cost ($/unit of production):								

*From mine mouth to first processing stage.

INDIGENOUS SUPPLY ASSESSMENT WORKSHEET

COAL

Country: U.S.A.

Coal:
Surface Mined X
Deep Mined ______

WAES SCENARIO CASE (1985): C

WAES SCENARIO CASE (2000): C-1 and C-2

Total Resource in Place: ______

Domestic Coal Price: ______

Technology Assumptions: ______

(Bil Tons)	1972	1975	1980	1985	1990	2000	2010	2025
(1) Ultimate Recoverable Resource:		137	137	137				
(2) Reserves Proved to Date:		123	123	123				
(3) Potential Production:		.340	.450	.589				
(4) Expected Annual Production:	.291	.340	.450	.589		1.548 C-1 1.120 C-2		
(5) Cumulative Production: From 1975		.340	2.304	4.960				
(6) Remaining Proved Reserves:		123	121	118				
(7) Reserve/Production Ratio:		362	53	24				
(8) Primary Transportation* —Capital Cost ($/unit of daily capacity): —Operating Cost ($/unit of daily capacity):								
(9) Capital Cost ($/unit of daily capacity):								
(10) Operating Cost ($/unit of production):								

*From mine mouth to first processing stage.

INDIGENOUS SUPPLY ASSESSMENT WORKSHEET

COAL

Country: U.S.A.

WAES SCENARIO CASE (1985): D

WAES SCENARIO CASE (2000): D-7 and D-8

Total Resource in Place:

Domestic Coal Price:

Technology Assumptions:

Coal:
Surface Mined X
Deep Mined

(Bil Tons)	1972	1975	1980	1985	1990	2000	2010	2025
(1) Ultimate Recoverable Resource:		137	137	137				
(2) Reserves Proved to Date:		123	123	123				
(3) Potential Production:		.340	.407	.496				
(4) Expected Annual Production:	.291	.340	.407	.496		1.162 D-7 / .854 D-8		
(5) Cumulative Production: From 1975		.340	2.222	4.517				
(6) Remaining Proved Reserves:		123	121	118				
(7) Reserve/Production Ratio:		362	54	26				
(8) Primary Transportation* —Capital Cost ($/unit of daily capacity): —Operating Cost ($/unit of daily capacity):								
(9) Capital Cost ($/unit of daily capacity):								
(10) Operating Cost ($/unit of production):								

*From mine mouth to first processing stage.

INDIGENOUS SUPPLY ASSESSMENT WORKSHEET

COAL

Country: U.S.A.

WAES SCENARIO CASE (1985): E

WAES SCENARIO CASE (2000): ________

Total Resource in Place: ________

Domestic Coal Price: ________

Technology Assumptions: ________

Coal:
Surface Mined X
Deep Mined ________

(Bil Tons)	1972	1975	1980	1985	1990	2000	2010	2025
(1) Ultimate Recoverable Resource:		137	137	137				
(2) Reserves Proved to Date:		123	123	123				
(3) Potential Production:		.340	.407	.496				
(4) Expected Annual Production:	.291	.340	.407	.496				
(5) Cumulative Production: From 1975		.340	2.222	4.517				
(6) Remaining Proved Reserves:		123	121	118				
(7) Reserve/Production Ratio:		362	54	26				
(8) Primary Transportation* —Capital Cost ($/unit of daily capacity): —Operating Cost ($/unit of daily capacity):								
(9) Capital Cost ($/unit of daily capacity):								
(10) Operating Cost ($/unit of production):								

*From mine mouth to first processing stage.

INDIGENOUS SUPPLY ASSESSMENT WORKSHEET

COAL

Country: U.S.A.

WAES SCENARIO CASE (1985): A

WAES SCENARIO CASE (2000): ______

Total Resource in Place: ______

Domestic Coal Price: ______

Technology Assumptions: ______

Coal:
Surface Mined ______
Deep Mined X

(Bil Tons)	1972	1975	1980	1985	1990	2000	2010	2025
(1) Ultimate Recoverable Resource:		300	300	300				
(2) Reserves Proved to Date:		150	150	150				
(3) Potential Production:		.300	.364	.395				
(4) Expected Annual Production:	.304	.300	.364	.395				
(5) Cumulative Production: From 1975		.300	1.948	3.863				
(6) Remaining Proved Reserves:		150	148	146				
(7) Reserve/Production Ratio:		500	76	38				
(8) Primary Transportation* —Capital Cost ($/unit of daily capacity): —Operating Cost ($/unit of daily capacity):								
(9) Capital Cost ($/unit of daily capacity):								
(10) Operating Cost ($/unit of production):								

*From mine mouth to first processing stage.

INDIGENOUS SUPPLY ASSESSMENT WORKSHEET

COAL

Country: U.S.A.

WAES SCENARIO CASE (1985): B

WAES SCENARIO CASE (2000): ______

Total Resource in Place: ______

Domestic Coal Price: ______

Technology Assumptions: ______

Coal:
Surface Mined ______
Deep Mined X

(Bil Tons)	1972	1975	1980	1985	1990	2000	2010	2025
(1) Ultimate Recoverable Resource:		300	300	300				
(2) Reserves Proved to Date:		150	150	150				
(3) Potential Production:		.300	.339	.361				
(4) Expected Annual Production:	.304	.300	.339	.361				
(5) Cumulative Production: From 1975		.300	1.900	3.666				
(6) Remaining Proved Reserves:		150	148	146				
(7) Reserve/Production Ratio:		500	78	40				
(8) Primary Transportation* —Capital Cost ($/unit of daily capacity):								
—Operating Cost ($/unit of daily capacity):								
(9) Capital Cost ($/unit of daily capacity):								
(10) Operating Cost ($/unit of production):								

*From mine mouth to first processing stage.

INDIGENOUS SUPPLY ASSESSMENT WORKSHEET

COAL

Country: U.S.A.

WAES SCENARIO CASE (1985): C

WAES SCENARIO CASE (2000): C1 and C2

Total Resource in Place: ______

Domestic Coal Price: ______

Technology Assumptions: ______

Coal:
Surface Mined ______
Deep Mined x

(Bil Tons)	1972	1975	1980	1985	1990	2000	2010	2025
(1) Ultimate Recoverable Resource:		300	300	300				
(2) Reserves Proved to Date:		150	150	150				
(3) Potential Production:		.300	.339	.361				
(4) Expected Annual Production:	.304	.300	.339	.361		.664 / .480	C1 / C2	
(5) Cumulative Production: From 1975		.300	1.900	3.666				
(6) Remaining Proved Reserves:		150	148	146				
(7) Reserve/Production Ratio:		500	78	40				
(8) Primary Transportation* —Capital Cost ($/unit of daily capacity): —Operating Cost ($/unit of daily capacity):								
(9) Capital Cost ($/unit of daily capacity):								
(10) Operating Cost ($/unit of production):								

*From mine mouth to first processing stage.

INDIGENOUS SUPPLY ASSESSMENT WORKSHEET

COAL

Country: U.S.A.

Coal:
Surface Mined
Deep Mined X

WAES SCENARIO CASE (1985): D

WAES SCENARIO CASE (2000): D7 and D8

Total Resource in Place: ________

Domestic Coal Price: ________

Technology Assumptions: ________

(Bil Tons)	1972	1975	1980	1985	1990	2000	2010	2025
(1) Ultimate Recoverable Resource:		300	300	300				
(2) Reserves Proved to Date:		150	150	150				
(3) Potential Production:		.300	.307	.304				
(4) Expected Annual Production:	.304	.300	.307	.304		.498 D7 / .366 D8		
(5) Cumulative Production: from 1975		.300	1.836	3.364				
(6) Remaining Proved Reserves:		150	148	147				
(7) Reserve/Production Ratio:		500	81	44				
(8) Primary Transportation* —Capital Cost ($/unit of daily capacity): —Operating Cost ($/unit of daily capacity):								
(9) Capital Cost ($/unit of daily capacity):								
(10) Operating Cost ($/unit of production):								

*From mine mouth to first processing stage.

INDIGENOUS SUPPLY ASSESSMENT WORKSHEET

COAL

Country: U.S.A.

WAES SCENARIO CASE (1985): E

WAES SCENARIO CASE (2000):

Total Resource in Place:

Domestic Coal Price:

Technology Assumptions:

Coal:
Surface Mined
Deep Mined X

(Bil Tons)	1972	1975	1980	1985	1990	2000	2010	2025
(1) Ultimate Recoverable Resource:		300	300	300				
(2) Reserves Proved to Date:		150	150	150				
(3) Potential Production:		.300	.307	.304				
(4) Expected Annual Production:	.304	.300	.307	.304				
(5) Cumulative Production: from 1975		.300	1.836	3.364				
(6) Remaining Proved Reserves:		150	148	147				
(7) Reserve/Production Ratio:		500	81	44				
(8) Primary Transportation* —Capital Cost ($/unit of daily capacity): —Operating Cost ($/unit of daily capacity):								
(9) Capital Cost ($/unit of daily capacity):								
(10) Operating Cost ($/unit of production):								

*From mine mouth to first processing stage.

NONFOSSIL SOURCES AND CONVERSION PROCESS WORKSHEET

Hydroelectric ______
Coal Gasification ______
Coal to Syncrude ______
Gas Liquifaction ______
Regasification ______
Other* (specify): X
Shale Oil

Country: U.S.A.

WAES SCENARIO CASE (1985): A, B & C

WAES SCENARIO CASE (2000): C1, C2

Total Resource in Place: ______

Technology Assumptions: 1976 Technology

	(Units)	1972	1975	1985	1990	2000	2010	2025
(1) Yield:								
(2) Maximum Potential Output:**		0	0	.4				A, B
	MMBD	0	0	.2		2.0		C1 and C2
(3) Capital Cost ($/unit of daily capacity):	$/BD			20,500		20,500		
(4) Cumulative Capital Investment:								
(5) Operating Cost ($/unit of production):	$/B			3.00		3.00		

*Specify other relevant non-fossil energy source or conversion process, i.e., geothermal, solar, wind, etc.

**"Maximum technological development" in GEMM.

NONFOSSIL SOURCES AND CONVERSION PROCESS WORKSHEET

Hydroelectric ______
Coal Gasification ______
Coal to Syncrude ______
Gas Liquifaction ______
Regasification ______
Other* (specify): X

Shale Oil

Country: U.S.A.

WAES SCENARIO CASE (1985): D & E

WAES SCENARIO CASE (2000): D7 & D8

Total Resource in Place: ______

Technology Assumptions: 1976 Technology

	(Units)	1972	1975	1985	1990	2000	2010	2025
(1) Yield:								
(2) Maximum Potential Output**	MMBD	0	0	0		0		
(3) Capital Cost ($/unit of daily capacity):	$/BD							
(4) Cumulative Capital Investment:								
(5) Operating Cost ($/unit of production):	$/B							

*Specify other relevant non-fossil energy source or conversion process, i.e., geothermal, solar, wind, etc.

**"Maximum technological development" in GEMM.

NONFOSSIL SOURCES AND CONVERSION PROCESS WORKSHEET

Hydroelectric ______
Coal Gasification ______
Coal to Syncrude X
Gas Liquifaction ______
Regasification ______
Other* (specify): ______

Country: U.S.A.

WAES SCENARIO CASE (1985): A,B,C,D&E

WAES SCENARIO CASE (2000): C1, C2

Total Resource in Place: ______

Technology Assumptions: 1976 Technology

	(Units)	1972	1975	1985	1990	2000	2010	2025
(1) Yield:								
(2) Maximum Potential Output:**		0	0	0		2.0		
(3) Capital Cost ($/unit of daily capacity):								
(4) Cumulative Capital Investment:								
(5) Operating Cost ($/unit of production):								

*Specify other relevant non-fossil energy source or conversion process, i.e., geothermal, solar, wind, etc.

**"Maximum technological development" in GEMM.

NONFOSSIL SOURCES AND CONVERSION PROCESS WORKSHEET

Hydroelectric ______
Coal Gasification ______
Coal to Syncrude X
Gas Liquifaction ______
Regasification ______
Other* (specify): ______

Country: U.S.A.

WAES SCENARIO CASE (1985): A,B,C,D&E

WAES SCENARIO CASE (2000): D7, D8

Total Resource in Place: ______

Technology Assumptions: 1976 Technology

	(Units)	1972	1975	1985	1990	2000	2010	2025
(1) Yield:								
(2) Maximum Potential Output	MMBD	0	0	0		0		
(3) Capital Cost ($/unit of daily capacity):								
(4) Cumulative Capital Investment:								
(5) Operating Cost ($/unit of production):								

*Specify other relevant non-fossil energy source or conversion process, i.e., geothermal, solar, wind, etc.

**"Maximum technological development" in GEMM.

NONFOSSIL SOURCES AND CONVERSION PROCESS WORKSHEET

Hydroelectric ______
Coal Gasification X
Coal to Syncrude ______
Gas Liquifaction ______
Regasification ______
Other* (specify): ______

Country: U.S.A.

WAES SCENARIO CASE (1985): A & B

WAES SCENARIO CASE (2000): ______

Total Resource in Place: ______

Technology Assumptions: 1976 Technology

	(Units)	1972	1975	1985	1990	2000	2010	2025
(1) Yield:								
(2) Maximum Potential Output	TCF	–	–	1.2				
(3) Capital Cost ($/unit of daily capacity)	$10^3$$/MMCFD			3750				
(4) Cumulative Capital Investment:								
(5) Operating Cost ($/unit of production):								

*Specify other relevant non-fossil energy source or conversion process, i.e., geothermal, solar, wind, etc.

**"Maximum technological development" in GEMM.

NONFOSSIL SOURCES AND CONVERSION PROCESS WORKSHEET

Hydroelectric ____
Coal Gasification X
Coal to Syncrude ____
Gas Liquifaction ____
Regasification ____
Other* (specify): ____

Country: U.S.A.

WAES SCENARIO CASE (1985): C

WAES SCENARIO CASE (2000): C1 and C2

Total Resource in Place: ____

Technology Assumptions: 1976 Technology

	(Units)	1972	1975	1985	1990	2000	2010	2025
(1) Yield:								
(2) Maximum Potential Output	TCF	-	-	.4		6.0		
(3) Capital Cost ($/unit of daily capacity):	$10^3$$/MMCFD			3750		3750		
(4) Cumulative Capital Investment:								
(5) Operating Cost ($/unit of production):								

*Specify other relevant non-fossil energy source or conversion process, i.e., geothermal, solar, wind, etc.

**"Maximum technological development" in GEMM.

NONFOSSIL SOURCES AND CONVERSION PROCESS WORKSHEET

Hydroelectric ______
Coal Gasification X
Coal to Syncrude ______
Gas Liquifaction ______
Regasification ______
Other* (specify): ______

Country: U.S.A.

WAES SCENARIO CASE (1985): D and E

WAES SCENARIO CASE (2000): D7

Total Resource in Place: ______

Technology Assumptions: 1976 Technology

	(Units)	1972	1975	1985	1990	2000	2010	2025
(1) Yield:								
(2) Maximum Potential Output	TCF	-	-	-		4.0		
(3) Capital Cost ($/unit of daily capacity):								
(4) Cumulative Capital Investment:								
(5) Operating Cost ($/unit of production):								

*Specify other relevant non-fossil energy source or conversion process, i.e., geothermal, solar, wind, etc.

**"Maximum technological development" in GEMM.

NONFOSSIL SOURCES AND CONVERSION PROCESS WORKSHEET

Hydroelectric ____
Coal Gasification X
Coal to Syncrude ____
Gas Liquifaction ____
Regasification ____
Other* (specify): ____

Country: U.S.A.

WAES SCENARIO CASE (1985): D & E

WAES SCENARIO CASE (2000): D8

Total Resource in Place: ____

Technology Assumptions: 1976 Technology

	(Units)	1972	1975	1985	1990	2000	2010	2025
(1) Yield:								
(2) Maximum Potential Output	TCF	-	-	-		-		
(3) Capital Cost ($/unit of daily capacity):								
(4) Cumulative Capital Investment:								
(5) Operating Cost ($/unit of production):								

*Specify other relevant non-fossil energy source or conversion process, i.e., geothermal, solar, wind, etc.

**"Maximum technological development" in GEMM.

NONFOSSIL SOURCES AND CONVERSION PROCESS WORKSHEET

Hydroelectric ______
Coal Gasification ______
Coal to Syncrude ______
Gas Liquifaction ______
Regasification ______
Other* (specify): X

Syngas from Liquids

Country: U.S.A.

WAES SCENARIO CASE (1985): A thru E

WAES SCENARIO CASE (2000): All

Total Resource in Place: ______

Technology Assumptions: 1976 Technology, Naphtha Gasification

	(Units)	1972	1975	1980	1985	2000	2010	2025
(1) Yield:			91	91	91	91		
(2) Maximum Potential Output	TCF		0.2	0.2	0.2	0		
(3) Capital Cost ($/unit of daily capacity):	10^3\$/MCFD		646	646	646	646		
(4) Cumulative Capital Investment:								
(5) Operating Cost ($/unit of production):								

*Specify other relevant non-fossil energy source or conversion process, i.e., geothermal, solar, wind, etc.

**"Maximum technological development" in GEMM.

NONFOSSIL SOURCES AND CONVERSION PROCESSES WORKSHEET

URANIUM AND NUCLEAR

Country: U.S.A.

WAES SCENARIO CASE (1985): A, B, & C

WAES SCENARIO CASE (2000): C-1

Total Resource in Place: ____________

Domestic Uranium Price: ____________

Technology Assumptions: 60% Load Factor in 1985
80% Load Factor in 2000

	(Units)	1972	1975	1980	1985	1990	1995	2000	2010
1. Ultimate Recoverable Uranium Resource:									
2. Reserve of Uranium Proved to Date:									
3. Potential Production of Uranium:									
4. Expected Production of Uranium:									
5. Reserve/Production Ratio:									
6. Maximum Installed Capacity (GWe)*:		15	40	70	166			380	
7. Capital Cost ($/KWe):			585	585	585			585	
8. Operating Cost (¢/Kwhr)**:									
9. Maximum Output	10^9 KWH	54	141	368	872			2663	

*Indicate assumed load factor and LWR or HWR.

**Excluding capital charges.

NONFOSSIL SOURCES AND CONVERSION PROCESSES WORKSHEET

URANIUM AND NUCLEAR

Country: U.S.A.

WAES SCENARIO CASE (1985): A,B,& C

WAES SCENARIO CASE (2000): C-2

Total Resource in Place:

Domestic Uranium Price:

Technology Assumptions: 60% Load Factor in 1985
80% Load Factor in 2000

	(Units)	1972	1975	1980	1985	1990	1995	2000	2010
1. Ultimate Recoverable Uranium Resource:									
2. Reserve of Uranium Proved to Date:									
3. Potential Production of Uranium:									
4. Expected Production of Uranium:									
5. Reserve/Production Ratio:									
6. Maximum Installed Capacity (GWe)*:		15	40	70	166			620	
7. Capital Cost ($/KWe):			585	585	585			585	
8. Operating Cost (¢/Kwhr)**:									
9. Maximum Output	10^9 KWH	54	141	368	872			4345	

*Indicate assumed load factor and LWR or HWR.

**Excluding capital charges.

NONFOSSIL SOURCES AND CONVERSION PROCESSES WORKSHEET

URANIUM AND NUCLEAR

Country: U.S.A.

WAES SCENARIO CASE (1985): D, E

WAES SCENARIO CASE (2000): D-7

Total Resource in Place:

Domestic Uranium Price:

Technology Assumptions: 60% Load Factor in 1985
80% Load Factor in 2000

	(Units)	1972	1975	1980	1985	1990	1995	2000	2010
1. Ultimate Recoverable Uranium Resource:									
2. Reserve of Uranium Proved to Date:									
3. Potential Production of Uranium:									
4. Expected Production of Uranium:									
5. Reserve/Production Ratio:									
6. Maximum Installed Capacity (GWe)*:		15	40	56	127			380	
7. Capital Cost ($/KWe):			585	585	585			585	
8. Operating Cost (¢/Kwhr)**:									
9. Maximum Output 10^9 KWH		54	141	294	668			2663	

*Indicate assumed load factor and LWR or HWR.

**Excluding capital charges.

NONFOSSIL SOURCES AND CONVERSION PROCESSES WORKSHEET

URANIUM AND NUCLEAR

Country: U.S.A.

WAES SCENARIO CASE (1985): D,E

WAES SCENARIO CASE (2000): D-8

Total Resource in Place: ______

Domestic Uranium Price: ______

Technology Assumptions: ______ 60% Load Factor in 1985
80% Load Factor in 2000

	(Units)	1972	1975	1980	1985	1990	1995	2000	2010
1. Ultimate Recoverable Uranium Resource:									
2. Reserve of Uranium Proved to Date:									
3. Potential Production of Uranium:									
4. Expected Production of Uranium:									
5. Reserve/Production Ratio:									
6. Maximum Installed Capacity (GWe)*:		15	40	56	127			620	
7. Capital Cost ($/KWe):			585	585	585			585	
8. Operating Cost (¢/Kwhr)**:									
9. Maximum Output 10^9 KWH		54	141	294	668			4345	

*Indicate assumed load factor and LWR or HWR.

**Excluding capital charges.

NONFOSSIL SOURCES AND CONVERSION PROCESS WORKSHEET

Hydroelectric X
Coal Gasification
Coal to Syncrude
Gas Liquifaction
Regasification
Other* (specify):

Country: U.S.A.

WAES SCENARIO CASE (1985): A, B & C

WAES SCENARIO CASE (2000): C-1, C-2

Total Resource in Place:

Technology Assumptions: 1976 Technology, 54% Load Factor

	(Units)	1972	1975	1980	1985	2000	2010	2025
(1) Installed Capacity	GWE	57	64	77	90	100		
(2) Maximum Potential Output	10^9 KWH	272	305	367	426	473		
(3) Capital Cost ($/unit of daily capacity):	$/KWE		520	520	520	520		
(4) Cumulative Capital Investment:								
(5) Operating Cost ($/unit of production):								

*Specify other relevant non-fossil energy source or conversion process, i.e., geothermal, solar, wind, etc.

**"Maximum technological development" in GEMM.

NONFOSSIL SOURCES AND CONVERSION PROCESS WORKSHEET

Hydroelectric X
Coal Gasification
Coal to Syncrude
Gas Liquifaction
Regasification
Other* (specify):

Country: U.S.A.

WAES SCENARIO CASE (1985): D, E

WAES SCENARIO CASE (2000): D-7, D-8

Total Resource in Place:

Technology Assumptions: 1976 Technology, 54% Load Factor

	(Units)	1972	1975	1980	1985	2000	2010	2025
(1) Installed Capacity	GWE	57	64	65	67	80		
(2) Maximum Potential Output	10^9 KWH	272	305	310	317	378		
(3) Capital Cost ($/unit of daily capacity):	$/KWE		520	520	520	520		
(4) Cumulative Capital Investment:								
(5) Operating Cost ($/unit of production):								

*Specify other relevant non-fossil energy source or conversion process, i.e., geothermal, solar, wind, etc.

**"Maximum technological development" in GEMM.

NONFOSSIL SOURCES AND CONVERSION PROCESS WORKSHEET

Hydroelectric ______
Coal Gasification ______
Coal to Syncrude ______
Gas Liquifaction ______
Regasification ______
Other* (specify): X
Geothermal

Country: U.S.A.

WAES SCENARIO CASE (1985): A, B & C

WAES SCENARIO CASE (2000): C-1, C-2

Total Resource in Place: ______

Technology Assumptions: Government R & D Support and Hot Water Technology Available. 70% Load Factor

	(Units)	1972	1975	1980	1985	2000	2010	2025
(1) Installed Capacity	GWE		.4	.9	10	50		
(2) Maximum Potential Output	10^9 KWH		2	5	61	307		
(3) Capital Cost ($/unit of daily capacity):	$/KWE		840	840	840	840		
(4) Cumulative Capital Investment:								
(5) Operating Cost ($/unit of production):								

*Specify other relevant non-fossil energy source or conversion process, i.e., geothermal, solar, wind, etc.

**"Maximum technological development" in GEMM.

NONFOSSIL SOURCES AND CONVERSION PROCESS WORKSHEET

Hydroelectric ______
Coal Gasification ______
Coal to Syncrude ______
Gas Liquifaction ______
Regasification ______
Other* (specify): X
Geothermal

Country: U.S.A.

WAES SCENARIO CASE (1985): D & E

WAES SCENARIO CASE (2000): D-7, D-8

Total Resource in Place: ______

Technology Assumptions: Hot Water Technology not Available until after 1985.
70% Load Factor

	(Units)	1972	1975	1980	1985	2000	2010	2025
(1) Installed Capacity	GWE		.4	.9	4	20		
(2) Maximum Potential Output	10^9 KWH		2	5	25	123		
(3) Capital Cost ($/unit of daily capacity):	$/KWE		840	840	840	840		
(4) Cumulative Capital Investment:								
(5) Operating Cost ($/unit of production):								

*Specify other relevant non-fossil energy source or conversion process, i.e., geothermal, solar, wind, etc.

**"Maximum technological development" in GEMM.